T0262371

Detecting Mineral Nutrient Deficiencies in Tropical and Temperate Crops

Westview Tropical Agriculture Series
Donald L. Plucknett, Series Editor

Sesbania *in Agriculture*, Dale O. Evans and Peter P. Rotar

Detecting Mineral Nutrient Deficiencies in Tropical and Temperate Crops, edited by Donald L. Plucknett and Howard B. Sprague

Detecting Mineral Nutrient Deficiencies in Tropical and Temperate Crops

edited by Donald L. Plucknett
and Howard B. Sprague

CRC Press
Taylor & Francis Group
Boca Raton London New York

CRC Press is an imprint of the
Taylor & Francis Group, an **informa** business

First published 1989 by Westview Press, Inc.

Published 2018 by CRC Press
Taylor & Francis Group
6000 Broken Sound Parkway NW, Suite 300
Boca Raton, FL 33487-2742

CRC Press is an imprint of the Taylor & Francis Group, an informa business

Copyright © 1989 Taylor & Francis Group LLC

No claim to original U.S. Government works

This book contains information obtained from authentic and highly regarded sources. Reasonable efforts have been made to publish reliable data and information, but the author and publisher cannot assume responsibility for the validity of all materials or the consequences of their use. The authors and publishers have attempted to trace the copyright holders of all material reproduced in this publication and apologize to copyright holders if permission to publish in this form has not been obtained. If any copyright material has not been acknowledged please write and let us know so we may rectify in any future reprint.

Except as permitted under U.S. Copyright Law, no part of this book may be reprinted, reproduced, transmitted, or utilized in any form by any electronic, mechanical, or other means, now known or hereafter invented, including photocopying, microfilming, and recording, or in any information storage or retrieval system, without written permission from the publishers.

For permission to photocopy or use material electronically from this work, please access www.copyright.com (http://www.copyright.com/) or contact the Copyright Clearance Center, Inc. (CCC), 222 Rosewood Drive, Danvers, MA 01923, 978-750-8400. CCC is a not-for-profit organization that provides licenses and registration for a variety of users. For organizations that have been granted a photocopy license by the CCC, a separate system of payment has been arranged.

Trademark Notice: Product or corporate names may be trademarks or registered trademarks, and are used only for identification and explanation without intent to infringe.

Visit the Taylor & Francis Web site at
http://www.taylorandfrancis.com

and the CRC Press Web site at
http://www.crcpress.com

Library of Congress Cataloging-in-Publication Data
Detecting mineral nutrient deficiencies in tropical and
    temperate crops.
    (Westview tropical agriculture series)
    1. Deficiency diseases in plants--Diagnosis.
I. Plucknett, Donald L., 1931-   . II. Sprague, Howard
Bennet. III. Series.
SB742.D47 1989   632'.3   88-28024

ISBN 13: 978-0-367-00539-9 (hbk)

ISBN 13: 978-0-367-15526-1 (pbk)

Contents

Preface

The ability to identify deficiencies of plant nutrients before they limit crop yields is a major need in modern agriculture. This ability requires using a combination of diagnostic methods that enable remedial steps to be taken before crop yields are severely reduced. Three main methods of diagnosis may be used: visual deficiency symptoms, soil analyses, and leaf tissue analyses. This book presents information on all three methods, and their integrated use, for managing most of the important tropical and temperate crops, including cereals, grain legumes, roots and tubers, sugar crops, vegetables, fruits and nuts, tropical industrial crops and forage crops. The book is written from a practical standpoint, providing soil and tissue analysis standards that are considered to be critical in plant nutrition along with color photographs of symptoms that are associated with specific nutrient deficiencies. Three introductory chapters are devoted to visual plant symptoms, soil testing, and plant and tissue testing. Individual chapters give background information on each crop, as well as details on how soil and tissue testing standards and visual symptoms can be used effectively in crop management. All chapters provide facts on nutrient requirements for each crop, as well as information on how to fertilize the crop for optimum production.

We are indebted to many persons who have helped to complete this book. Special thanks go to Audrey Mitchell, Noemie Del Marr, and Janette Abdul-Ghani, who typed or prepared the final manuscript. Mary Horne provided valuable editorial assistance.

Donald L. Plucknett
Washington, D.C.

Howard B. Sprague
University Park,
Pennsylvania

1
Visual Plant Symptoms as Indicators of Mineral Nutrient Deficiencies

J. Fielding Reed

INTRODUCTION

For generations, the appearance of a plant has been used by scientists and by laymen, by gardeners and by farmers as an indication of the health of the plant. The plant speaks through distress signals. The message may tell us that there is simply a shortage of water or that there is too much water and, hence, a shortage of air. Or the signals may tell of a disease caused by an organism such as a virus, fungus or bacteria. Or, the plant may be complaining because it is being attacked by insects or rodents, either below or above ground. And if there is an excess of some chemical that can be toxic to the plant, this condition is also indicated by the appearance of the plant.

Finally, if one or more of the essential mineral nutrients is not present in an adequate amount, the plant will do its best to tell us this. So, we have learned to observe the plant's appearance and to use visual plant symptoms as indicators of mineral nutrient deficiencies. At the same time we must realize that visual symptoms tell many stories in addition to that of mineral deficiencies, and it is not always an easy matter to know just what the plant is trying to tell us. One of the purposes of this book is to help us to find out what is limiting production of a plant and, especially, to learn how to tell one symptom from another.

This calls for great patience and study and, most of all, experience. It requires a knowledge of what the healthy plant should look like. It means using the other diagnostic tools to help identify or confirm the visual symptoms. This has been brought out in the following chapters on soil testing and plant analysis. The modern farmer or scientist will use a soil test or plant analy-

---

[1]Retired President of Potash Phosphate Institute, in Atlanta, Georgia; Current address -- 175 Lull Water Road, Athens, GA 30606, USA.

sis, or both, to help identify the symptom. This is especially necessary when one is just beginning to study visual symptoms.

Then, as one learns the symptoms of nutrient deficiencies, he must be aware of the hazards in relying on his knowledge and must be aware of the possibilities of errors. This involves going beyond the field of agronomy and plant nutrition to include the areas of plant diseases, insect damage, and weather and environmental factors. A good diagnostician must be broad in his knowledge and meticulous in his approach. But when all of this is done, it provides a very valuable aid to profitable crop production.

Another very important point should be kept in mind. Often a plant will border on deficiency of a plant nutrient and yet not show any symptoms. This condition is frequently referred to as hidden hunger. When there are no visual symptoms, the plant is not producing at its capacity. This is one of the dangers of relying on symptoms. When the plant reaches the level where a symptom appears, the yield has already been reduced. Too many areas of the world are being farmed in the hidden hunger zone.

In the chapters that follow, the detailed symptoms of various factors that limit production will be described for many important crops. This chapter deals with overall principles of visual symptoms.

The Plant Nutrient

Scientists have determined that 16 elements are necessary for plant growth. Some of these come from the air, some from water, and others from the soil or from fertilizers and lime. They are needed in different amounts, from as much as several million kilograms of water to very small traces of some of the elements. A general guide is shown in Table 1.1.

These amounts of nutrients are broad approximations and depend on the crop being grown and, especially, on the yield of the crop. For example:

Crop	Yield kg/ha	Kg per ha contained in the crop plant		
		N	P_2O_5	K_2O
Corn (maize)	7,500	170	64	170
	12,000	292	125	292
Wheat	2,200	74	30	90
	4,400	150	60	180

Thus, when we look for deficiency symptoms, we must keep in mind the high requirements for certain elements if high yields are the goal.

TABLE 1.1
Elements essential for plant growth

Element/Material	Chemical symbol	Kilograms per hectare
Supplied by air and water:		
Hydrogen (as water)	H_2O	2,200,000 to 6,600,000 kg/ha
Oxygen	O_2	5,500 to 8,800 kg/ha
Carbon (as carbon dioxide)	CO_2	16,500 to 28,000 kg/ha
Primary/major elements supplied by soil and fertilizers:		
Nitrogen	N	60 to 300 kg/ha
Phosphorus	P	10 to 200 kg/ha
Potassium	K	20 to 400 kg/ha
Secondary elements supplied by soil, fertilizer or lime:		
Calcium	Ca	20 to 400 kg/ha
Magnesium	Mg	20 to 400 kg/ha
Sulfur	S	10 to 200 kg/ha
Micronutrients needed in small amounts:		
Iron	Fe	1 to 5 kg/ha
Manganese	Mn	0.5 to 5 kg/ha
Boron	B	Trace* ppm
Zinc	Zn	Trace* ppm
Copper	Cu	Trace* ppm
Molybdenum	Mo	Trace* ppm

*Usually measured in parts per million.

Nutrient Balance

Even more important is the interrelationship of one element to another. A high quantity of phosphorus in the soil or in the plant may result in a deficiency of zinc. A high amount of potassium may result in a deficiency of magnesium. And then there is the need for more of one element as more of another is added. When we add more nitrogen, we create the need for more potassium because the yield is greater and the plant's needs increase. A farmer might see no potassium deficiencies in one field where he has added only a small amount of nitrogen; but on an adjoining area where he has applied more nitrogen, he might see potassium deficiencies.

This whole science of interrelationships, of the effect of one element upon another, is a complex one. It is mentioned here so that the grower can be aware of the pitfalls that are present when one uses visual symptoms alone. This does not suggest that we should not learn to use symptoms. But it should warn the user that the symptom simply tells us what the limiting factor is at that time. When this limiting factor is corrected, another may then turn up, so one must be aware of these interrelationships.

FACTORS AFFECTING SYMPTOMS

Why do symptoms of a plant nutrient deficiency occur? Because there is not enough of that nutrient present in a form so that the plant can take up what it needs. Often this is simply because the soil is infertile and not enough of that nutrient was added. But we must recognize that there are various other factors that affect uptake of nutrients and hence lead to the appearance of symptoms.

Root Zones

Plants differ a great deal in the extent of their root systems. And even the same plant will vary in the amount of roots, depending upon the environment in which it grows. Since some of the plant nutrients do not move very far in the soil, the extent of the root system will determine whether the plant gets enough of a nutrient. Indeed, if conditions are such that root growth is rather shallow, a plant may show a deficiency symptom when the soil actually contains a rather good supply of that nutrient.

Root growth is affected by the physical condition of the soil, by a compact layer, by too little or too much water, by tilth and cultivation practices, by the amount of organic matter present in the soil, and by many other factors.

When a deficiency symptom is noted, one should always examine the root zone to see if a greatly restricted root zone may be contributing to the deficiency.

Temperature

Many growers have seen a visual symptom when the plants are young, only to see the plant "grow out" of this symptom as the season progresses. Often this is caused by a combination of temperature and root growth. If the soils are cold and the air temperature is cold, the plant just does not grow, root systems are small, and plant nutrient uptake is reduced.

Also, when air temperatures are too low or too high, photosynthesis and plant respiration rates are lowered. For example, if temperatures are too high at night, respiration continues at a high rate, burning up plant sugars, and of course photosynthesis stops at night. Thus, temperature could create visual symptoms, and one should always look into the day and night temperatures when considering the cause of a visual symptom.

Acidity or Alkalinity

When a visual symptom is noted, one should look especially into the degree of acidity or alkalinity of the soil upon which the plant is growing. Very often this is closely related to the cause of the symptom.

The solubility and the availability of many plant nutrients are dependent upon the soil pH.[1] When the pH value of the soil goes above pH 6.0-6.5, elements such as iron (Fe), manganese (Mn), zinc (Zn) and boron (B) decrease in solubility, often to the point where the plant is deficient and shows symptoms. And in contrast, when the soil pH is on the acid side (pH below 6.0), molybdenum (Mo) becomes less soluble and a deficiency of this element may be evident. Liming the soil may correct this deficiency.

Because of the importance of soil pH upon plant nutrient availability, it is essential to find out the liming history of a field, or even a small area, where a symptom is observed. Over-liming is easy to do on acid, sandy soils. When these soils are over-limed, deficiency symptoms of iron, manganese, or zinc are likely to appear.

In some parts of the world the soils are not acid, but are naturally alkaline. This is often the case in arid regions where alkaline salts have accumulated or even in areas of higher rainfall where there are outcroppings of limestone and the soil pH is above 7.0. In such soils, it is not uncommon to see symptoms of deficiencies of iron, manganese, or zinc.

Where soils have been limed in excess or where the soil is naturally alkaline, it is often difficult to correct a deficiency of iron or manganese by simply adding these materials to the soil. When these are added, they are made unavailable by the excess of alkalinity. So, often it is best on such soils to correct the deficiency by adding these elements in a spray form on the foliage.

Variety--genetic (heritable) factors

Sometimes a visual symptom of a deficiency may be noticed in one variety of corn (maize) but not in another. This is not uncommon, but it is often overlooked. Difference in heredity makeup may affect the ability to take up and utilize plant nutrients. One variety may show symptoms of magnesium deficiency while another vari-

[1]pH is a chemical measure of acidity or alkalinity. pH values below 7 are acid; lower pH values indicate increasing acidity; and values higher than pH 7 indicate increasing alkalinity.

ety growing beside it may not show the symptom. One should therefore watch for this possibility, for it often helps to explain why nutrient deficiencies appear.

State of Maturity

As a plant nears maturity, it shows signs of "old age." This may be a reddening or browning, or tip or edge "burn." But it does often look like a nutrient deficiency symptom and the grower may mistake it for such. In fact, there is a relationship, because as a plant grows older, it may "run out" of nitrogen or potassium, and therefore may "mature" before it has reached its full yield potential. In many crops there is a delicate balance between the amount of nitrogen required for full yield and the amount that may be too little or too much. Visual symptoms help us to recognize the correct amount.

Distinction of Deficiency Symptoms from Other Symptoms

The necessity for study and experience has been stressed in learning to detect symptoms due to plant nutrient deficiencies. It is especially important to learn to distinguish between these symptoms and other visual symptoms that may be present.

Herbicide[1] Injury

With the introduction of many new herbicides, some growers are not aware of their properties, and injury to plants may result from misuse of certain herbicides. These may be mistaken for symptoms of nutrient deficiency. This can be avoided by a good review of the history of the field treatment.

Diseases and Insects

If the symptom is one of a deficiency that is well known to the examiner, then there may be no problem. But often when a plant shows distress signals, the average person has a hard time deciding. Is it a deficiency or is it a disease? If it is a disease, is deficiency partly responsible for the disease? Could it be nematodes? Or maybe it's none of these but a matter of high air temperature, too much water, or too little water.
Frequently these questions are difficult for the inexperienced grower to answer. And that's why we do not rely on visual symptoms alone but use every tool that is available. First look for insects by examining the roots, leaves, and stems. At the same time use a small

---

[1]Herbicides are special chemicals developed to kill weeds, without injury to the crop. Apply only the particular herbicide that will not harm the crop plant.

hand lens and look for evidence of disease. Of course, to be sure, plants should be taken to a plant pathologist or an entomologist for positive diagnosis. And plant analysis, tissue tests and soil tests are very helpful in arriving at the answer. This whole approach may be summarized.

Be a complete crop diagnostician. To be a complete diagnostician, you must look beyond fertility problems. Know your plant environmental conditions. Such knowledge may help you pinpoint a problem that is inducing, or magnifying, apparent nutrient shortage. Look at all factors that influence crop growth, response to fertilization and yield.

1. Root zone. The soil must be granular and permeable enough for roots to expand and feed extensively. A crop will develop a root system 2 m or more deep on some soils to get water and nutrients. A shallow or compacted soil does not offer this root feeding zone. Wet or poorly drained soils result in shallow root systems.

2. Temperature. Cool soil temperatures reduce organic matter decomposition. This slows the release of nitrogen (N) and other nutrients. Nutrients are less soluble in cool soils, increasing deficiency potential. Phosphate (P) and potash (K) diffuse more slowly in cool soils. Root activity is decreased in cool soils.

3. Soil pH. Acid soil conditions reduce the availability of Ca, Mg, S, K, P, and Mo... and increase the availability of Fe, Mn, B, Cu, and Zn. Nitrogen is most available between pH 6.0 - 7.0.

4. Insects. Don't mistake insect damage for deficiency symptoms. Examine roots, leaves and stems for insect damage that may look like a nutrient deficiency.

5. Diseases. Close study will show the difference between plant disease and nutrient deficiency. A disease can often be detected with a small hand lens.

6. Moisture conditions. Dry soil conditions may create deficiencies. Boron, copper and potash are good examples. This is why crops respond so well to such nutrients when they are available in dry periods. Drought slows movement of nutrients to roots.

7. <u>Soil salinity problems</u>. Soluble salts and alkali are problems in some areas, particularly with low rainfall. These conditions may occur in just part of the field--usually where a high water table exists or where poor quality water has been used for irrigation.

8. <u>Weed identification</u>. Herbicides and mechanical controls are more important today than ever before. Weeds rob crop plants of water, air, light, and nutrients. Some weeds may even release substances that inhibit crop growth. Learn to identify weeds and to know the herbicide materials used to control them.

9. <u>Herbicide damage</u>. Under certain conditions plants may suffer from carryover herbicides from previous crops or those applied the current year. Learn the symptoms.

10. <u>Tillage practices</u>. Some soils develop hard pans and require deep plowing. This calls for more phosphorus and potassium to build up fertility. In conservation tillage much of the fertilizer is broadcast and is on or near the surface. Here more P and K may be needed to build fertility. Also in some cases it is desirable to know the fertility level of the subsoil.

11. <u>Plant spacing</u>. Row width, spacing of plants in the row and number of plants per hectare are important in yields.

12. <u>Water management</u>. Adequate drainage, either surface or tile, is the key. With irrigation, time and amount of watering are of prime importance in good crop growth. Learn what the irrigation program has been. In regions of limited rainfall, soil management should be planned to retain all rainfall on the field, for soil storage.

13. <u>Date of planting</u>. This will affect crop appearance and optimum growth. Get the information.

14. <u>Fertilizer placement</u>. Under many conditions a small amount of fertilizer near the roots is important for a fast start. The fertilizer may have been broadcast on the soil or placed too deep for prompt uptake by roots.

Importance of Cultural Practices

Knowing what has been done in a field before you go into that field can be one of the most important diagnostic techniques you will develop.

Get the facts on field cropping histories; on planting dates; on seeding rates; on varieties; on row widths; on tillage practices; on depth and method of planting; on past fertilizer and liming practices; on past weather conditions, if you can.

Remember, the more you know about a field before you go into it, the better you may diagnose its problems.

Get the facts systematically. And record them! A checklist will keep you from forgetting key information.

General Features of Symptoms

The chapters on individual crops that follow will be of great value in identifying the symptoms of plant nutrient deficiencies for each crop. While symptoms differ for different crops, there are some general clues for nutrient deficiencies and some specific symptoms for the individual element deficiencies that can serve as a guide for all crops (See Figure 1.1). A nutrient should be suspected when these conditions occur:

1. Very poor growth at seedling stage;
2. Plants badly stunted in early growth;
3. Root growth restricted or abnormal;
4. Internal discolorations or abnormalities;
5. Matures too soon or too late;
6. Difference in growth from adjacent crops, even without leaf symptoms;
7. Poor quality crops -- appearance, taste, firmness, moisture content;
8. Specific leaf symptoms that may appear at different times during growth.

GENERAL KEY TO DEFICIENCY SYMPTOMS

Nitrogen (N). Plants are light green and growth is stunted. The lower leaves may be affected first, but other leaves follow, with later yellowing and drying up or firing and final shedding of the lower leaves.

In tree crops, the leaves are often small, pale in color and may be abnormal on any part or all of the plant.

Phosphorus (P). Plants are often small and growth is stunted, but in many crops the leaves are darker green than normal. The leaves and sometimes the stems may develop a reddish-purplish cast especially during early stages of growth.

Maturity is delayed and fibrous root development restricted. Petioles, leaves, and leaf margins may take an upward direction. Frequently the only symptoms may be smaller plants.

Potassium (K). Scorching or firing along leaf margins is the most common symptom. This usually appears first on the older leaves. Plants grow slowly, have poorly developed root systems. The stalks are weak and lodging is common. Seed and fruit are small and shriveled. Plants possess low resistance to disease.

In the case of legume crops (beans, peas, alfalfa, clover, etc.) the first signs of potassium deficiency are small white spots or yellowish dots around the outer edges of the leaves. Later the edges turn yellow and die.

Calcium (Ca). Calcium deficiencies are not often seen in the field because secondary deficiency effects associated with high soil acidity limit growth first.

Leaves may be cup-shaped and crinkled, and the terminal buds deteriorate with some breakdown of petioles (leaf stems).

Fruits may break down at the blossom end. Calcium deficiency is known to be associated with "blossom-end rot" in tomatoes and other crops.

Magnesium (Mg). Magnesium deficiency symptoms appear first on lower (older) leaves. They appear first as a light, yellowish, faded discoloration with the veins remaining green. In crops such as corn (maize) the leaves are yellowish or very light-green-striped while veins remain green. In some crops, as the deficiency progresses, a reddish-purplish color develops with green veins.

The pattern is distinct and characteristic and can usually be identified after some experience in observation.

Sulphur (S). Plants are pale green, and look very much like nitrogen deficient plants. The symptoms generally appear first on the upper leaves while nitrogen starved plants generally show up first on the lower leaves; however, in sulfur deficiency the entire plant can take on a pale green appearance.

Leaves tend to shrivel as the deficiency progresses and plant stems grow thin and woody. Sulfur deficiency occurs most often on sandy soils low in organic matter and in areas of moderate to heavy rainfall. It occurs early in the season and the symptoms may disappear as roots penetrate the subsoil and into areas of higher sulfur content.

Corn: Effect of boron deficiency on ears. (Potash and Phosphate Institute)

NPK   −N   −P   −K

Citrus: Nitrogen-deficient trees show leaf chlorosis with small puckered leaves. Phosphorus-deficient citrus usually shows no specific symptoms. (Potash and Phosphate Institute)

Stylosanthes capitata: On acid soils and soils low in calcium, this deficiency somewhat resembles potassium deficiency. Leaf firing and dieback are evident. (Jose G. Salinas)

Cassava: Interveinal chlorosis of the lower leaves, commencing at the tips and margins and spreading inward toward the midrib. (R. Howeler)

Wheat: Bronzing and dieback from the tips with some leaf spotting. Copper deficiency is difficult to diagnose. (G. W. Wallingford)

Cassava: A characteristic interveinal chlorosis of the younger leaves. Small white or light yellow chlorotic patches develop between the veins. The leaf nodes become narrow, and the margins curl upward. (R. Howeler)

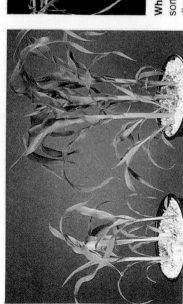

Corn: Sulphur deficiency symptoms. (Sulphur Institute)

MANGANESE STARVED

Soybeans: Manganese deficiency. (Potash and Phosphate Institute)

From left: complete, -S, -P, -N

From left: complete, -Mn, -Mg, -K

Maize: Nitrogen (N) deficiency shows up as a yellowing of the tips of lower leaves, then follows up the midrib. In contrast, potassium (K) deficiency is characterized by a bronze to yellow discoloration along the edges of the older, lower leaves. Phosphorus (P) deficiency causes no obvious leaf symptoms in older plants, but younger plants may show purpling or bronzing of leaves. Magnesium (Mg) deficiency causes a yellow discoloration or striping between the veins, usually first in the lower leaves. Later, the edges become reddish purple. Manganese (Mn) deficiency in maize is very difficult to diagnose in the field, so it must be related to plant tissue tests. Usually only a slight stunting of growth and a slight loss of color between the veins of upper leaves may be seen, but these are not specific symptoms. Sulphur (S) deficiency shows up as a general yellowing of the foliage, sometimes mistaken for nitrogen deficiency. However, usually both upper and lower leaves are affected and there is a pale, almost white, interveinal area in severe cases. (D. Plucknett)

Squash: Potassium deficiency causes marginal chlorosis and sometimes necrosis of the leaves. (Rutgers University)

Corn: Iron deficiency. Note interveinal yellowish chlorosis of the younger leaves. (Potash and Phosphate Institute)

Minor Nutrient Element Deficiencies

Zinc (Zn). Symptoms of zinc deficiency appear first on the younger leaves and other plant parts. Some plants are much more likely to show the symptoms, and they have been defined on these plants. In corn (maize), the deficiency is called "white bud" because the young bud may turn white or light yellow while the leaves show bleached bands.

Other symptoms include bronzing of rice; rosette of pecans; "little leaf" of fruit trees; brown spots with yellowing leaf tissues in legumes; and small, pointed, yellow mottled leaves in citrus.

Manganese (Mn). Symptoms first appear in younger leaves, with yellowing between the veins--and sometimes brownish-black specks.

The deficiency is sometimes confused with magnesium deficiency; however, it usually appears first on the newer (upper) leaves while magnesium occurs on older or all leaves. Best way to distinguish is to check the soil properties. Manganese is more likely if the soil pH is above the neutral point of 7 and in soils higher in organic matter during cool spring months when soils are waterlogged. Liming history is important, but pH values will indicate any excess use of lime.

Iron (Fe). Iron deficiency shows up as a very light pale leaf color with veins remaining green, usually first appearing on younger leaves; but severe deficiency may result in the entire plant showing such symptoms.

It can easily be mistaken for manganese and also occurs on high pH (alkaline) soils. The interveinal chlorosis in iron deficiency is often whiter than that of manganese.

It really takes experience and other tests to distinguish between these two.

Copper (Cu). Organic soils are most likely to be copper deficient, since copper is fixed in unavailable forms in these soils.

Common symptoms of copper deficiency include dieback in citrus, and blasting of onions and vegetable crops. Many vegetable crops show copper hunger with leaves that lose turgor and develop a bluish-green shade before becoming chlorotic and curling. Also the plants may fail to flower, and there is often excessive leaf shedding.

Boron (B). Boron deficiency generally stunts plant growth--the growing point and the lower leaves are affected first.

In many crops the symptoms of boron deficiency are well defined and quite specific, such as crooked and cracked stem in celery, corky core in apples, black heart

in beets, hollow heart in peanuts, and ringed or banded
leaf petioles in cotton.

Alfalfa is especially susceptible to boron deficiency, which is shown by rosetting, yellop top, and death of
the terminal bud. With experience, this can be distinguished from potash deficiency or leafhopper damage.

<u>Molybdenum (Mo)</u>.   Molybdenum deficiency symptoms
show up as general yellowing and stunting of the plant.
In fact, this deficiency can cause nitrogen deficiency in
legumes because the soil bacteria on legumes must have
molybdenum to help fix nitrogen from the air.

A soil test helps because molybdenum becomes more
available as the soil pH becomes more alkaline. So liming may often correct the deficiency.

Molybdenum is not an easy deficiency to identify
just from visual symptoms without a soil or plant test
and a history of field treatment.

2
Soil Testing as a Guide to Productive Crop Yields

J. W. Fitts

The soil in which crop plants grow must provide all of the mineral nutrients that each crop requires for satisfactory yields. There is widespread hidden hunger due to nutrient deficiencies that seriously depress yields of crops and the vigor of forage plants on pastures and rangeland. Soil testing is the most effective means of coping with hidden hunger by identifying it before the crop is planted, and undertaking corrective measures by application of suitable fertilizers and soil amendments prior to planting.

The soil is also the reservoir for water that the crop absorbs to sustain growth. The structure of the soil also must be permeable for root penetration and occupation, and to permit soil aeration sufficient to favor root functioning. The present state of soil fertility, as it relates to its nutrient supplying power for crops, is crucial to high crop yields. Modern technology has made soil testing an indispensable method of rapidly and effectively measuring the status of mineral nutrient supply and thereby providing information on the most profitable use of fertilizers and soil amendments. By this means, the maximum crop production is made possible to utilize available soil moisture and desirable soil aeration on each land area.

ESSENTIAL MINERAL NUTRIENTS

All crops, pasture and range plants require mineral nutrients for satisfactory growth. The major nutrients (and their chemical symbols) are:

Nitrogen (N)
Phosphorus (P)
Potassium (K)

Magnesium (Mg)
Sulfur (S)

---

[1]Dr. J.W. Fitts is President of Agro-Services International, Inc., 215 East Michigan Ave., Orange City, Florida 32763, USA.

The _minor nutrients_, essential for plant growth, but in very small amounts, are:

Boron (B)	Manganese (Mn)
Copper (Cu)	Molybdenum (Mo)
Iron (Fe)	Zinc (Zn)

It should be noted that these mineral nutrients are assimilated exclusively by the roots of all plants. Also, it is important to know that a serious deficiency in supply of any single nutrient element will prevent plant response to any abundance for all the other elements. These deficiencies must be identified and corrected to permit vigorous plant growth. Such corrections must be suited to the needs of each type of crop or forage plant.

RAPID SOIL TESTING

Rapid soil testing to determine the present supply of these essential nutrients in representative soil samples from designated crop fields, orchards, pastures, and other grazing lands, is a powerful tool available to farmers and herders. The soil testing laboratory may be planned to test many samples per day, at relatively low cost. Trained technicians will use appropriate methods for making liquid extracts from the soil samples, and expressing the contents of each mineral nutrient so that any deficiencies may be noted, to guide in the economic use of fertilizers and soil amendments.

Soil tests are also highly useful for determining whether soils are acid or alkaline, and the strength of this characteristic. Soil pH (soil reaction) is vitally important in determining the kind of crop best suited to a soil, as well as the application of soil amendments to meet specific crop needs, such as application of agricultural lime to correct excessive soil acidity.

Soil tests also should reveal any toxicities, such as excessive soluble aluminum (Al), manganese (Mn), or iron (Fe), that are harmful to many important crops. Such tests also may reveal excessive salinity due mainly to excessive sodium (Na) in some low-rainfall areas or irrigated soils with imperfect internal drainage. Since various crops have different tolerances for such unfavorable conditions, the soil test results should be useful in selecting the crop species most likely to yield well in spite of the unfavorable soil conditions.

SOIL TESTING IN RELATION TO EFFECTIVE CULTURAL PRACTICES

It should be noted that soil test results may be used to guide other cultural practices to obtain the highest crop yields or forage growth. These practices

may include the following: (i) the particular improved
variety of the crop selected, capable of responding to
soil treatment; (ii) the best season of planting to utilize rainfall and sunshine; (iii) the plant population (of
tilled crops) to respond fully to fertilizers and soil
amendments; (iv) pest control to minimize weed competition, insect pest damage, plant diseases, nematode damage, and even bird damage; and (v) land management to
control rainfall run-off losses and soil erosion.

Fertilization and use of soil amendments to correct
nutrient deficiencies of soils, as revealed by appropriate soil tests, should not be considered as corrections
for unwise management practices. Instead, correction of
nutrient deficiencies is the means of obtaining the highest yields that are economically possible for each field
or pasture, under the prevailing conditions of climate
and local hazards, and for those soil conditions that
cannot be readily changed.

The supplemental information (noted later in this
chapter) is of great importance in making recommendations
based on the results of soil tests. The proposed use of
the field sampled is an essential consideration in making
recommendations to the farmer or herder. When soil samples on specific fields are submitted well in advance of
planting dates, soil treatments may be made in time to
affect crop vigor.

Even when the crop has already been planted, and is
lacking in vigor, soil test information may be useful in
stimulating the crop by making sidedressing of necessary
fertilizers or soil amendments. If the soil test indicates that other factors than nutrient deficiencies are
stunting the crop, careful examination is needed to determine whether insect pests or diseases, weeds, drought,
poor soil drainage, or other unfavorable conditions are
responsible.

SOIL IMPROVEMENT

Yearly sampling of arable soils to determine their
nutrient supplying power is an excellent method of measuring progressive soil improvement under a consistent
management system. It is rare indeed for soils to remain
at high levels of productivity -- in terms of nutrient
supply, status of soil acidity or alkalinity, soil organic matter, and so on -- under sustained cropping without
suitable fertilizers or soil amendments. The effectiveness of the land management system being followed may be
progressively improved, using regular soil testing as a
guide. The more productive agricultural regions of the
world make use of long-range management systems to identify and correct any deficiencies in mineral nutrient
supplying power of soils.

SOIL REACTION

The acidity or alkalinity of soils is quite important, both to suit the requirements of different crop or pasture plants, and in affecting the availability for plant uptake of nutrient elements added in fertilizers or soil amendments. The best measure of soil reaction is by the pH scale. This is shown in the following chart (Figure 2.1).

Effects of Soil Reaction on Nutrient Availability to Plants

The current level of soil acidity or alkalinity has a substantial effect on the availability to plants of the several mineral nutrients. Figure 2.2 indicates this effect.

The width of the bar for each nutrient shows the relative availability of each element at different soil reactions.

In general, the six elements required by plants in greatest amounts (nitrogen, phosphorus, potassium, sulfur, calcium and magnesium) have the highest degree of availability in the range from slightly acid to neutral to slightly alkaline.

The trace elements that are needed in very small amounts (iron, manganese, boron, copper and zinc) become excessively soluble in strongly acid soils and quite toxic to the plant roots. Correction of excess soil acidity is necessary to avoid such toxicity. Molybdenum increases in availability as acidity is reduced.

SAMPLING FIELD SOILS

The modern soil testing laboratory will require very small amounts of soil in each sample, but the sample must be representative of entire fields. Where the field to be sampled includes both sloping and level areas, areas of different soil color, or of texture (sandy, loamy, or clayey), it is essential that separate sampling be made for distinct areas. To mingle samples from different kinds of soils will not reveal the true nutrient status of any area.

When to Sample

The best time is well in advance of the next planting season for annual crops. For grazing lands, the sampling may just precede the season of normal growth of forage plants.

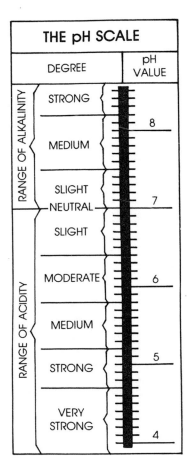

Figure 2.1. The pH scale showing the ranges of soil acidity and alkalinity at which plants grow (Courtesy, National Lime Association).

HOW SOIL pH AFFECTS AVAILABILITY OF PLANT NUTRIENTS

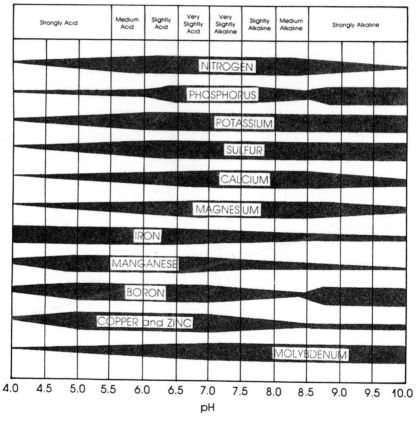

Figure 2.2. Effects of soil reaction on availability of soil nutrients to plants. The width of the bar indicates the relative availability of each element with a change in soil reaction (pH) (Courtesy, National Plant Food Institute).

Method of Sampling

The goal is to take 10 to 20 small slices of soil to a depth of 15 cm, using a spade, trowel or auger. Scrape off any surface cover of plant material, and cut a thin soil slice to the desired depth. (Note the presence of any impervious soil layer in this sample, or in the subsoil.)

1. Place these soil slices in a wooden bucket, or plastic, or enameled container. Do not use a metal container, since the trace of metal will contaminate the soil and invalidate testing for the minor essential nutrients.

2. Mix the slices from 10 to 20 locations, very thoroughly, place on wooden surface to air-dry if the soil is very moist. Do not dry with heat, since this affects measures of nutrient availability.

3. After remixing to get a representative sample, place about 500 gm of this soil in a plastic container and close the container to prevent soil loss or contamination. Mark clearly the field from which the sample was taken, the date, and the name and location of the farmer or herder.

4. Provide the additional information that will provide the laboratory technician with a guide to the test(s) desired, and indicate to the soils specialist those recommendations sought to correct any nutrient deficiencies disclosed by the test. The information should include: (i) the previous crop grown, and approximate yield; (ii) the crop species to be grown; (iii) any recent applications of agricultural lime; (iv) any animal manures added in growing the previous crop; (v) normal depth of tillage in preparation for planting; (vi) nature of soil; color, texture, impervious layers, etc; (vii) soil drainage; (viii) what crop yield levels are sought; (ix) will irrigation be practiced; (x) what unhealthy plant symptoms were observed in the previous crop; and (xi) what method of fertilization is planned: broadcast, followed by tillage; band placement near row; or sidedressing of growing crop.

GUIDANCE ON EFFECTIVE USE OF FERTILIZERS AND SOIL AMENDMENTS

The value of soil tests as an aid to satisfactory crop yields is greatest when the farmer or herder has ready access to a competent soil testing laboratory that can make the evaluations promptly and provide the results without delay to the man in the field. Since a competent laboratory can analyze a large number of samples per day, the timeliness of soil testing should be determined by field sampling well in advance of the planting season, and a prompt report by the soils specialist as to recommendations for the farmer.

Where excessive soil acidity is to be corrected by use of agricultural lime, the recommendations should specify the amounts to be applied, the method of incorporation in the soil, and the time that must elapse for correction of acidity.

The farmer should have separate guidance on proper use of fertilizers. It should be noted that laboratory tests for nitrogen needs are only of general value. More emphasis must be placed on the inclusion of leguminous crops in the rotation, and the application of animal manures. In general, recommendations of nitrogen fertilizers should specify amounts required to achieve the desired crop yields, in the current season.

FIELD EVALUATION OF SOIL TEST DATA

Although most soil testing procedures have been thoroughly evaluated, each of these has been specifically designed to be used under a particular set of conditions. Thus, a different extracting solution is used on alkaline soils as a better indicator of the nutrients that plant roots can take up from such soils. Other forms of extracting solutions are found superior on moderately acid or neutral soils, or on soils with unusual properties.

Since there are a good many major kinds of soils in agricultural areas of the world, in a wide range of climates, producing many kinds of crops, the soil testing practices for each region must be calibrated to suit the conditions of that region. The soil test data are usually recorded as parts per million (ppm) in the extracted solution for the particular nutrient element being measured. These values are usually expressed as high, medium, low, or very low. However, such qualitative terms should be proven by actual field trials on local or regional soils, using typical crops of that region or ecological zone. The field trials should include measurements of crop yields or pasture production without fertilization, and with several levels of fertilizer applications. When such field trials are compared with the soil test data, the ppm test values may be reliably assigned such qualitative values as high, medium, or low, for each of the mineral nutrients that are generally important in the region.

This calibration is not immediately available for a soil testing laboratory operating in a new region. However, such a new laboratory is justified in borrowing evaluation standards from an established laboratory in a similar region elsewhere, until such time as local calibration is made. When local field trials have been made to ensure that the meaning of soil test information is known, very wide application of soil test results may be made with confidence. Soil testing to determine the soil's supplying power for essential plant nutrients is firmly science-based, but the interpretation is largely dependent on the professional judgment of the crop and soils specialists who are familiar with the geographic region in which agriculture is practiced.

SUMMARY

Successful, profitable agricultural production in any region may be greatly supported by effective soil testing, combined with appropriate use of fertilizers and soil amendments to make full use of soils, climate, and the farmer's managerial skill in growing crops and forage plants.

3
Plant and Tissue Testing Principles
Relating to the Identification
of Nutrient Deficiencies

Werner L. Nelson

Plant analysis for mineral nutrient elements is based on the concept that the concentration of an essential nutrient element in a plant, or part of the plant, is an indication of the supply of that nutrient. As such it is directly related to the quantity in the soil that is available to the plant. A second aspect is that, to a certain point, as the percentage content of a nutrient in the plant increases, yield increases.

Plant analysis has been used as a diagnostic tool for many years, and there is now a renewed interest and activity in such methods. There are several reasons for the renewed interest; greatly improved techniques for making plant analysis, increased amount of calibration information, and greater demand from farmers and their advisors on nutrient supplies needed for high yields.

Plant analysis refers to the quantitative analysis for the total amount of essential elements in plant tissue. It should be distinguished from rapid plant tissue tests, which will be discussed later.

There are many diagnostic tools to use. In particular, plant analysis and soil testing go hand in hand.

REASONS FOR USING PLANT ANALYSES

1. <u>To diagnose or confirm diagnoses of visible symptoms</u>. Symptoms are often difficult to identify because a number of factors may cause symptoms that may look alike. Often analyses are used to compare normal and abnormal plants.

---

[1]Senior Vice President, Potash and Phosphate Institute, 402 Northwestern Avenue, West Lafayette, Indiana 47906, USA.

2. To identify "hidden hunger". Sometim...
    be suffering from a deficiency but show...
    A plant analysis looks beyond the app...
    crop.

3. To indicate if applied nutrients entered...
    If no response was obtained to applied nu...
    might be concluded that the nutrients
    lacking.

4. To indicate interactions or antagonism among...
    nutrients. Sometimes the addition of one...
    will affect the amount of another taken up...
    plant. For example, zinc (Zn) uptake may be...
    with high application rates of phosphoru...
    fertilizers.

5. To study trends during the year or over the y...
    Periodic sampling during the season may help to...
    termine if a nutrient is becoming deficient. Sa...
    ling a crop over the years monitors trends in b...
    level of fertility in the soil. Most of the mo...
    common nutrient deficiencies are the result of lon...
    term improper lime and fertilizer practices. Plan...
    nutrient deficiencies or excesses can be detecte...
    before they appear as visual symptoms or reduce...
    yields and quality.

6. To suggest additional tests or studies to identify
    problems in a field. Analyses can be useful in lo-
    cating areas for fertility trials on farm lands.

Comparing Crop Yields with Nutrient Supplies in the Soil

Of prime importance is how the plant analyses relate
to available nutrient levels in the soil and/or to mineral nutrients applied. Studies involving rate of fertilizers applied at various levels to the soil are essential
in defining the relationship of nutrient supply to crop
yields. While many such studies have been conducted,
yield levels have often been relatively low in earlier
studies. There is an urgent need for carefully controlled experiments at modern yield levels, and up to the
maximum yield for a given soil and environment.

Yield level may have a distinct effect on what is
considered to be an adequate level of a nutrient in the
plant. A prime example is potassium (K) in alfalfa. A
number of years ago 1.25% K found in plant tissues was
considered adequate for good yields. As yields moved up,
the value was changed to 1.5, 1.75, 2.0 and 2.5%. Now,
some think 3.0% or more is needed in plant tissues for
sustained high yields and quality, and to maintain the
alfalfa stand. Examples might be cited for other crops
such as maize and soybeans.

At a low yield level, factors other than plant nutrients may be limiting yields and the content of a
nutrient in the plant may mean little.

Ranges in concentration. Ranges have been developed
giving the ranges for deficiency, low, sufficiency or adequate, high and toxic for many crops. This concept is
usually more useful than the critical level. In Georgia
(USA), the concentration of each element is reported as
less than, greater than, or within the sufficiency range
(Plank, 1979). In using plant analysis as a diagnostic
tool, an excessively high amount of an element may be as
important as a deficiency.

FACTORS AFFECTING NUTRIENT CONCENTRATIONS IN PLANTS

The soil test level (see Chapter 2) and amounts of
nutrients added in fertilizers, manures, crop residues
and soil amendments are key factors. However, the concentration of a nutrient within a plant is the integrated
value of all the factors that interacted to affect plant
growth. When one considers the multiplicity of factors
that influence growth and resulting crop yields, it is
surprising that plant analysis relationships hold as well
as they do. Some of these factors are discussed in the
following paragraphs.

Soil moisture. This is a key factor. With low soil
moisture it is more difficult for plants to absorb nutrients and the content of a nutrient will be lower.

TABLE 3.1
Influence of applied N, P and K* and moisture stress on
percent N, P and K in maize leaves (Voss, 1970)

Nutrients Applied (kg/ha)			NPK Concentration	
			No. of stress days	Maximum stress
N	P	K		
			%N	
0	78	47	2.0	1.5
179	78	47	2.9	2.2
			%P	
179	0	47	0.26	0.12
179	78	47	0.32	0.18
			%K	
179	39	0	1.1	0.7
179	39	93	1.6	1.2

* NPK - nitrogen, phosphorus, potassium.

This is illustrated in Table 3.1. With moisture stress NPK in the maize plant was reduced. With application of these nutrients, more NPK moved into the plant, but concentrations in stress periods were still below the optimum. It is important to add enough nutrients to help insure against seasonal variations in need.

Temperature. Low temperature reduces the uptake of a number of elements including N, P, K, zinc (Zn), sulfur (S), and magnesium (Mg). In cooler areas it has been found that a higher soil test level, or rate of applied nutrients, must be used to achieve plant nutrient concentrations comparable to those found in warmer climates. Under cooler conditions root growth is slower and plant uptake processes are slowed. Also, release of such elements as S, P and N from organic matter through mineralization in the soil is slowed by lower temperature.

Soil pH. This influences the availability of many nutrients. For example, a higher pH tends to reduce iron (Fe), aluminum (Al), zinc (Zn), manganese (Mn), and boron (B) and increases molybdenum (Mo) in plants. A lower pH (greater acidity) makes it more difficult for the plant to absorb Mg and P, but more Mn and Al are absorbed from acid soils.

Tillage and placement. Conservation tillage practices may cause a reduced uptake of nutrients, particularly K, in drier, cooler areas. An example from a medium level of available soil potassium in Wisconsin is shown in Table 3.2. The K concentration in the maize ear leaf from the unplowed area was lower than that from the plowed, with or without potassium (K). This is in part due to position availability of the fertilizer in the soil, because much of the fertilizer is broadcast and remains on or near the surface. Band placement of nutrients near the seed or deeper in the soil improves uptake by the plant.

TABLE 3.2
Effect of tillage and added potassium (K) on maize ear leaf (Schulte, 1979)

K applied Annually 1972 - 1976	% K in ear leaf of corn on:	
	Plowed land	Unplowed land
(kilograms per hectare)		
0	0.73	0.59
66	1.40	1.04
133	1.71	1.42

Placement. Side band placement is effective for increasing uptake of plant nutrients in early growth stages, and may influence yield.

Hybrid or variety. The yield of a crop is the result of its inherited capability, and the environment. Hybrids or varieties vary greatly in their yield capability. For example, in an experiment in 1980 in New Jersey one maize hybrid yielded 17,500 kg per hectare and another 12,700 kg per hectare under exactly the same environment. Obviously the total nutrient uptake was much different for the two hybrids.

Also, varieties or hybrids may vary widely in concentration of nutrients, and the greater the inherent capacity, the greater the yield. As yet, however, little information is available on critical or sufficiency levels of nutrients for specific varieties and hybrids. Management practices appear to be of greater significance in controlling nutrient concentrations than hybrid or variety (Munson, 1969).

Interactions. High concentrations of one element may cause imbalances or deficiencies of other elements. The relationship between P and Zn is one example. A high amount of phosphorus (P) may reduce the amount of zinc (Zn) in a plant. Also, under a marginal magnesium (Mg) supply in the soil, a high application of potassium (K) may reduce Mg to the deficient point in the plant. The concentration of K in the plant may be reduced also by a high rate of ammonium nitrogen (NH_4-N).

Stage of growth. The concentration of an element considered to be adequate changes as the plant grows and matures (Table 3.3). Hence, it is important that plants be sampled at comparable and recognizable growth stages.

TABLE 3.3
Effect of stage of growth of maize on sufficiency ranges for N, P, and K (Plank, 1979)

| | Plant Tissue Content | | |
| | Sufficiency Range % | | |
	Nitrogen (N)	Phosphorus (P)	Potassium (K)
Whole plants, less than 30 cm tall	3.5-5.0	0.3 -0.5	2.5 -4.0
Leaf below the whorl, before tasseling	3.0-3.5	0.25-0.45	2.0 -2.5
Ear leaf at tasseling before silks turn brown	2.75-3.2	0.25-0.45	1.75-2.25

TAKING PLANT SAMPLES FOR ANALYSES OF NUTRIENT CONTENT

1. Mailing kit. Many laboratories have plant analysis mailing kits. Instructions for sampling and submitting samples should be followed specifically.

2. <u>What to sample</u>. The sampling procedures for collecting leaf and plant tissue for analysis is shown in Table 3.4. When no specific sampling instructions are given for a particular crop, the general rule of thumb is to sample the upper, recently matured leaves. The recommended stage of growth to sample is just prior to the beginning of the reproductive stage for many plants. Roots from total plant sample should be removed. It is well to take a soil sample from the same area at the same time.

3. <u>Comparison samples</u>. Where a deficiency is suspected, take samples from normal plants in an adjacent area as well as from the affected plants. Take a soil sample from each area also.

4. <u>Washing to remove contaminants</u>. Dusty plants should be avoided but if dust is present, brushing or wiping with a clean damp cloth may be sufficient. If not, rinse briefly in running water while the material is still fresh.

5. <u>What not to sample</u>.

    - Diseased or dead plant material, damaged by insects, or mechanically injured.

    - Plants stressed severely by cold, heat, moisture deficiency or excess.

    - If roots are damaged by nematodes, insects or diseases.

6. <u>The questionnaire</u>. This is the means of communication between the sample and the laboratory. Completion of the questionnaire is important if the interpreter is to evaluate properly the analysis and make a recommendation.

7. <u>Packaging the plant tissue</u>. Partially air dry and put in a clean <u>paper bag</u> or envelope. Do <u>not</u> put in polyethylene bags or tightly sealed containers, since this permits molding.

POSSIBILITIES FOR ERRORS

It is well to keep in mind that plant analyses are not foolproof. There are many possibilities for errors: (i) collecting the sample--plant part, stage of growth, environment; (ii) analysis in laboratory; (iii) interpretation; (iv) "sufficiency ranges" may not apply to modern yield goals; (v) improperly completed questionnaire; (vi) very acid or alkaline soil so that plant growth is adversely affected; (vii) amounts of other nutrients: a defi-

TABLE 3.4
Sampling procedures for collecting leaf and plant tissue
for a plant analysis (Plank, 1979)

Stage of Growth	Plant Part to Sample	Number of Plants to Sample
FIELD CROPS		
Maize*		
(1) Seedling stage (less than 30 cm)	All the above ground portion.	20-30
(2) Prior to tasseling	The entire leaf blade fully developed below the whorl.	15-25
or		
(3) From tasseling and shooting to silking	The entire leaf blade at the ear node (or immediately above or below it).	15-25

* Sampling after silking occurs is not recommended.

Soybeans or Other Beans*		
(1) Seedling stage (less than 30 cm)	All the above ground portion.	20-30
or		
(2) Prior to or during initial flowering	The leaflets on 2 or 3 fully developed leaves at the top of the plant.	20-30

* Sampling after pods begin to set not recommended.

Small Grains (including rice)*		
(1) Seedling stage (less than 30 cm)	All the above ground portion.	50-100
or		
(2) Prior to heading	The 4 uppermost leaf blades.	

* Sampling after heading not recommended.

(continued)

Stage of Growth	Plant Part to Sample	Number of Plants to Sample
Hay Pasture, or Forage Grasses		
Prior to seed head emergence or at the optimum stage for best quality forage	The 4 uppermost leaf blades	40-50
Alfalfa		
Prior to or at 1/10 bloom stage	Mature leaf blades taken about 1/3 of the way down the plant.	40-50
Clover and Other Legumes		
Prior to bloom	Mature leaf blades taken about 1/3 of the way down from top of the plant.	40-50
Cotton		
Prior to or at first bloom or when first squares appear	Youngest fully mature leaf blades on main stem.	30-40
Tobacco		
Before bloom	Uppermost fully developed leaf blade.	8-12
Sorghum		
Prior to or at heading	2nd leaf blades from top of plant.	15-25
Peanuts (Groundnuts)		
Prior to or at bloom stage	Leaflets of mature leaves from both the main stem and either cotyledon lateral branch.	40-50
VEGETABLE CROPS		
Potato		
Prior to or during early bloom	3rd to 6th leaf blade from growing tip.	20-30

Stage of Growth	Plant Part to Sample	Number of Plants to Sample
Head crops (cabbage, etc.)		
Prior to heading	First mature leaves from center of whorl, without stem.	10-20
Tomato (field)		
Prior to or during early bloom stage	3rd or 4th leaf from growing tip, without stem.	20-25
Tomato (greenhouse)		
Prior to or during fruit set	(1) Young plants: leaves adjacent to 2nd and 3rd clusters.	20-25
	(2) Older plants: leaves from 4th to 6th clusters.	20-25
Beans		
(1) Seedling stage less than 30 cm)	All the above ground portion.	20-30
(2) Prior to or during initial flowering	Leaflets from 2 or 3 fully developed leaves at the top of the plant.	20-30
Root Crops (carrots, onions, beets, etc.)		
Prior to root or bulb enlargement	Center mature leaves.	20-30
Leaf Crops (lettuce, spinach, etc.)		
Mid growth	Youngest mature leaf.	35-55
Peas		
Prior to or during initial flowering	Leaves from the 3rd node down from the top of the plant.	30-60
Sweet Corn		
(1) Prior to tasseling	The entire fully mature leaf blade below the whorl.	

(continued)

Stage of Growth	Plant Part to Sample	Number of Plants to Sample
(2) At tasseling	The entire leaf at the ear node.	20-30

Melons (water, cucumber, muskmelon)

Early stages of growth prior to fruit set	Mature leaf blades near the base portion of plant on main stem.	20-30

FRUITS AND NUTS

Apple, Apricot, Almond, Prune, Peach, Pear, Cherry

Mid season	Leaf blade near base of current year's growth or from spurs.	50-100

Strawberry

Mid season	Youngest fully expanded mature leaves, without petioles.	50-75

Pecan

6 to 8 weeks after bloom	Leaflets from terminal shoots, (taking the pairs from the middle of the leaf).	30-45

Walnut

6 to 8 weeks after bloom	Middle leaflet pairs from mature shoots.	30-35

Grapes

End of bloom period	Petioles from leaves adjacent to fruit clusters.	60-100

Raspberry

Mid season	Youngest mature leaves on laterals or "primo" canes (without leaf petioles).	20-40

Stage of Growth	Plant Part to Sample	Number of Plants to Sample
ORNAMENTALS AND FLOWERS		
Ornamental Trees		
Current year's growth	Fully developed leaves (without petioles).	30-100
Ornamental Shrubs		
Current year's growth	Fully developed leaves.	30-100
Turf		
During normal growing season	Leaf blades. Clip by hand to avoid contamination with soil or other material.	½ liter of material
Roses		
During flower production	Upper leaves on the flowering stem, without petioles.	20-30
Chrysanthemums		
Prior to or at flowering	Upper leaves on flowering stem.	20-30
Carnations		
(1) Unpinched plants	4th or 5th leaf pairs from base of plant.	20-30
(2) Pinched plants	5th or 6th leaf pairs from top of primary laterals.	20-30
Poinsettias		
Prior to or at flowering	Most recently mature, fully expanded leaf blades.	15-20

ciency of nutrient A may cause nutrient B to accumulate. However, when nutrient A is supplied in adequate amounts, nutrient B may be deficient; (viii) additions of a nutrient may increase yield but not increase the percentage content in the plant, however, because of a larger plant the total content of the nutrient in the total plant will be greater; (ix) the sufficiency or adequate range at a given location may vary from one year to the next because of climatic conditions; and (x) environmental factors affect nutrient levels. Factors that reduce concentrations are shown in Table 3.5.

TABLE 3.5
Environmental factors affecting nutrient levels in plant tissue (Sparr, 1981)

Environmental factor	May reduce tissue concentration of:
Acid soil	N, P, K, Ca, Mg, Mo
Alkaline soil	Zn, Mn, B, P, K
Low organic matter in soil	B, Cu, Zn, S
Cold soils	P, Zn, Mn, B
Drought	N, B, Mn
Excess moisture	N, P
Compacted soils	All elements
Disease	Mg, N
High light intensity	B
Low light intensity	Mn
Sampling plants too young	Ca, Mg, Zn, B, Cu
Sampling plants too old	N, P, K
Sampling too high on a plant	Ca, Mg, Zn, B, Cu
Sampling too low on a plant	N, P, K

USE OF PLANT ANALYSIS

With most field crops, if analyses are made as the reproduction stage begins, they are too late to be of much value for corrective applications the current year. Adjustments can be made in lime and fertilizer practices for next year's crop when used in conjunction with soil tests.

However, with N, and particularly under irrigation, plant analyses for N before the reproductive stage may indicate a deficiency of N which might be corrected in the current season for crops such as maize, cotton and tomatoes. Increased use of foliar (leaf) applications, greater effort towards top profit yields, and more irrigation; any or all of these will increase efforts for supplying corrective applications of N fertilizers during growth of the current crop, as well as certain other elements such as potassium (K). Some micronutrients such as manganese (Mn) are readily applied as a foliar (leaf) spray, if found to be deficient by plant analyses.

For many years plant analysis has been used for tree crops such as citrus, peaches, apples, pecans and other nuts and fruits. Because of the perennial nature of these crops and their extensive root systems, plant analysis is especially useful for determining their nutrient status and possibly corrective applications of leaf sprays during the current year.

Tables are available to provide a guide for interpreting plant analyses and to relate the plant analysis results to probable causes for elemental concentrations falling outside the sufficiency range. The tables in University of Georgia Bulletin 735 (Plank, 1979) are an example.

It must be emphasized that interpretations are not valid for crops infested with nematodes, insects, diseases and weeds, damaged by chemicals or mechanically injured. Also, it is difficult to interpret results from crops after a severe drought. Hence, it is important to examine the plants carefully for such problems and to be aware of rainfall records.

DRIS APPROACH (Diagnosis and Recommendation Integrated System)

The critical value and sufficiency range approaches have limitations. An important limitation is that the stage of growth greatly influences the values, and unless the sample for diagnosis is taken at exactly the right time, the critical value approach may become insensitive to the insufficiency to be diagnosed. For example, if the leaf sample is taken earlier than it should have been, the nutrient content may be higher than given critical values when, in fact, the plant has a deficiency. On the other hand, if the sample is taken later than it should have been, nutrients which were not insufficient may be diagnosed as deficient.

The Diagnosis and Recommendation Integrated System (DRIS) has several unique features which help overcome this problem (Sumner, 1979). It is possible to make diagnoses at different ages of the crop and to classify the order in which nutrients limit yield.

In the DRIS approach, indices or norms are available for various crops. The index for a crop in a given growth situation is determined from plant analyses for such elements as N, P, K, Ca, and Mg. The more negative an index value for a nutrient the more deficient that nutrient is relative to the others. Because the indices rank the nutrients in order of limiting importance they automatically incorporate the concept of balance into the system.

The DRIS approach is not widely used as yet but it is an approach which should be kept in mind.

TOTAL NUTRIENT NEEDS

Another use for plant analysis is to determine the amount of nutrients in the total above-ground plant, or in the harvested grain. When these data are given for several yield levels the increasing nutrient needs for higher yields are emphasized. While the fertilizer needs cannot be predicted from removal data, the nutrients used must come from the soil or the fertilizer.

Approximate amounts of certain nutrients in the above ground portion for various crops at three yield levels are shown in Table 3.6. Actual amounts will vary with variety, soil and climate.

There is some emphasis on fertilizing just once in a rotation. For example, in a maize-soybean rotation the P and K may be applied ahead of the maize. This is satisfactory providing adequate quantities are applied to take care of both crops. The high amount of nutrients removed in forage is often emphasized. However, high amounts of nutrients are also removed just in the grain of high yielding crops (below).

Grain Yield (kg/ha)	N	P_2O_5	K_2O	Mg	S
			(kg/ha)		
Maize - 11,200	168	97	64	20	15
Soybeans - 3,600	269*	54	95	19	12
Total	437	151	159	39	27

* Legumes can get most of their N from the air.

Analysis of the total plant at various growth stages throughout the season gives an indication of the total amounts of nutrients needed and when. An example is shown for wheat in Table 3.7.

TISSUE TESTING

A tissue test is the determination of the amount of a plant nutrient in the sap of the plant. This is a semi-quantitative measurement of the unassimilated, soluble contents of the plant sap.

Green plant tissue can be tested in the field, and this helps a person to verify or predict the nutritional situation while still in the field. The results can be read as very low, low, medium or high. The purpose is not to split hairs but assess general levels. Several elements can be tested for: nitrate-N(NO_3), P, K and sometimes Mg, Mn and Fe.

The tissue test is a great help in verifying plant deficiency symptoms and in detecting deficiencies before symptoms appear. A good approach is to compare healthy

TABLE 3.6
Amounts (in kg/ha) of N, P_2O_5*, K_2O*, Mg and S in the above ground portions of four crops (Griffith, 1979)

| | Yields in kilograms per hectare | | | | | | | | Yields in metric tons | | |
| | Maize | | | Soybeans | | | Wheat | | | Alfalfa | | |
	7,000	11,200	14,000	2,400	3,600	4,800	2,400	3,600	6,000	7	14	18
N	155	266	350	216	324	432	67	134	168	255	450	600
P_2O_5	58	144	150	43	64	85	27	54	68	40	80	120
K_2O	165	266	350	95	142	189	81	162	203	200	480	600
Mg	41	65	81	18	27	36	12	24	30	20	40	53
S	21	33	41	17	25	33	10	20	25	20	40	51

*Phosphoric acid and potash, respectively, P_2O_5 x 0.44 = P, K_2O x 0.83 = K.

plants along with poor ones. Quick pH soil tests are helpful in interpreting tests.

TABLE 3.7
Total N, P_2O_5 and K_2O that the grain in a wheat crop of 6,500 kg/ha takes up while it grows (Koehler, 1976)

	Kilograms per hectare, plant nutrient content				
	0-30 days Tillering Stage	31-73 days Boot Stage	74-93 days Milk Stage	94-114 days Mature Stage	Season Total
N	38	70	50	-17	158
P_2O_5	15	49	35	- 6	99
K_2O	20	108	16	-39	144
Total Dry Weight	790	5,570	4,490	1,790	12,640

	Percent of maximum uptake				Straw	Grain
N	24	68	100	89	22	67
P_2O_5	15	65	100	94	14	80
K_2O	14	89	100	73	56	17
Total Dry Weight	6	50	86	100	49	51

The concentration of the nutrients in the cell sap from the green tissue is usually a good indication how well the plant is supplied at the time of testing. The cell sap of conducting tissues might be compared with the conveyor belts in a factory. If the factory is to operate at full capacity, all the belts bringing in raw materials must be running on schedule. If one raw material is short, its belt will run empty, the other raw materials will pile up, and production will be drastically reduced.

An alert factory manager will make sure there are no shortages. Likewise, an alert farmer or advisor will make certain that no nutrient is limiting crop growth and that the supplies are in proper balance.

In using tissue tests we are looking for one element such as N, P, or K that may be limiting crop yields. If one element is very low this may allow others to accumulate in the sap because plant growth has been reduced. If the deficiency had been corrected and the plant had grown vigorously, other elements then may not have been in sufficient supply for top yields.

Materials needed for tissue tests are simple and easy to carry. The kit may include some tissue test papers, pliers, two solutions, a knife and nitrate powder. Plant Check, 4200 Woodville Pike, Urbana, Ohio 43078 or Urbana Laboratories, P.O. Box 399, Urbana, Illinois 61801 can supply kits to test for NPK.

It takes much practice and care in running the tests and interpreting the results. The kits will have instructions. Also, the Potash and Phosphate Institute, 2801 Buford Hwy., N.E., Suite 401, Atlanta, Georgia 30329 has a slide set and script "Field diagnosis and tissue testing," which covers techniques, interpretation, and possibilities for errors.

It is important to consider: (i) The ammonium molybdate solution may deteriorate. Hence, it must be checked frequently according to instructions. (ii) The potash papers deteriorate with time. They should be kept cool and a new supply obtained every three months. (iii) Plants damaged for any reason will give misleading results. (iv) Time of sampling and plant part, as for plant analysis, is important. (v) Because of all these considerations, a training course in making determinations and interpreting is almost essential.

Tissue tests have not been widely used because most people would rather send in a plant sample and a soil sample to a laboratory rather than develop the skill to run and interpret the tests properly. Yet properly used, they fit in nicely with soil tests and plant analysis as an important diagnostic tool.

SUMMARY

Many factors affect uptake of nutrients. Plant analysis or tissue tests give an indication what the plant was able to absorb from the soil or from foliar application. Variations from one field to another or among years on the same field are sometimes difficult to interpret. Hence a key point is continued use of plant analysis to pick up nutrient trends over the years.

Plant analysis and tissue testing must be used in conjunction with other diagnostic tools. For example, soil testing, plant analysis and tissue testing go hand in hand. One is not a substitute for the other. All are useful tools in diagnosis and many good farmers and advisors use both programs. An accurate information sheet on past management practices, careful interpretation and being aware of the possibilities for errors are essential for most effective use of plant analysis and tissue testing.

Demand for plant analysis will continue to increase as farmers strive for higher, more profitable yields. The challenge will be to determine the levels of nutrients needed in the plant for top profit yield, particularly when we consider that top profit yields will keep moving up over the years.

SUPPLEMENTAL READING

GRIFFITH, W.K. 1979. Stepping up the yield ladder increases plant food needs. Better Crops with Plant Food. 63 (Summer): 4-6.

KOEHLER, F.E. 1976. Wheat takes up much plant food. Better Crops with Plant Food. 60 (1): 16-18.

MUNSON, R.D. 1969, Plant analysis: Varietal and other considerations. p. 85-104. In: F. Greer (ed.). Proceedings of Symposium of Plant Analysis. International Minerals & Chemical Corp., Skokie, IL.

PLANK, OWEN C. 1979. Plant Analysis Handbook for Georgia. Bulletin 735, Georgia Extension Service, University of Georgia.

SCHULTE, E.E. 1979. Buildup soil K levels before shifting to minimum tillage. Better Crops with Plant Food 63 (fall): 25-27.

SPARR, M.C. 1981. Using plant analysis. Technical Service Department, Crops Division. Indiana Farm Bureau Coop Assoc. Indianapolis, IN.

SUMNER, M.E. 1979. Interpretation of foliar analyses for diagnostic purpose. Agronomy Journal 71:343-8.

VOSS, R.D. 1970. Proceedings 22 Annual Fertilizer Agricultural Chemical Dealers, 14:1, Iowa State University.

4
RICE

S. K. De Datta

Rice (Oryza sativa) is one of the most important food crops of the world. There are 111 rice-growing countries in the world covering all Asian countries, most countries of West and North Africa, some countries in Central and Fast Africa, most of the South and Central American countries, Australia, a few countries in Europe and four states in the USA. China contributes 36% of the world's rice production on 23% of the world's rice area; India contributes 19% of production on 28% of the area. The other countries of South, Southeast and Fast Asia contribute 35% of world production on 38% of the area; Latin American countries produce 4% of the total on 6% of the area; and African countries south of the Sahara produce less than 2% of the world's total on 3% of the area.

Modern semi-dwarf rices are fertilizer responsive and photoperiod insensitive. They cover about 40% of the rice areas in 11 important rice-growing countries in Asia, excluding China and Japan. In China and Japan, the entire rice-growing areas are planted to modern varieties including 6 million hectares under hybrid rice in China.

Rice provides major caloric intake and an important protein source in Asia. Rice production provides employment to the largest sector of the rural population in most of Asia.

ECOLOGICAL AND PRODUCTION REQUIREMENTS

Rice is a variable crop which can be grown from non-flooded dryland conditions (upland rice) to water depths up to 6 meters (floating rice). Rice grows best and produces most under irrigation with continuous flooding of about 5 cm. More than 50% of the world's rice area is rainfed lowland and another 10% of the area is rainfed upland. In rainfed rice, variability of rainfall is the

---

[1]Head, Department of Agronomy, International Rice Research Institute, P.O. Box 933, Manila, Philippines.

most critical production constraint that limits yield stability.

In the monsoon tropics, solar radiation seldom exceeds 350 cal/m^2 per day during the ripening period in the wet season when the most rice is grown. Temperatures in the tropics are high and constant, allowing year-round rice production. In some temperate countries such as southern Australia and the USA, one rice crop can be harvested, and solar radiation during the ripening period can average 700 cal/cm^2 per day. Most of the year, however, low temperature will limit rice growth and development in many temperate countries. Temperature extremes, both low and high, result in a high percentage of unfilled spikelets and low yields.

Rice is grown from the equator to 50°N and from sea level to 2500 m. Rice will grow under appropriate temperature regimes wherever there is enough water to sustain a crop. Rice is grown on a variety of soils ranging from waterlogged and poorly drained to well drained. The crop is grown on all 10 soil orders, as classified under the modern soil taxonomy. Most lowland rice soils in tropical Asia are considered either Entisols or Inceptisols. Rice is grown on problem soils such as saline, alkaline, and acid sulfate soils and high organic matter soils (Histosols). The productivity of lowland rice soils is heavily dependent on the fertility and chemical nature of the soil. Soil tilth, which is considered highly important for an upland crop, is generally considered unimportant for lowland rice; soil physical properties are considered relatively unimportant as long as sufficient water is available.

MINERAL NUTRITION

It is important to understand the unique properties of flooded soils in order to manage soil, fertilizers, and moisture regimes, and to maximize rice production in a given environment.

There are a number of chemical and electrochemical changes when a soil is flooded; these are: (i) depletion of molecular oxygen; (ii) chemical reduction of the soil, or a decrease in redox potential; (iii) increase in pH of acid soils and decrease in pH of calcareous and sodic soils; (iv) increase in specific conductance; (v) reduction of Fe (III) to Fe (II) and Mn (IV) to Mn (II); (vi) reduction of NO_3^- and NO_2^- to N_2 and N_2O; (vii) reduction of SO_4^{2-} to S^{2-}; (viii) increase in supply and availability of nitrogen, silicon, and molybdenum; (ix) decrease in concentrations of water-soluble zinc and copper; and (x) generation of carbon dioxide, methane, and toxic reduction products such as organic acids and hydrogen sulfide.

The stabilization of the soil pH after submergence has several effects on rice growth: (i) adverse effects

of low or high pH per season are minimized; (ii) excess
aluminum and manganese in acid soils are rendered harm-
less; (iii) iron toxicity in acid soils is lessened; (iv)
availability of phosphorus, molybdenum, and silicon is
increased; (v) mineralization of organic nitrogen is
favored; (vi) organic acids are decomposed; and (vii)
lime is seldom necessary.

Oxygen Transport in Rice

Submergence creates a unique environment for rice
growth and nutrition. There is direct and indirect evidence to conclude that oxygen transport takes place from
shoot to root. The rice plant develops air spaces in the
culm even in upland conditions. The rice plant grown in
flooded soils, however, develops more and larger air
spaces. Air enters the rice plant through the leaf blade
and leaf sheath stomates and moves downward to the nodes
at the plant base. Finally, the air diffuses outward
from the plant roots into the surrounding soil.

Nutritional Requirements

The major elements C, H, O, N, P, K, Ca, Mg, and S
are needed by plants in relatively larger amounts than
the minor elements, Fe, Mn, Cu, Zn, Mo, B, and Cl.
All of these 16 essential elements must be present
in optimum amounts and in forms usable by rice plants. A
rice crop that produced 9.8 t/ha rough rice (with hull)
and 8.3 t/ha of straw had the following mineral concentrations and amounts of nutrients removed (Table 4.1).

DIAGNOSIS OF NUTRIENT DEFICIENCY AND TOXICITY

Nutrient deficiency symptoms in the rice plant are
seen in color of the leaves, stems, and roots, in plant
height and tillering habit, and in the development of
root systems..
The best time to observe deficiency or toxicity
symptoms in rice is in the early stages of symptom development. For example, zinc deficiency in lowland rice
usually appears within 2-3 weeks after transplanting,
after which the crop apparently begins to recover when
the deficiency is moderate. The symptoms of iron toxicity may appear in 1-2 weeks, or not until 1-2 months
after transplanting.

TABLE 4.1
Nutrient removal by a rice crop (variety IR36)[a] yielding
9.8 t/ha rice[b] in a farmer's field in Calsuan, Laguna
Province, Philippines, 1983 dry season

Nutrient Element	Mineral Concentration (%) in		Amount of Mineral Removed at Harvest (kg)		
	Straw	Grain	Straw	Grain	Total
N	0.90	1.46	75	143	218
P	0.06	0.26	5.0	25.5	30.5
K	2.80	0.27	232	26	258
Ca	0.32	0.01	27	1.0	28
Mg	0.16	0.10	13	10	23
S	0.04	0.06	3.3	5.9	9.2
Fe	0.018	0.02	1.7	2.0	3.7
Mn	0.037	0.006	3.1	0.6	3.7
Zn	0.002	0.002	0.2	0.20	0.40
Cu	0.0002	0.0025	0.02	0.24	0.26
B	0.0019	0.0016	0.16	0.16	0.32
Si	10.6	2.1	879	206	1086
Cl	0.65	0.42	54	41	95

[a] 174N-17P-33K kg/ha fertilizer treatment.

[b] straw yield = 8.3 t/ha.

Mineral Deficiency Symptoms

Nitrogen (N). Stunted plants with limited number of
tillers; narrow and short leaves which are erect and become yellowish green as they age (young leaves remain
greener); and old leaves become light-straw-colored and
die.

Phosphorus (P). Stunted plants with limited number
of tillers; narrow and short leaves that are erect and
dirty dark green; and young leaves remain healthier than
older leaves, which turn brown and die.

Potassium (K). Stunted plants, and tillering slightly reduced; short, droopy, and dark green leaves; yellowing at the interveins, on lower leaves, starting from the
tip, and eventually drying to a light brown color: brown
spots sometimes develop on dark green leaves; irregular
necrotic spots may develop on the panicles; long thin panicles form; and some symptoms of wilting when there is
excessive imbalance with nitrogen (low K:N ratios in
plant).

Calcium (Ca). Deficiency causes little change in
general appearance of the plant, except in cases of acute
deficiency; the tips of the upper growing leaves become
white, rolled, and curled; and in an extreme case, the
plant is stunted and the growing point dies.

Magnesium (Mg). With moderate deficiency, height and tiller number are little affected; wavy and droopy leaves due to expansion of the angle between the leaf blade and the leaf sheath; and interveinal chlorosis characterized by an orangish yellow color on lower leaves.

Sulfur (S). Symptoms are similar to those of nitrogen deficiency, which makes it impossible to distinguish between the two deficiencies by visual symptoms alone. Symptoms appear initially on leaf sheaths, which become yellowish, proceeding to the leaf blades, with the whole plant chlorotic at the tillering stage; reduced plant height and tiller number; and fewer panicles, shorter panicles, and reduced number of spikelets per panicle at maturity.

Zinc (Zn). The midribs of the younger leaves, especially the base, become chlorotic; brown blotches and streaks in lower leaves appear, followed by stunted growth, although tillering may continue; reduced size of the leaf blade but with the leaf sheath little affected; and uneven growth and delayed maturity in the field.

Iron (Fe). Entire leaves become chlorotic and then whitish; and the newly emerging leaf becomes chlorotic if iron supply is cut suddenly.

Manganese (Mn). Stunted plants with normal tiller number, and interveinal chlorosis on the leaves.

Boron (B). Reduced plant height; the tips of emerging leaves become white and rolled as in the case of calcium deficiency; and the growing points may die in severe cases, but new tillers continue to be produced.

Copper (Cu). Bluish-green leaves, which become chlorotic near the tips; development of chlorosis downward along both sides of the midrib, followed by dark brown necrosis of the tips; new leaves fail to unroll and maintain a needlelike appearance of the entire leaf, or occasionally of half the leaf, with the basal portion developing normally.

Chlorine (Cl). Deficiency symptoms have not been described in rice.

Silicon (Si). Deficient plants become soft and droopy.

Mineral Toxicity Symptoms

Iron (Fe). Toxicity symptoms are: tiny brown spots on lower leaves, starting from the tips and spreading toward the bases (the spots are generally combined on interveins); leaves usually remain green; and in severe cases, the entire leaf becomes purplish-brown.

Aluminum (Al). Toxicity symptoms are orangish-yellow interveinal chlorosis, which may become necrotic in serious cases.

Manganese (Mn). Toxicity symptoms are stunted plants and limited tillering; brown spots on the veins of the leaf blade and the leaf sheath, especially on the lower leaves.

Boron (B). Toxicity symptoms include chlorosis at the tips of the older leaves, especially along the margins, followed by the appearance of large, dark brown elliptical spots in the affected parts, which ultimately turn brown and dry up.

Salt injury. Salt toxicity in rice is seen as stunted growth, reduced tillering, and whitish leaf tips; frequently some parts of the leaves become chlorotic.

Soil Analysis

The measurement of soil pH is the simplest and most informative analytical technique in the diagnosis of nutrient deficiency or toxicity in rice. The following pH values and their corresponding possible nutritional problems are often cited.

pH*	Possible Nutritional Problem
< 4.0	acid sulfate soil
4.0 - 5.0	Fe toxicity and P deficiency in lowland rice; Al and Mn toxicity and P deficiency in upland rice
6.5 - 8.5	Zn deficiency in lowland rice; Fe deficiency in upland rice
> 8.5	Alkali soil

*< = less than, > = greater than

The best correlation with the response of rice to a given element has been with determination of:

(i) available nitrogen by waterlogged incubation and alkaline permanganese methods;

(ii)   available potassium by exchangeable potassium;

(iii)   available phosphorus by Olsen and Bray P_1 methods;

(iv)   available sulphur extraction by $Ca(H_2PO_4)_2H_2O$;

(v)   available zinc by extraction with buffered chelating agents or weak acids; and

(vi)   available silicon by extraction with sodium acetate.

Critical limits using soil analysis:

Total N %	N need
< 0.1	high
0.1 - 0.2	moderate
> 0.2	low

Available N (ppm)	N need
50 - 100	high
100 - 200	moderate
> 200	low

Available P (Olsen P, ppm)	P need
< 5	high
5 - 10	moderate
> 10	low

Exchangeable K (me/100 g)	K need
> 0.2	low or nil

Available Zn (ppm)	Zn need
> 1.0	low or nil

If pH is >6.8, Zn deficiency is likely to occur, particularly if Zn-inefficient varieties are grown.

Available S Methods Used	Critical S Concentration (ppm)
Calcium phosphate	9
Lithium chloride	25
Ammonium acetate	30
Hydrochloric acid	5

TABLE 4.2
Critical deficiency levels of some micronutrients in rice
soils

Element	Method	Critical Level (ppm)
B	Hot H_2O	0.1 - 0.7
Cu	DTPA + $CaCl_2$ (pH 7.3)	0.2
Fe	DTPA + $CaCl_2$ (pH 7.3) $NH_4C_2H_3O_2$ (pH 4.8)	2.5 -4.5
Mn	DTPA + $CaCl_2$ (pH 7.3) 0.1 $\underline{N}$ H_3PO_4 and 3 $\underline{N}$ $\overline{N}H_4H_2PO_4$	1.0 15 - 20
Mo	$(NH_4)_2(C_2O_4)$ (pH 3.3)	0.04-0.2
Zn	0.05 $\underline{N}$ HCl Dithi$\bar{z}$one + $NH_4C_2H_3O_2$ EDTA + $(NH_4)_2CO_3$ DTPA + $CaCl_2$ (pH 7.3)	1.5 0.3 - 2.2 1.5 0.5 - 0.8

TABLE 4.3
Deficiency and toxicity critical concentrations of
various elements in the rice plant[a]

Element	(D)ef. or (T)ox.	Critical concentration	Plant part analyzed	Growth stage[b]
N	D	2.5%	Leaf blade	Til
P	D	0.1%	Leaf blade	Til
K	D	1.0%	Straw	Mat
	D	1.0%	Leaf blade	Til
Ca	D	0.15%	Straw	Mat
Mg	D	0.10%	Straw	Mat
S	D	0.10%	Straw	Mat
Si	D	5.0%	Straw	Mat
Fe	D	70 ppm	Leaf blade	Til
	T	300 ppm	Leaf Blade	Til
Zn	D	10 ppm	Shoot	Til
	T	1500 ppm	Straw	Mat
Mn	D	20 ppm	Shoot	Til
	T	>2500 ppm	Shoot	Til
B	D	3.4 ppm	Straw	Mat
	T	100 ppm	Straw	Mat
Cu	D	<6 ppm	Straw	Mat
	T	30 ppm	Straw	Mat
Al	T	300 ppm	Shoot	Til

[a]Figures for critical concentrations collected from
various references but adjusted to round figures
[b]Mat = maturity; Til = tillering

Plant analysis

In general, nutrient deficiency or toxicity symptoms appear when plants are young, which allows whole plants (excluding roots) to be sampled for chemical analyses. Plant samples are washed, dried and ground. Table 4.3 gives information on the critical concentrations of nutrient elements in the rice plant.

It is difficult to delineate precisely the critical concentration for zinc deficiency; however, the following guidelines are helpful:

Zinc Concentration in the Whole Shoot	Diagnosis of Deficiency
<10 ppm	Definite
10 - 15 ppm	Very likely
15 - 25 ppm	Likely
>25 ppm	Unlikely

Similarly for sulfur, plant critical levels vary with physiological age and plant part.

Critical Total S Level in the Plant	S Content (%)	Critical N:S Ratio
Shoot at maximum tillering	0.11	15
Straw at maturity	0.055	14
Grain at maturity	0.065	26

FERTILIZATION

The steadily increasing cost of fertilizer has resulted in more emphasis on improved management of fertilizer for efficient use in rice. One method to increase efficiency is to avoid fertilizers where they are not needed and then apply the fertilizers saved to soils where substantial responses are likely. In many instances, soil tests and plant analysis provide reasonably good guidelines.

In lowland rice, losses of N take place through: (i) volatilization; (ii) denitrification; (iii) leaching; and (iv) run-off. Ammonium fixation by clay minerals and immobilization also cause temporary losses and unavailability of N.

The recovery of fertilizer N applied to rice is seldom more than 30-40%. Even with the best agronomic practices, recovery seldom exceeds 60-65%.

Nitrogen response is determined by variety grown, soil characteristics (organic matter, total N, soil texture and others), water control, and cultural practices. In order to maximize fertilizer N use efficiency in lowland rice, the following alternative concepts appear promising: (i) varietal differences in N utilization efficiency; (ii) improved timing of N application; (iii) deep placement of N fertilizer; (iv) controlled-release N fertilizers; and (v) use of nitrification and urease inhibitors.

In order to minimize gaseous losses, it is critical to study floodwater chemistry following fertilizer N application. Basal fertilizer N applied into standing water with or without incorporation is subject to volatilization loss.

Among the alternative practices for increasing fertilizer N efficiency, proper timing of applications of N fertilizers and proper water management at the time of basal fertilizer application appear most relevant. Deep point-placement, with or without applicators, appears highly promising in most soils, including calcareous soils. Furthermore, the following general practices help to improve fertilizer use efficiency: (i) prepare the land thoroughly and level it well; (ii) prepare and plaster the bunds to minimize loss of water and nutrients; (iii) plant modern disease- and insect-resistant rice varieties suited to the area; (iv) wherever possible, grow an upland crop such as mungbean or cowpea or keep the soil flooded between two lowland rice crops; and (v) exploit varietal tolerance to nutritional deficiencies or toxicities such as zinc and phosphorus deficiencies, salinity, alkalinity, acidity, and iron toxicity in lowland rice and aluminum toxicity in acid upland soils.

Management of phosphorus fertilizers is largely dependent on soil characteristics such as soil pH, degree of weathering, kind of clay minerals, water regime, cropping intensity, and cropping pattern. Superphosphate (single and triple) is the most common fertilizer for rice. Finely ground rock phosphate with high citrate solubility is often considered to be as effective as superphosphate in acid soils. Also, residual effects should be considered in determining the efficiency of different phosphorus sources.

Potassium fertilizer management is simple; K should be applied during the final land preparation. In coarse textured soils benefits to split application of K have been recorded. Potassium response in lowland rice is usually recorded only at very high yield levels and under intensive cropping of rice and other crops. Irrigation water, rice stubble and soil potassium often fulfill the K requirement of marginally-K-deficient soils at a 3-5 t/ha grain yield level. At high yield levels (6-10 t/ha), K fertilizer should be applied where responses are expected.

A possible solution to zinc deficiency is to drain the field and thereby increase zinc solubility. However, draining is often not possible or even desirable considering other benefits of continuous flooding. Dipping rice seedlings in a 1-2% solution of zinc oxide is the cheapest method of correcting zinc deficiency if labor is

inexpensive. Otherwise, 5-10 kg/ha $ZnSO_4$ should be applied to the fields if zinc deficiency symptoms are apparent. In severe zinc-deficient areas (soil pH 8.0 or
above), 25-30 kg $ZnSO_4$/ha should be applied.

Sandy soils low in organic matter and located in the
interior of humid regions may need sulfur. The simplest
remedy is to use ammonium sulfate instead of urea or
single superphosphate instead of triple superphosphate.
Gypsum is available in many countries and should be applied as a source of sulfur.

Use of Straw and Green Manures

Five tons of rice straw contain 30 kg N, 5 kg each
of P and S, 75 kg K, 250 kg Si, and 4 tons of organic
matter. Spread the straw in the field after threshing.
Incorporate it at land preparation for the next rice crop
or use it as a mulch for the upland crop. Straw incorporation increases yield and improves soil fertility.

Wherever feasible, green manures, azolla, or compost
should complement inorganic fertilizer, not only to supply nutrient elements but to improve soil fertility and
to improve physical properties and water-holding
capacity.

SELECTED REFERENCES

DE DATTA, S.K. 1981. Principles and Practices of Rice
Production. John Wiley & Sons, New York. 641 pp.

DE DATTA, S.K., I.R.P. FILLERY, and E.T. CRASWELL.
1983. Results from recent studies on nitrogen
fertilizer efficiency in wetland rice. Outlook on
Agriculture 12(3):125-134.

HERDT, R.W. and C. CAPULE. 1983. Adoption, spread, and
production impact of modern rice varieties in Asia.
International Rice Research Institute, Los Banos,
Philippines. 54 p.

YOSHIDA, S. 1981. Fundamentals of Rice Crop Science.
International Rice Research Institute, Los Banos,
Philippines. 269 pp.

5
WHEAT

R. A. Olson

The importance of wheat in world agriculture cannot be overestimated. Its cultivation probably began as early as 10,000 B.C., it continues to exceed other cereal crops in total production, and it ranks essentially equivalent to rice as the foremost crop for the nutrition of mankind. Wheat is the major food for about 35% of the world's people, is the staple of some 43 countries of the world, and remains the largest single food item of the USA in per capita consumption.

Preeminence of wheat for human consumption is due in considerable part to the unique properties of wheat proteins to form gluten that permits the dough from mixed flour and water to retain gas needed for bread production. The protein component serves an invaluable role in the diets of many people. Virtually all of the countries having less than 60 grams daily protein intake per capita are found in tropical regions where wheat is not the staple food crop. Among the three major cereals, wheat diets would require no protein supplement, while modest protein supplementation is needed for rice and maize if all the daily energy requirements were provided by a cereal grain.

Wheat, during historical times at least, has been regarded as a premium human food, thereby limiting its use as an animal feed grain. Efforts were made from very early times to grind the grain in a way that would separate the endosperm from the germ and outer bran portion of the kernel. The latter was not used for human food and rather became a fuel source or food for animals. With growing world demand for food grain the use of wheat for animal feed remains limited, although it is quite acceptable for this purpose.

---

[1]Professor of Agronomy, University of Nebraska, Lincoln, Nebraska 68503, USA.

ECOLOGICAL ADAPTATION

Wheat is produced under a wide diversity of environmental conditions, probably exceeding that of any other major crop. There is, however, variation in wheat types related to climatic conditions. Most wheat is of the bread-making type (<u>Triticum</u> <u>aestivum</u>), hard or soft in texture, red or white in color, chiefly classed as "bread wheats." Lesser amounts of a durum (pasta) type (<u>Triticum</u> <u>durum</u>) are grown as "macaroni" wheats which may be white or red, and of generally flinty character.

Climatic adaptation determines regions where spring rather than winter wheats are grown. Cool temperatures are required during a portion of the life cycle for "vernalization" of winter types, which will flower and produce grain only when subjected to temperatures in the range of 0-8°C or lower during early vegetative development. There are, however, limits to the coldness that winter wheats will endure; for example, little is grown above 42°N latitude in the USA, where the spring types take over. Spring wheats are produced in some quite cold regions, such as elevations as high as 3300 m in Tibet where frost can occur at any time, but for the most part winter wheats are grown where winters are dry and cold with limited snow cover.

Most wheat is grown in temperate regions between 30-55° North latitudes and 25-40° South latitudes, within a precipitation range of 300-1100 mm per annum.

REGIONS OF LIMITED RAINFALL

Much wheat production is confined to generally dry climates. Total annual precipitation is no certain indicator of moisture adequacy or deficiency for the crop since distribution and timeliness of moisture received can be equally important as the amount. Nonetheless, few situations exist where continuous wheat cropping is successful with less than 380-500 mm annual rainfall. To meet the moisture needs of wheat in dry regions, summer fallowing is often practiced in vast areas in an attempt to store a portion of the moisture received in one season to complement that received during the subsequent growing season.

HUMID REGIONS

Because of diseases like mildew, rust, scab and various leaf spots, there is limited production of wheat and other small grains other than rice in very humid regions. High moisture levels also promote excessive vegetative growth and coincident lodging and result in difficulties in land preparation, seeding, and harvesting of the crop.

SOIL ADAPTATION

Wheat is tolerant of a wide range of soil conditions. Optimum yields require sufficient soil moisture
for adequate nutrient supplying capacity of the soil. A
depth of 90 cm or more of friable soil without restric-
tive horizons is essential for root development and for
meeting water and nutrient needs. Medium to fine textured soil types are superior to coarse textured soils,
which are better used to grow crops such as rye and
triticale. An inherent granular or induced crumb soil
structure produced by tillage is desirable to ensure
germination and plant establishment. Given favorable
soil structure, aeration and moisture supply, roots will
readily penetrate 2 m or more in the soil. Poorly
drained soil is quite intolerable, especially for winter
wheat in cool regions, because of heaving damage.
From a chemical standpoint, soils would ideally be
in near-neutral pH range to maximize nutrient absorption
and to prevent toxicities associated with high Al, Mn and
Fe contents. Some modest level of soil organic matter
would be desirable for assisting stability of soil
structure, for storage of such essential nutrients as N,
S, and P, and for raising soil cation exchange capacity.
The above desirable attributes notwithstanding, good
yields of wheat are obtained on soils with a pH range of
5.5 to 8.5. The worldwide average for International
Winter Wheat Performance Nursery plots in 1975/76 was pH
7.6, with free lime content of 0 to 25% and with soil organic matter content of 0.7 to 8%.

APPROPRIATE PLACE IN THE FARMING SYSTEM

In dry temperate regions where much of the annual
precipitation is received in winter and spring, wheat is
best adapted of all major crops and is commonly grown in
monoculture, albeit with fallowing in semi-arid regions.
In less dry areas where continuous cropping is possible,
wheat of either spring or winter types will fit into
cropping sequences in whatever position that tillage and
harvesting operations demand. Where moisture supply allows multiple cropping, wheat fits best as a fall planting with late spring harvest followed by an irrigated
summer crop like rice, as practiced in vast areas of
China.

MINERAL NUTRIENT REQUIREMENTS

Highest yields of wheat are produced in central and
northern Europe where the climate is generally favorable
for soft red winter types. A part of the high yields
must be attributed to the fact that larger quantities of
fertilizer are used there than in any other region of the
world. Record yields of 14 mt/ha have been obtained in

the Pacific Northwest of the USA with recently developed short, stiff stemmed, tillering white varieties. Few yields in excess of 6 mt/ha have been recorded with the hard red winter wheats in drier regions.

AVERAGE YIELDS

Worldwide average yield of wheat was in the order of 1800 kg/ha during 1977-79, reflecting especially the dry conditions under which much of the crop is grown (Table 5.1). This yield represents a 20% improvement over results in 1969-71. Highest average yields are being realized in Europe followed by North and Central America. Yields in France exceed the averages achieved by all the major wheat producing countries. Lowest regional yields exist in Africa, associated with the very dry climate of the North African countries that produce wheat.

IMPROVED VARIETIES

Major advances in wheat production potentials were made in the 1960s with release of the semidwarf winter wheat variety "Gaines" in 1961 by Washington State University and the semidwarf spring wheats Sonora 64 and Lerma Rojo by CIMMYT in Mexico. Yields upwards of 10 tons grain/ha suddenly became possible when the necessary packages of fertilizers, proper watering, and pest control were adopted. Actually, a large part of the value derived from the semidwarf wheat is associated with the capability of the short, stiff-stemmed plant to withstand very high N treatments without plant lodging, a major limitation of older varieties.

SOIL TESTS

The soil test values in Table 5.2 apply to conditions in midwestern USA, and perhaps poorly to other regions with different clay mineral type or on soils that are strongly acid or strongly alkaline. Certainly there would be no harm in having higher levels than those indicated for most nutrients, although substantial excesses of some are known to have deleterious interaction with others in crop nutrition as P on Zn, and Fe; K on Mg; Zn on Fe; and so on.

Advancing technologies in wheat cultivars, moisture efficiency and improved cultural practices would likely call for higher levels of the more mobile nutrients such as N, S and boron (B). It is for these mobile nutrients that testing of the entire rooting profile is desirable whenever soils are deep, well drained, and where evaporation of free water and the plant's use of water during the growing season exceeds the moisture (rainfall) received. Deep profile testing information is useful for less mobile nutrients as well, particularly P, in predicting response to fertilizers.

TABLE 5.1
Wheat production and fertilizer use by region and select-
ed countries (FAO Production Yearbook, Vol. 33, 1979 and
FAO Fertilizer Yearbook, 1978)

Region[a]	Average Yield[b] 1969-71	1977-79	Average Fertilizer Use on Arable Land 1977
	kg/ha		kg/ha $N+P_2O_5+K_2O$
World	1500	1800	11+6+5
North Africa	1000	1000	1+1+.5
North America	2100	2100	18+9+9
South America	1200	1300	2+3+2
Asia	1100	1400	14+5+2
Central Europe	2600	3400	56+38+36
Australia	1200	1300	.4+2+.2
U.S.S.R	1400	1600	12+8+9
Average for developed countries	1800	2100	16+11+10
Average for developing countries	1100	1400	6+3+1

[a]Statistics given for each region and for the two major
producing countries in each region.

[b]Yields rounded to nearest 100 kg/ha.

PLANT TISSUE TESTING

   Precision in plant tissue testing of most crops has
not advanced to the level realized with soil testing.
This is the particular consequence of varied nutrient
concentrations found in plant parts, different stages of
growth, environmental conditions just prior to sampling
(wet-dry, hot-cold, etc.), and concentrations of other
nutrients. Keeping these conditions in mind, the values
presented in Table 5.3 can nonetheless be helpful in
diagnosing problem situations that affect the wheat crop.

VISUAL SYMPTOMS

   Nitrogen (N). The most readily recognized
deficiency symptom during the vegetative stage of wheat
is the stunted and overall yellowish color of the crop,
with green clumps arising from animal urine spots, that
proclaims a shortage of N. The chlorotic yellowing progresses from leaf tips inward, and from older to younger
leaves, with the former dying prematurely when shortage
is severe.

TABLE 5.2
Soil test levels above which no yield response to fertilizer occurs under optimum wheat production in Midwest USA

Element	Test Method of Soil Extraction	Soil Sampling Depth	Optimum Value
Nitrogen (N)	Residual NO_3-N	Profile to 180 cm depth	*> 150 kg/ha
Phosphorus (P)	Bray & Kurtz #1	Plow layer	> 25 ppm P for slightly alkaline soils
	$NaHCO_3$ extractable P	Plow layer	> 15 ppm P for highly calcareous soils
Potassium (K)	Exchangeable K	Plow layer	> 100 ppm K
Calcium (Ca)	pH	Plow layer	> 6.2 pH value
Magnesium (Mg)	0.1 N HCl	Plow layer	> 30 ppm
Sulfur (S)	$Ca(H_2PO_4)_2$ extract	Profile	> 6 ppm SO_4-S
Zinc (Zn)	DTPA extract	Plow layer	> 1 ppm
Iron (Fe)	DTPA extract	Plow layer	> 4 ppm
Copper (Cu)	DTPA extract	Plow layer	> 0.2 ppm
Manganese (Mn)	DPTA extract	Plow layer	> 1 ppm
Boron (B)	Hot water extract	Plow layer	> 0.5 ppm

* > = greater than.

TABLE 5.3
Nutrient concentrations in the tops of wheat during rapid
vegetative stage expressing varied nutritional levels of
the crop[1]

Nutrient	Nutrient Content of Growing Wheat Tops		
	Deficient	Low	Sufficient
N, %	< 2.50	2.50 - 3.00	> 3.00%
P, %	< .15	.15 - .25	> .25%
K, %	< 1.00	1.00 - 1.50	> 1.50%
Ca, %	< .15	.15 - .30	> .30%
Mg, %	< .10	.10 - .15	> .15%
S, %	< .15	.15 - .20	> .20%
Fe, ppm	< 5.00	5.00 - 10.00	> 10.00 ppm
Mn, ppm	< 10.00	10.00 - 15.00	> 15.00 ppm
Cu, ppm	< 2.00	2.00 - 5.00	> 5.00 ppm
Zn, ppm	< 10.00	10.00 - 15.00	> 15.00 ppm
B, ppm	< 2.00	2.00 - 5.00	> 5.00 ppm
Mo, ppm	< .02	.02 - .50	> .50 ppm

[1]Total concentrations following complete digestion of
plant sample.

< = less than.
> = greater than.
ppm = parts per million.

Phosphorus (P). Deficient plants are stunted, slow
growing and tiller poorly. Yellowing of leaf tips may be
observed with severe shortage. Delayed heading is certain to be noted if there is adjacent wheat experiencing
adequate P nutrition for comparison.

Potassium (K). As with N deficiency will show up
first in older leaves because of K translocation to growing tissues. Leaves become chlorotic beginning at the
tip and progressing to the base, eventually giving a
scorched appearance. K deficient plants may have weak
stems with a tendency to lodge.

Manganese (Mn). The so-called "grey speck" of wheat
grown on calcareous and organic soils is due to Mn deficiency, and is accentuated by cool, dry weather. On
merging of spots leaves often drop at the base. Plants
are lighter green than normal and are slow to mature.

Copper (Cu). Organic soils in Europe and northern
USA have expressed Cu shortage by chlorotic (light colored) leaves that curl and twist at the tips and margins.
Heads are bleached in appearance (white heads) and with
severe Cu shortage are poorly filled.

ROLE OF NUTRIENT REQUIREMENTS IN CULTURE OF WHEAT

Animal and human wastes, and legume residues, have been used from the dawn of history for satisfying crop needs, long before the discovery of the elements and their functions in crop growth. In most countries, however, perhaps no more than 10% of crop nutrient requirements could be met from this source. Even the best organic farmers, the Chinese, are looking to non-organic sources of crop nutrients for meeting their food needs. World food grain production has increased approximately 60% in the past 25 years, and these gains can be attributed in considerable part to the correction of crop nutrient deficiencies by the use of mineral fertilizers, from about 20 million metric tons $N + P_2O_5 + K_2O$ in 1955 to over 100 million tons in 1980.

EFFECTIVE USE OF FERTILIZERS AND AMENDMENTS

Judicious fertilizer use will enhance water use efficiency by wheat, and may help the crop to better utilize moisture stored at depth in the soil, but can in no way substitute for water. Accordingly, much of the world wheat production effort must be made in balancing fertilizer practice against available moisture supply.

Soil Amendments. Provision of Ca and Mg where needed is most commonly accomplished by liming, with the primary objectives of correcting soil acidity or neutralizing exchangeable Al. The main purpose of liming in subtropical/tropical areas is to achieve neutralization of soluble aluminum (Al); and rates of lime beyond the amount required for that purpose are likely to have little value. Raising pH by liming in temperate regions normally enhances availability of molybdenum (Mo), Ca, and P, and also Mg if the lime used is dolomitic. Further, the soil's capacity to nitrify N from native organic sources is enhanced.

Nitrogen (N). Wheat has by far the greatest need for N among all nutrients and the greatest yield response occurs from its judicious application. In humid and irrigated regions where legumes can be included in a cropping sequence, the associated symbiotic N fixation can reduce the amount of fertilizer N needed for optimum production, but higher levels of P, K and Ca are required to grow the legume effectively in such rotations. Similarly, where significant quantities of animal manures, human wastes and similar organic residues are returned to the soil, the need for inorganic N supplements is lessened.

Quantity of fertilizer applied per hectare is the foremost consideration in N fertilizer management, since the crop cannot reach its yield potential where N is in short supply. Excess N can be responsible for prolonged and excessive vegetative growth in many cultivars, resulting in tall, weak-stemmed plants that lodge readily. Overstimulated and prolonged vegetative growth delays maturity and is especially harmful if summer drought occurs in the mid-latitudes. When fertilizer use begins in a given region, a series of fertilizer trials will tell the appropriate rate of N fertilizer likely to give best results on local soils. After a few years of N fertilizer use under different soil and fertilizer management practices, however, the dominant factor determining most efficient N rate is likely to be the amount of residual mineral N in the crop rooting zone at the beginning of the growing season.

The best time for N fertilizer application is governed by the climate, the kind of N carrier employed, and the type of wheat grown. In very humid regions where rainfall during the growing season substantially exceeds evaporation and plant water use, it may be necessary to apply some N at planting, with the remainder being supplied later in the season to minimize leaching and denitrification losses. In dry regions in other than sandy soils, preplant or planting time applications will usually give good results, especially if a carrier of somewhat delayed availability is employed such as anhydrous ammonia placed at moderate depth in the soil. From the standpoint of yield alone, in the Central Great Plains of the USA fairly comparable results have been obtained from fall and spring applications of N for winter wheat, but maximum grain protein yield almost always accompanies spring treatments. Opportunity is afforded by spring topdressing to evaluate crop stand and moisture availability at a relatively late stage in the crop season and thereby judge probability of profitable response to fertilizer N. This avoids the possibility of winter and spring N leaching losses from the soil, and of excessive fall vegetative growth and moisture use at the expense of grain yield the following summer.

Nitrogen placement requirements are governed chiefly by the kind of fertilizer used. Anhydrous ammonia must be placed at sufficient soil depth under suitable soil texture and moisture conditions to assure an adequate soil capacity for holding the ammonia against loss as a vapor. Urea and ammonium sulfate should be incorporated in the soil, particularly on alkaline soils to reduce ammonia losses. Soil incorporation is not so necessary for nitrate forms such as ammonium nitrate.

Phosphorus (P). Wheat is one of the more responsive crops to P fertilization, with critical levels of soil P being essentially twice that required for maize and three times that for soybeans. Since most wheat is grown in relatively dry regions where soils are commonly calcareous, P proves to be nearly as great a limitation for production as N, since calcareous soils are usually inherently low in available P. Fertilizer P rarely presents the over stimulation problem commonly encountered with N in dry farming. When added to correct a deficiency, fertilizer P will induce added tillering and more vigorous plant growth, but does not stimulate excessive vegetative growth if substantially more than required is applied. The predominant factor controlling optimum rate of fertilizer P is the level of soil solution P maintained in balance with solid phase P throughout the soil rooting profile. The level of other soil nutrients and the capacity of soil to convert fertilizer P to less available forms are lesser contributing factors. Strongly acid soils containing substantial iron and aluminum oxides in the clay fraction, and highly calcareous soils, especially where both types are low in organic matter, are likely to require heaviest P rates.

More than any other nutrient, phosphorus promotes tillering and establishment of an effective root system. In achieving these results the plant must have a plentiful supply of available P from the earliest stages of growth. Phosphorus also has limited mobility in the soil, and this necessitates its placement in the rooting zone at or before planting. For wheat, the advantage for row placement of P over other application methods has long been known. There are situations, however, where no differences existed between row and broadcast applications, and also where banding was as much as four times as effective. Many factors could be responsible for this anomaly, including availability of soil moisture and amount of P fixation by the soil, but the predominant one appears to be the level of available soil P; the lower the soil P the greater the benefit from row application.

Potassium (K). In drier regions where most of the wheat crop is produced responses to K have been much less common than to N and P. Because there is little K in the grain, K is not removed from the soil in appreciable quantity where the straw is left on the land. This limited removal plus inherently high levels of exchangeable soil K account for the slow depletion of K in most dry farming regions. On the other hand, where straw is continuously removed for fuel, fodder or bedding, the depletion rate is greatly accelerated.

Because of low K mobility in soils K fertilizer must be applied at or before planting. There does not appear to be any special advantage in either row application, or broadcasting with incorporation by plowing or disking. The rate of K needed for optimum yield is determined by the level of soil K, and by potential yield.

Other Nutrients. Wheat yields are restricted by a shortage of sulfur (S) in local areas, especially on soils that are sandy, acid and low in organic matter. Most medium to fine textured soils of drier regions have large S reserves (as gypsum) in their subsoils, which coupled with aerial fallout in industrial regions adequately serves current crop needs. Thus S shortage is much less common than that of N, but where it is limiting, applied S does not express itself fully in yield of protein content, while various soluble non-protein forms of S increase in tops and roots. Sulfur and N metabolism is closely related, with one part available S being needed for each 12-16 parts of available N for maximum grain yield and protein production. The sulfate form of S is moderately mobile in most soils. Sulfur carriers are best applied at planting time.

Zinc (Zn). Zinc deficiency has been identified locally in wheat, most commonly on calcareous soils of high pH. Since Zn is a highly immobile element in soils, it must be placed in the rooting zone at or before planting (as with P and K), and for this reason is generally applied as part of the basal or starter fertilizer treatment either broadcast, incorporated in the soil, or row placed. A small amount of N placed in immediate contact with the Zn carrier in row application facilitates plant uptake of Zn.

Copper (Cu). Shortage of Cu for efficient wheat production has been recognized on organic soils of northeastern USA, eastern Canada and northern countries of Europe. As with Zn, low mobility in soil necessitates application of Cu compounds at or before planting in the seedling root zone. Foliar treatment (leaf spray) as with Zn can be regarded only as supplemental in the correction of a recognized deficiency, and the emphasis rather should be on soil treatment.

TIME OF PLANTING AND PLANT POPULATION IN RELATION TO NUTRIENT SUPPLY

Soil nutrient supply in relation to planting time has its greatest impact with spring wheat planted in the cooler regions. Given the short growing season of high elevation areas, and above 45° N or S latitude, the early stimulation afforded by a high level of nutrition with early planting can make the difference in bringing the crop safely through to maturity. Limited soil moisture during the vegetative stage also can make a good supply of readily available nutrients more necessary, due to retarded active transport and diffusion of nutrients to the root system. Given these stress conditions during the growing season and the possibility of frost damage before harvest, early planting and a high level of nutrition go hand-in-hand in the cooler regions.

Where length of growing season is not a major limitation, as in mid-temperate zone winter wheat, earliness in maturity is commonly desirable as well because of possible crop desiccation by summer drought. An adequate level of P nutrition especially may make the difference between good and poor filling of the wheat kernel, and excessive N is quite intolerable. On those occasions where excessive moisture or dryness, or where the potential for Hessian fly attack delays autumn planting of winter wheat, effective use of fertilizer can assist stand establishment to the extent of reducing or eliminating winter killing. But the combination of early planting and high nutrient status should be avoided in the dry farming areas of winter wheat production because of unnecessary moisture consumption by excessive growth in the fall.

Unlike most row crops, plant population does not have the impact on yield potential of wheat, since limited number of seeds per unit area is usually compensated for by tillering. It is necessary, however, that the wheat plant should have a good level of nutrition for effective tillering to occur, and adequacy of P is particularly important in this respect.

PEST CONTROL IN RELATION TO NUTRIENT SUPPLY

Pest and disease attacks are usually most harmful to starved and weak plants with limited capacity to recover from damage. Root-rot organisms (Pythium) are especially damaging under P deficiency. Heavy N fertilization, however, produces an increased density of rank, succulent plants that are more subject to damage from various pathogens causing powdery mildew, Septoria leaf- and glume-blotch, or "Take-all" disease.

Attacks by insects such as aphids and Hessian fly are usually most damaging to plants suffering from nutritional deficiencies. It is not always clear whether the insect is repelled by a well-fed plant, or whether the nutritional adequacy assists the plant to recover more readily from inflicted damage.

Fertilizer practice is one of the better weed control measures available to the farmer. Appropriate timing, rate and placement of fertilizers that increase the competitiveness of wheat produces an early crop canopy that shades out weed seedlings.

BENEFITS FROM ADEQUATE MINERAL NUTRITION

Individual states within the USA and individual countries elsewhere have reported a doubling or tripling of average wheat yields with increased use of mineral fertilizers during the past 30 years.

During the period 1961-77 some 250,000 fertilizer trials and demonstrations have been conducted under the FFHC Fertilizer Program of FAO (UN Food and Agriculture Organization) in 40 developing countries, and economic analysis has been made on all of the resulting yield data. Wheat was the crop investigated in approximately one-tenth of these trials and demonstrations. As noted in Table 5.4 average crop response to fertilizer and value/cost ratios were lowest in Morocco, Syria and Tunisia, three of the driest countries among those included. It is further evident that a generally lower economic return was derived from fertilizer treatments throughout the Near East and North African Region compared with the much more favorable responses in the Far East and the Central and South American regions where rainfall is generally higher. FAO has found that farmers of less developed countries accept fertilizer use readily when the value/cost ratio exceeds 2.0, but are not inclined to take the economic risk below that level.

TABLE 5.4
Production and economic response of wheat to fertilizer treatment in FFHC Fertilizer Program trials of FAO (1961-77)

Location	Country Trials[1] (No.)	Average Response to Best Treatment Yield Increase (%)	Value/Cost Ratio (V/C)
Near East and North Africa			
Ethiopia	52	74	3.6
Lebanon	87	98	3.0
Morocco	35	71	1.9
Syria	46	36	1.0
Tunisia	28	39	1.8
Turkey	137	94	3.6
Far East			
Afghanistan	638	294	4.1
Nepal	365	180	3.1
Central and South America			
Colombia	15	73	4.4
Ecuador	66	94	2.0
Guatemala	48	166	3.5
Peru	124	102	4.5

[1]Thousands of fertilizer demonstrations without replication were conducted during the interval for which data are not presented. Trials contained rate variables of NPK with internal replication, thereby supplying more reliable data.

6
MAIZE

John J. Hanway

    Maize, wheat and rice are the primary cereal grain
crops in the world and approximately equal quantities of
each are produced. Total world production of maize in
1975 was 323 million metric tons.

    Maize is used predominantly as a feed for animals,
but it is used extensively for human food and in industry. The grain is used most generally, but the entire
above-ground part of the plant is used as silage or fodder for animal feeding in many areas. The grain is used
for human food as whole meal for bread or porridge, as
popped corn, as tortillas, or it may be processed to produce starch, oil, sugar syrups, and so on. Maize enters
into industrial use as paper products, construction
materials, textiles, metal castings, pharmaceutics, ce-
ramics, paints, and explosives.

ECOLOGICAL ADAPTATION

    Approximately half of the world production of maize
is in the USA but it is an important crop in Europe,
Asia, South America, and Africa (Table 6.1). Most maize
producing areas are in the 30 to 55° latitude range. Only two areas not in that range -- Brazil near 30° and Mexico at high altitudes -- produce significant amounts.

    Most maize production is in regions which have a
freeze-free season of 120 to 180 days and the warmest
months have average temperatures between 21°C (70°F) and
27°C (80°F). Maize is grown in areas where the annual
precipitation ranges from 250 to over 5000 mm. Summer
rainfall of at least 150 mm is required if the crop is
not irrigated, and yields can be very high with
irrigation.

---

[1]Professor of Soil Fertility, Iowa State University,
Ames, Iowa 50011, USA.

TABLE 6.1
Maize Grain Production - 1975[a]

Region	Area 10^d hectares	Production 10^d metric tons	Average yield Kg/ha
N. America	38	161	4250
Oceania	0.1	0.4	4230
Europe	13	46	3840
Asia	28	54	1920
S. America	17	28	1710
Africa	17	25	1430
World Total	115	323	2820

[a]World Crop Statistics. 1975. FAO, Rome.

Although maize can be grown in successive seasons for a long time, better yields are generally obtained and the crop is commonly grown in cropping systems that include other crops. Alternating crops in the cropping system aids in weed and disease control and in maintaining soil fertility. Other crops in the cropping system include soybeans, small grains, and legume and grass forage crops.

MINERAL NUTRIENT REQUIREMENTS

Maize grain yields greater than 10,000 kg/ha have been produced under optimum conditions. The world average yield is now about 2800 kg/ha (Table 6.1). Average yields vary from 1430 to 1920 kg/ha in Africa, South America, and Asia and from 3840 to 4250 kg/ha in Europe, Oceania, and North America. Average yields now being produced are much less than potential yields. Approximately one-half of the total above-ground plant weight at maturity is in the grain.

To increase yields in any region it will be necessary to: (i) use hybrids that are adapted to the region, (ii) use cultural practices appropriate to the soil and the region; (iii) use effective methods for control of weeds, diseases, and insects; and (iv) use soil fertilization practices as needed.

Goals of Mineral Nutrients for Acceptable Yields

Nutrient uptake by the plants is essential throughout the growing season, but the nutrients required and the plant's ability to take up nutrients from the soil vary at different times during the season.

Soil tests, plant tissue tests, and visual observations of the growing plants are used as guides for determining optimum fertilization practices. For any of these to be useful in a region, it is necessary to conduct research to develop a relationship between the test results or visual observations and the crop yield response obtained from fertilizer applications in the field. The goal in soil fertilization will vary among regions depending on the cost of the fertilizer used and the value of the crop produced. Where high crop yields are possible and fertilizer costs are reasonable, the goal may be to raise the nutrient availability to a high level where further increases would result in no additional yield increase. But, where crop yields are limited by some uncontrollable factor or where fertilizer costs are high in relation to the value of the crop, the goal in soil fertilization may be a much lower level of nutrient availability in the soil.

Soil tests. Soil tests for acidity (pH) and for P and K availability are now used most commonly. Soil tests for other nutrient elements have been shown to be useful in limited areas, but are not generally accepted. Soil acidity is usually considered to be optimum at pH values between 6 and 7. Three tests most commonly used for P are: (i) 0.025N HCL-0.03N NH$_4$F(Bray-1); (ii) 0.025 N H$_2$SO$_4$ - 0.05N HCL; and (iii) 0.05M NaHCO$_3$(Olsen). High test values for these methods are >30, >38, and >10 ppm, respectively. Potassium tests most commonly use 1N NH$_4$OAc at pH 7 as the extractant and exchangeable K values greater than 80 ppm (for sands) to 125 ppm (for clay loam) are considered high.

Plant Tissue Tests. Chemical analyses of maize leaves from plants growing in the field provide very useful information concerning the availability of different nutrients to the plants. The results of these analyses can serve as guides for fertilizing succeeding crops. Although much research remains to be done, the following concentrations of different nutrient elements in dried samples of the ear leaf of maize plants at the tasseling-silking stage may serve as indices of sufficiency levels:

N	3.0%	Mn	15	ppm
P	0.30%	Zn	15	ppm
K	2.0%	Cu	5	ppm
Ca	0.40%	Fe	15	ppm
Mg	0.25%	B	10	ppm

Visual Symptoms. Deficiencies of most of the nutrient elements result in reduced growth (stunting) of the plants, but definite symptoms characteristic of specific nutrients often are not visually apparent.

Deficiencies of N or S do result in a characteristic general loss of green color and development of a yellowish color in the leaves. This is more severe in lower than in upper leaves. Deficiency of K results in death (firing) of the leaf margins, especially in the lower leaves.

IMPROVED VARIETIES OR HYBRIDS IN RELATION TO NUTRIENT REQUIREMENTS

Yield differences among maize hybrids at any one fertility level or resulting from fertilizer applications are often observed. Differences in nutrient contents of seeds and leaves of different hybrids have been reported. The differences in yield at a given soil fertility level are generally considered to be due to climatic adaptation at that site and are used in selection of hybrids to be grown there. Differences among hybrids or varieties in their ability to absorb or to respond in yield to applications of a specific nutrient have not been found for maize as has been done for some other crops.

ROLE OF NUTRIENT REQUIREMENTS IN THE CULTURE OF MAIZE

To Correct Deficiencies

Many of the cultural practices that are or have been followed in crop production influence nutrient availability to the crop being grown. Applications of animal manure, commercial fertilizers, and other soil amendments such as lime are often used to increase the supply and availability of nutrients for maize. Effective use of legumes in the crop rotation and of the crop residues left in the field following grain crops should also be considered as an essential part of the cultural practices necessary to provide the essential nutrients.

Animal manures are an effective source of many nutrient elements and should be used whenever they are available. However, the amounts required to provide optimum levels of nutrient availability are seldom available or the costs to handle them may make manure more expensive than other fertilizers. Grazing animals deposit much of their manure on fields being grazed but the manure is not well distributed.

Commercial fertilizers are used extensively primarily as sources of N, P, and/or K and also to supply certain minor elements where they are needed. Nitrogen is seldom adequate in soils without fertilization, and maize will respond to P and/or K fertilization on many soils.

Applications of lime to adjust soil acidity often result in increased availability of several soil nutrients, especially phosphorus.

Including legumes such as clover or alfalfa in a rotation with maize is an excellent method to increase the N available for maize. Crop residues left in the field after harvest of maize or other cereals provide certain nutrient elements, especially K, but usually are very low in N and immobilize available soil N.

As Affected by Fertilizer Rates and Placement

Effective rates and placements of fertilizers vary greatly with different soils and conditions. Field fertilizer trials conducted on soils in the region provide the best guides for effective fertilizer use.

Generally, it is desirable to place fertilizer in the soil where it will remain moist and well-aerated and where it will be encountered by the growing roots. Plant roots are not attracted to zones of high fertility in the soil and do not grow into or absorb nutrients effectively from dry soil. However, roots that encounter areas in the soil that are moist and well-aerated and that have a high level of nutrient availability do grow and proliferate profusely.

Fertilizer nutrients, except N, do not move extensively in most soils so nutrient availability generally is greatest in or very near the zone of fertilizer placement. Because plant residues that return nutrients to the soil fall to the ground, nutrient availability generally is highest near the soil surface, unless deep plowing is practiced. Moisture conditions are often limiting for root growth and nutrient uptake at the soil surface. Nitrogen applied to the soil in organic or ammonium form will be nitrified to nitrate which is water-soluble and moves with soil water to deeper depths.

In Relation to Time of Planting and Plant Population

The ability of plants to absorb nutrients is markedly influenced by temperature. Maize generally is planted in the early spring when the soil is cool or cold. Therefore, applications of relatively low rates (100 kg/ha) of a complete NPK fertilizer (starter fertilizer) where it will be available to the very young plants are often beneficial. Later in the season after the soil is warm, additional fertilizer may not be needed.

Increasing the number of maize plants per unit area reduces the amount of soil available to each plant. Where the moisture available is adequate, the amount of fertilizer required to avoid nutrient deficiencies in the plants increases as the plant population increases. This is especially true for N.

In Relation to Weed Control, Insect Pests, and Diseases

Weeds absorb and accumulate nutrients similar to crop plants and therefore compete with crops for available nutrients. Insect pests and diseases reduce nutrient absorption to maize or otherwise reduce crop growth and yield. Cultural practices to control weeds, insects, and diseases are essential for effective maize production.

CROP MATURATION AND HARVEST

Maize matures about 40 to 50 days after silking and pollination. Mineral nutrient deficiencies have little influence on the length of this period. Appreciable amounts of several nutrient elements are retranslocated from the leaves and other vegetative plant parts to the developing seeds, but uptake of all nutrients, except K, continues throughout the seed-filling period. The amount and quality of the grain produced are influenced by the nutrients available to the plants during this period. Nutrient deficiencies result in small or poorly-filled grain that is low in some nutrients. For example, N deficiency results in grain with a low protein content. The low nutrient content of the grain produced may reduce its value as a feed or reduce the vigor of the seedlings if the seed is planted.

SUGGESTED REFERENCE

G.F. SPRAGUE, (ed.) Corn and Corn Improvement. 1976, 2nd Ed. American Society of Agronomy, Inc. Madison, WI, USA.

7
SORGHUM

B. A. Krantz

Sorghum is an important crop in both temperate and tropical regions. In some tropical areas of Africa and Asia, sorghum constitutes more than 70% of the total calories and much of the protein in the diet. Most of the sorghum food grains in South Asia and Africa are grown as a rainfed crop in the semi-arid tropics (SAT) where infertile soils and erratic and undependable rainfall are major constraints. Due to the lack of research, especially in soil and water management, yields have shown very little increase over the past four decades, even though improved varieties have been developed. In 1979, about 61% of the total sorghum-growing area of the world was in South Asia (33%) and Africa (28%), producing about 16% and 15%, respectively, of the world's total production. Although the combined production was only about 31% of the world's total production of 67.3 million metric tons (mmt), it was very important as most was used as human food. Sorghum is also used primarily as human food in the northern part of China and Manchuria, as well as in some West Asian countries.

In contrast to the SAT, sorghum production in the temperate climatic regions, especially in the USA, has increased dramatically. Production in the USA increased about 14-fold from the late thirties to the mid-sixties. In 1979, the total USA production increased to 20.7 mmt with 79% of the production being in the states of Texas, Kansas and Nebraska.

Sorghum yields in the temperate regions of developed countries have increased dramatically due to research in hybridization, accompanied by research in soil, water, and crop management. In the more developed countries, grain sorghum is used almost exclusively for livestock

[1]Emeritus Soils Specialist, Cooperative Extension Service, University of California at Davis, Davis, California 95616, USA. Formerly Head of Farming Systems Program, International Crops Research Institute for the Semi-Arid Tropics (ICRISAT); Hyderabad, India.

feed, with less than 2% of the USA production being used for food and industrial purposes. With the recent increase in export of feed grains, there has been a large increase in sorghum grain export. During 1979, about 43% of the USA sorghum production was exported.

Sorghum is the third largest food grain in the world, after wheat and rice. In area planted and total cereal grain production, sorghum ranks fifth after wheat, rice, maize, and barley. Because of the importance of sorghum as a food grain to the people of the semi-arid tropics, potential increases in yields/ha made possible by recent research on soil and water management and cultivar improvements offer great hope for improving world food resources.

ECOLOGICAL ADAPTATION

Sorghums are grown over a wide range of ecological conditions and can produce better than maize under extreme conditions of either temporary drought or waterlogging. Sorghum is grown on all six continents in regions where the growing season temperature is above 20°C and the frost-free period exceeds 125 days. Most sorghum cultivars tend to become partially dormant during hot, dry periods and will resume growth when soil water becomes available. This drought "escape" characteristic makes sorghum better adapted than maize to semi-arid regions with erratic rainfall.

Sorghum has a greater ability than maize to sustain grain yields under water stress. From the available evidence, Hsiao et al. (1976) concluded that "differences between the root systems of the two crops were minor and were insufficient to explain the differing yield behavior under short water supply." With increasing water stress, stomatal (leaf pores) closure and reduction in carbon dioxide assimilation from the air occurred sooner with maize than with sorghum. This observation helps to explain the different reaction of these two crops to water stress. However, the biggest difference between sorghum and maize is the greater ability of sorghum to form new tillers or branched heads after a water stress period is relieved by rain. Even though sorghum can endure drought conditions, it is also very responsive to irrigation.

Sorghum is grown successfully over a wide range of soils, from sands to clays. Sorghum also has some tolerance to saline conditions and Al toxicity of acid soils. Although it can endure adverse soil conditions and low fertility, sorghum is very responsive to optimum soil fertility and management.

The place of sorghum in the farming system is quite flexible. In warm, temperate climates sorghum is often grown as a second crop after the early summer harvest of a crop such as wheat, barley or sugar beets. In cool, temperate climates of low rainfall where it is grown as

the only summer crop, its yields are usually more dependable than maize.

In the semi-arid tropics of Africa, it is often
grown in intercropping systems with pearl millet, cowpeas, groundnuts, and other crops. In India, sorghum is
most often intercropped with pigeonpeas. Since pigeonpea
is a very slow-starting, long-seasoned (180-200 days)
crop, it is ideal as an intercrop with cereals such as
sorghum, maize or pearl millet which normally mature in
90-110 days. The slow-starting pigeonpea does not normally reduce yields of the fast-growing cereal. If the
cereal grains and stalks are harvested as soon as they
are physiologically mature, the intercropped pigeonpea
yield reduction is minimized.

The basic concept of intercropping is that two or
more species when intercropped can exploit the environment better than either species grown separately. Although intercropping is not adapted to mechanical harvesting, it is at present usually advantageous to intercrop in the semi-arid tropics where most crops are harvested by hand. In traditional farming, sorghum is often
grown as a broadcast mixed crop with many other species
of cereals, pulses, oil seeds, and vegetables. Mixed
cropping is a risk-evasion practice often used on uneven
land surfaces where depression areas are favored in a low
rainfall year and the hummocks are favored in a wet
year. This inefficient mixed cropping system is gradually being replaced by intercropping in which crops are
sown in separate rows. Under this system, inter-row cultivation, appropriate fertilization, timely cultural operations, and harvest of each component crop can be
accomplished.

Recent research in the semi-arid tropics has shown
substantial yield increases of sorghum and associated
crops by use of a broadbed and furrow system compared to
traditional flat cultivation. The broadbed and furrow
system laid out on a graded contour of about 0.5% slope
increases early season rainwater infiltration, reduces
erosion, provides crop drainage, and facilitates supplemental irrigation, where feasible.

NUTRIENT REQUIREMENTS

Sorghum is very responsive to fertilization and irrigation, and yields under optimum moisture and nutrient
conditions are generally about 90% of maize yields.

In the semi-arid countries of Africa and South Asia
little or no chemical fertilizer is used and only minimal
amounts of animal manures or residues are returned to the
soil. Sorghum production in the semi-arid countries is
almost entirely rainfed (non-irrigated). Because of erratic rainfall and ever-present risk of drought, the
semi-arid farmers with their low capital base have been
reluctant to use fertilizers and other costly yield-

increasing inputs. Thus, the yields in these semi-arid countries are very low (600-700 kg/ha) and have shown no consistent increases.

In developed countries the fertilization and yield levels under rainfed conditions depend upon the average rainfall in various regions. Under irrigated conditions, high levels of fertilization are used and grain yields of 11,000 kg/ha may be obtained. In 1979, Spain had the highest national average yield (5326 kg/ha) of any country. However, its total production was only 0.2 million metric tons. The USA with 31% of the world's total production reported an average national yield of 3947 kg/ha. China, which produces 17% of the world production, had a national average yield of 1323 kg/ha, while the average world yield was 1322 kg/ha.

SOIL TESTS

Sorghum sensitivity to N, P, K, S, and Zn deficiencies in soils is similar to that of maize. However, the critical soil test values, especially for Zn and P, are lower for sorghum than for maize (Table 7.1). Sorghum responds markedly to P application.

TABLE 7.1
Critical and adequate levels of soil nutrients for sorghum

Nutrient	Critical level ppm (less than)	Adequate level ppm (greater than)	Method of analysis
P	< 4	> 7	Olsen-0.5M, NaHCO
K	< 40	> 60	Pratt-1N Nh$_4$-Acetate
S	< 5	> 10	Johnson and Nishita-M, LiCl
Zn	< 0.3	> 0.6	Lindsay-DTPA
Fe	< 4	> --	Lindsay-DTPA

> = greater than.
< = less than.

The Bray Powder Method is the most commonly used procedure of rapid plant tests for sorghum. This method can be used to diagnose N deficiency before symptoms appear, which permits timely correction by N fertilization. Sorghum is much more sensitive to Fe deficiency than most other annual crops, including maize. Iron deficiency symptoms on sorghum are very distinctive, and sorghum is an excellent indicator crop for Fe deficiency in soils. Since Fe is a relatively immobile element in the plant, the symptoms of interveinal chlorosis show first on young leaves on the upper part of the plant. Iron deficiency is usually found where highly calcareous subsoils have been exposed by land leveling or soil erosion. Foliage sprays of 3% ferrous sulfate have been found to be the most satisfactory method to correct Fe deficiency on sorghum.

IMPROVED HYBRIDS AND VARIETIES

The discovery of cytoplasmic male sterility in sorghum in 1954 made the production of F_1 (first generation) hybrids feasible on a commercial scale. Since this discovery, sorghum yields in the USA have more than doubled.

In developed countries, where combined harvesting is used, low-stature dwarf plant types are used. However, in the semi-arid tropics of Africa and Asia tall sorghums are preferred because of multipurpose uses of the stalk, including feed, fencing, thatch, and fuel. Tall late-maturing sorghums have a low harvest index (grain to straw ratio) and a low capability for grain production. Thus, breeders in the semi-arid tropics are developing sorghum varieties and hybrids of medium height and life period which have high grain-yielding capacity and moderately high fodder production. Major emphasis is also placed on development of varieties with disease resistance, insect and drought tolerance, and the ability to perform well under low fertility as well as high fertility conditions.

ROLE OF NUTRIENT REQUIREMENTS IN SORGHUM PRODUCTION

In the semi-arid tropics of Asia and Africa, manure, household refuse or compost are the main fertilizers used in traditional agriculture. The composition varies and the rate of application used ranges from two to five tons/ha. Because of shortages of materials, applications are normally made every second year. The amounts used are usually not adequate and nutrient deficiencies, especially N, are widespread. Chemical fertilizers are very expensive in West Africa and are not normally used by farmers. In India more chemical fertilizers--mainly N with P--are being used, especially with the introduction of new high-yielding, fertilizer-responsive varieties. Nitrogen application at medium levels, when combined with other good practices, gives profitable yield increase in most situations. Recommended rates and methods of application vary with the rainfall pattern and moisture holding capacity of the soils. In low P soils, N-P fertilizer is recommended as a starter fertilizer at planting time.

In the USA prior to 1950 relatively little fertilizer was used on sorghum. With the development of high-yielding hybrids and expansion of sorghum into irrigated areas or higher rainfall areas, the fertilizer use has increased dramatically. However, in drier areas moisture is the limiting factor and fertilizer usage is very limited. Normally the grain is harvested and removed from the field and all stalk and leaf residues are returned to the soil. A 6700 kg/ha grain crop would remove about 121, 37, and 27 kg/ha of N, P_2O_5, and K_2O, respectively. Under irrigated conditions, much higher yields are normally obtained and nutrient removal is proportionately higher.

In most soils sorghum responds to application of N and P but K deficiency is less widespread. Zinc deficiency may occur, especially in areas where Zn deficient subsoils are exposed by land leveling for irrigation. Zinc deficiency can be corrected by annual spraying of Zn compounds dissolved in water, or by incorporation of about 10 kg Zn/ha as sulfate or other compounds in soil during seed bed preparation. The soil application normally lasts 4 or more years. Recommended fertilization of irrigated sorghum in California, USA is about 20-40-40 banded at planting and about 140 kg N/ha, applied as a side dressing in each growth stage of the crop.

Since sorghum is a tropical crop, planting is normally delayed until soil temperatures reach 16°C. In areas where chinch bugs or mites are potential problems, early planting is advised. In the semi-arid tropics of India, where shootfly damage can be very serious, early planting at the onset of monsoon rain is recommended. The shootfly population is greatly reduced during the hot dry period before monsoon and early planting and the use of starter fertilizer banded near the crop row gets the sorghum plants beyond the vulnerable seedling stage before buildup of shootfly populations takes place. Recent breeding programs show substantial promise for the development of shootfly tolerance in sorghum.

An inadequately fertilized sorghum crop may cause non-uniform maturation and delayed harvest. Mature sorghum grain is vulnerable to damage by migratory birds, such as the Quelea of Africa, by insects, head molds, and grain weathering, especially in wet, tropical conditions. Therefore adequate fertilization and other factors which promote uniform maturation and timely harvest can enhance grain quality and yield.

BENEFITS FROM ADEQUATE MINERAL NUTRITION

Sorghum yields vary greatly. In 1979 the yield in the semi-arid countries of Niger, Upper Volta, India, Tanzania, and Nigeria were 427, 600, 645, 686, and 730 kg/ha, respectively. Because of the ever-present risks of drought, farmers have been reluctant to use fertilizers. However, with recent potentials shown by water, soil, crop and fertility management research using improved varieties, the possibilities for increased and more stable yields are encouraging. Since most of the sorghum grain in the semi-arid regions of Asia and Africa is used for food, the sound adaptation of this research, along with the availability of moderately priced fertilizer, could have a substantial effect on the human welfare of these areas.

SUPPLEMENTARY READING

FAO. 1979. Production Yearbook. Food and Agriculture Organization, Vol. 33. FAO, Rome.

HAHN, R. R. 1970. Dry milling and products of grain sorghum. pp. 573-601. In: J. S. Wall and W. M. Ross (eds.) Sorghum Production and Utilization. The AVI Publishing Co., Westport, Connecticut.

HSIAO, T. C., E. FERRES, E. ACEVEDO, and D. W. HENDERSON. 1976. Water stress and dynamics of growth and yield of crop plants, pp. 281-305. In: O.L. Lange, L. Kappen, and E.D. Schulze (eds.) Ecological Studies Analysis and Synthesis. vol. 19. Springer-Verlag, Berlin, Heidelberg, New York.

International Crops Research Institute for the Semi-arid Tropics. 1974, 1979, 1980. ICRISAT Annual Reports for 1973-74, 1977-78, and 1978-79. ICRISAT, Patancheru, Andhra Pradesh 502324, India.

KRANTZ, B. A. 1981. Water conservation, management, and utilization in semi-arid lands. pp. 339-378. In: J. T. Manassah and E. J. Briskey (eds.) Advances in Food Producing Systems for Arid and Semi-arid Lands. Academic Press, Inc., New York.

RACHIE, K. O. 1970. Sorghum in Asia. pp. 328-381. In J. S. Wall and W. M. Ross (eds.) Sorghum Production and Utilization. The AVI Publishing Co., Westport, Connecticut.

USDA. 1980. USDA Agricultural Statistics, Washington, D.C.

8
BARLEY

William C. Dahnke

Barley (<u>Hordeum</u> <u>vulgare</u>) is one of the oldest domesticated grain crops. It is grown throughout the temperate zones, in many subtropical zones and in high altitude
areas of the tropical zones of both hemispheres.

World barley production for the 1970-80 period aver-
aged 158 million mt per year. Approximately 83 million
ha are used for barley production each year for a worldwide average yield of 1952 kg/ha. The principal producers of barley in 1980 are shown in Table 8.1.

TABLE 8.1
Principal barley growing countries, 1980 (FAO, 1980)

Country	Area Harvested	Yield	Production
	1,000 Ha	Kg/Ha	1,000 mt
Morocco	2,180	1,015	2,212
Canada	4,837	2,282	11,041
USA	2,927	2,667	7,806
China	4,401	1,091	4,800
Turkey	2,900	1,897	5,500
Czechoslovakia	921	4,234	3,900
Denmark	1,588	3,840	6,098
France	2,648	4,440	11,758
German DR	976	3,615	3,528
German FR	2,002	4,409	8,826
Poland	1,318	2,580	3,400
Romania	860	2,791	2,400
Spain	3,425	2,500	8,561
Sweden	658	3,780	2,486
United Kingdom	2,338	4,406	10,300
Australia	2,608	1,108	2,890
USSR	32,151	1,384	44,500
World	83,195	1,952	162,402

[1]Department of Soil Science, North Dakota State University, Fargo, North Dakota 58105, USA.

Barley is utilized as a feed grain for livestock, a food grain, malt for beer, hay, silage, and as a cover or green manure crop. In North America barley is used mainly as a feed grain and for malted products with only a small amount being used as food. In other areas of the world, a major part of the crop is used directly as human food. It is generally used as flour in flatbreads and as pearled barley in soups or cooked with other seeds.

Barley may have been the first cereal used in making beverages. Malting is a controlled germination during which enzymes change complex compounds in the grain to simpler compounds. After the grain has germinated for 5 to 7 days, the green malt is dried down to a moisture content of 4 to 7 percent to stop germination.

BIOLOGICAL ADAPTATION

The effects of environment and man's activities have resulted in the development of many diverse types of barley. This can be seen from the fact that barley is grown from north of the Arctic Circle in Norway to near the equator in Ethiopia. In mountainous regions, barley cultivation reaches higher elevations than any other grain crop (between 3,600 and 4,500 m in Tibet).

This wide adaptation and the short growing season needed make barley man's most dependable grain crop. The length of the growing period of barley varieties varies considerably but is generally from 55 to 111 days. Length of growing period of a particular variety is influenced by temperature, light and soil moisture conditions. The emergence-to-headed period of a variety is shortened as one proceeds northward (longer days), but this will also be influenced by temperature conditions.

GROWTH

The period from sowing to emergence fluctuates considerably but takes 5 to 7 days on the average. Seeds begin to germinate at about 2° C but germination is very slow at this temperature. The optimum temperature for emergence is in the range of 15 to 20° C. Young barley plants have considerable tolerance to low temperature and have been reported to survive temperatures as low as -9°C.

Tillering starts after the appearance of the third leaf, which is usually 10 to 15 days after emergence. The rudimentary spike, with nodes and internodes, forms during tillering. Flowering and fertilization of spring barley take place mainly before the spike has emerged from the boot. With winter barley, heading generally occurs earlier during cooler weather and flowering does not begin until after the heads emerge.

The period from shooting until shortly after heading is considered to be the growth period during which cereals are most sensitive to environmental stress. The actual temperature or moisture conditions under which stress occurs vary greatly among varieties. Varieties developed for cooler environments will often do poorly when grown in a warmer climate, and vice versa.

Of the major small grain crops, barley has the lowest transpiration ratio. Under average soil fertility and a favorable growing season, barley will require 300 to 455 kg of water to produce one kg of dry matter (grain and straw). Water is used more efficiently at higher fertility levels and good growing conditions.

SOIL REQUIREMENTS

Barley grows best on a well-drained, medium-textured soil with a pH of 7-8. Barley grown on high N soils will often lodge and/or be too high in protein for malting. Some varieties have a high salt tolerance, an important factor in stand establishment and grain yield in areas with slightly saline soils. While barley is more tolerant of soil salinity and alkalinity, it is more sensitive to soil acidity than other cereals.

ACTUAL AND POTENTIAL YIELD OF BARLEY

Barley can be grown any place in a rotation but may lodge due to excess N when grown after fallow, legumes, or on land that has been in grass for many years. These are also situations where malting barley may end up with a protein level too high for malt grade.

Average yields of barley for a particular environment are usually only 30 to 60% of possible yields with present technology and varieties. Some of the highest average yields (1969-71) for several countries are shown in Table 8.2. The potential yield of barley under ideal conditions of moisture, temperature, soil physical conditions and nutrients should be about 15,000 kg/ha (280 bu/acre).

MINERAL NUTRITION

The average yield of barley for various countries is in large part a reflection of available moisture supply and level of mineral nutrition.

TABLE 8.2
Average barley yields for several countries for the years
1969-71 (FAO, 1980)

Country	Yield	
	Kg/ha	Bu/acre
Zimbabwe	3714	69
Yemen Democratic Republic	3236	60
Austria	3334	62
Belgium	3480	65
Denmark	3856	71
Germany FR	3586	66
Ireland	3955	73
Netherlands	3627	67
Switzerland	3775	70
United Kingdom	3564	66
New Zealand	3312	61

SOIL TESTING

Many different soil testing procedures are used
throughout the world, so no attempts will be made to give
the interpretation of these tests for barley. In general, soil test levels for P and K should be maintained in
the medium category or above to avoid mineral deficiencies. In many areas N recommendations are based on the
amount of residual nitrate-nitrogen in the top 60 cm of
the soil profile. In other areas the amount of organic
matter in the surface soil is used, or general N recommendations are made without the aid of any soil test.

When a residual nitrate-nitrogen test is used, it is
commonly assumed that residual nitrate-nitrogen in the
top 60 cm of soil has the same availability as fertilizer
N. A common N recommendation would be to supply a total
of approximately 3 kg of residual soil plus fertilizer N
per 100 kg of expected yield of barley per ha (assuming
an N mineralization of 60 kg N/ha during the growing season). If N mineralization is less than this, more residual plus mineral fertilizer would be needed; if mineralization is more than 60 kg N/ha, less than 3 kg of N per
100 kg of barley would need to be supplied.

In addition to N, P and K, barley will respond to
any of the other essential elements when the level in the
soil is deficient. Deficiencies of the secondary and
trace elements in barley are uncommon.

TISSUE TESTING

Plant tissue testing can be used to determine the
nutrient content of barley, but because of its short
growing season tissue tests are not too useful to correct

current problems. Tissue tests are probably more useful in evaluating the present fertility program and updating plans for next year.

The rate of nutrient uptake and hence the nutrient concentration in plants is influenced by environmental conditions. Some soil factors that are known to influence plant nutrient concentration are pH, temperature, moisture and salinity. In addition, plant nutrient concentrations will vary between varieties of the same species. For plant analysis to be of aid in a fertilization program, the influence of all these factors must be taken into consideration. Some general guidelines on the amounts of several nutrients to be found in the above-ground portion of barley plants at the early heading stage are given in Table 8.3. At the heading stage, the preferred plant part to sample is either the whole plant (above ground) or the upper leaves. The specific plant part used is not as important with barley as is consistently using a certain part for analysis.

TABLE 8.3
Interpretation of plant analyses for barley based on the above-ground portion of plants collected as the head emerges from the boot (stage 10.1 on the Feekes scale, Large, 1954)

	Nutrient Concentration in Dry Tissue			
	Deficient	Low	Sufficient	High
			%	
Nutrient				
(winter barley)	<1.25	1.25-1.74	1.75-3.00	>3.00
(spring barley)	<1.50	1.50-1.99	2.00-3.00	>3.00
Phosphorus	<0.15	0.15-0.19	0.20-0.50	>0.50
Potassium	<1.25	1.25-1.49	1.50-3.00	>3.00
Calcium		<0.30	0.30-1.20	>1.20
Magnesium		<0.15	0.15-0.50	>0.50
Sulfur		<0.15	0.15-0.40	>0.40
			ppm	
Manganese	<5	5-24	25-100	>100
Zinc		<15	15-70	>70
Copper		<5	5-25	>25

> = more than.
< = less than.

VISUAL DEFICIENCY SYMPTOMS

Nitrogen (N). Deficiency is characterized by spindly growth and yellow or light green foliage. Nitrogen deficiency appears somewhat similar to S deficiency. The main difference between these two deficiencies is that in the case of N the yellowing appears first on the older leaves and proceeds upward on the plant while in the case of S the yellowing is more marked on the younger leaves.

Phosphorus (P). Deficiency is usually evident on very young plants as browning of the leaf tips and eventual death of the tissue starting with the oldest leaves.

Potassium (K). The usual yellow discoloration at leaf tips and margins indicating K deficiency in small grains is preceded by purplish-brown spots in the case of barley. These symptoms begin on the older lower leaves.

Minor Elements. Deficiencies of Zn, Fe, Mn, Cu, Mo and B occur infrequently on barley and are difficult to diagnose visually.

FERTILIZATION OF BARLEY

Nutrient requirements can be met with animal manures, composts or mineral fertilizer. Barley responds well to fertilizer on soils found to be low in nutrients by soil test. Since both barley protein content and yield can be increased by N, N applications must be controlled to grow barley of malting quality. North American malting and brewing companies prefer barley that has 11.5-13.0% protein. Barley with more than 13.5% protein is used as feed.

Time of N application relative to the growth stage of barley has a large influence on protein content. For example, it was found that when 22, 45 or 67 kg N/ha were applied to barley in early growth stages, increased vegetative growth and grain yield resulted without an increase in grain protein content. However, when the same rates of N were applied at flag leaf emergence, yields did not increase but grain protein content increased.

Likewise, the date of planting will influence protein content. In the North Central USA, barley planted before the middle of May will usually have protein levels low enough to be acceptable for malting, but barley planted after the middle of May is more likely to have high protein.

The supply of K and P in the soil also affects the quality and yield of barley. When these nutrients are applied to deficient soils, they tend to improve straw strength, increase kernel plumpness, increase yields and reduce grain protein content.

In general, barley is planted at the rate of 80 to 135 kg seed/ha. Where rainfall and soil nutrients are adequate for high yields, higher seeding rates should be used. Although small grains have the ability to tiller when stands are low, it is risky to expect tillering to occur. High soil temperatures during the normal tillering period will greatly reduce the number of tillers that develop.

FERTILIZER PLACEMENT

When soil fertility is high and a maintenance program is followed, there is usually very little yield difference between band and broadcast applications. On
soils testing low in P and K, a given amount of fertilizer placed near the seed will often give more yield response than the same rate broadcast.

Overall fertilizer philosophy, therefore, influences
the best placement method for P and K. If a farmer is
interested in building up his soil fertility to the medium or high soil test range, a broadcast application is
probably best. This would be true even though more fertilizer would be needed to obtain a given yield in the
year applied. Fertilizer that is not removed by the crop
in the year of application will raise the overall soil
fertility and be available to future crops. Soils that
are medium to high in fertility allow much greater flexibility in method and time of application of P and K fertilizers. In the long term the savings realized because
of this greater flexibility in fertilizer application
will be equal to or greater than the investment needed to
increase the fertility of infertile soils.

Nitrogen in the form of the nitrate ion is subject
to leaching and denitrification losses and must be managed on a year to year basis. Nitrogen must be applied
at near recommended rates or serious yield reductions
will likely occur. The most important factor in the
method of placement of N fertilizer is the chemical form
of N fertilizer. For example, highly volatile forms of N
cannot be broadcast on the soil surface.

BENEFITS FROM ADEQUATE MINERAL NUTRITION

In countries that use fertilizers intensively it is
generally agreed that 50 to 70% of long-term yield is due
to the application of fertilizer. Yield increases from
manure and mineral fertilizer are usually in close agreement when the amounts of nutrients supplied are the
same. Using current grain and mineral fertilizer prices,
a return of up to 5 to 1 will be obtained on a long-term
basis from the application of NPK fertilizer, assuming
all three nutrients are needed. There are many soils in
barley growing areas that are naturally very high in K
and/or P, and crops can be grown on these soils for many
years without the addition of these nutrients. In these
cases returns mainly from the application of N will be
much greater than 5 to 1.

88

SUPPLEMENTAL READING

ANONYMOUS. 1973. Malting barley-protein content. Malting Barley Improvement Association, 828 N. Broadway, Milwaukee, WI.

ATKINS, R. E., GEORGE STANFORD, and LLOYD DUMENIL. 1955. Effects of nitrogen and phosphorus fertilizers on yield and malting quality of barley. Agriculture and Food Chemistry 3:609-614.

BAUER, A. and E. H. VASEY. 1964. Potassium fertilizer for barley. North Dakota Farm Research 23:19-22.

BRIGGS, D. E. 1978. Barley. Chapman and Hall, London.

DAHNKE, W. C., J. C. ZUBRISKI, and E. H. VASEY. 1981. Fertilizing malting and feed barley. Circular SF-723, Cooperative Extension Service, North Dakota State University, Fargo, ND 58105.

GATELY, T. F. 1968. The effects of different levels of N, P, and K on yields, nitrogen content, and kernel weights of malting barley (var. Proctor). Journal of Agricultural Science, Cambridge 70:361-367.

LARGE, E. C. 1954. Growth stages in cereals:illustration of the Feekes Scale. Plant Pathology 3:128-129.

LEONARD, WARREN H. and JOHN H. MARTIN. 1963. Cereal Crops. The Macmillan Company, New York.

MCBEATH, D. K. and J. A. TOOGOOD. 1960. The effect of nitrogen top-dressing on yield and protein content of nitrogen deficient cereals. Canadian Journal of Soil Science 40:2, pp. 130-135.

NUTTONSON, M. Y. 1957. Barley-climate relationships and the use of phenology in ascertaining the thermal and photo-thermal requirements of barley. American Institute of Crop Ecology, Washington, D.C.

PENDLETON, J. W., A. L. LANG, and G. H. DUNGAN. 1953. Response of spring barley varieties to different fertilizer treatments and seasonal growing conditions. Agronomy Journal 45:529-532.

RUFFING, B. J., D. T. WESTERMANN and M. E. JENSEN. 1980. Nitrogen management for malting barley. Communications in Soil Science and Plant Analysis 11:889-894.

WARD, R. C., D. A. WHITNEY, and D. G. WESTFALL. 1973. Plant analysis as an aid in fertilizing small grains. In: Leo M. Walsh and James D. Beaton (eds.). Soil Testing and Plant Analysis. Soil Science Society of America Inc., Madison, WI.

ZUBRISKI, J. C., E. H. VASEY, and E. B. NORUM. 1970. Influence of nitrogen and potassium fertilizers and dates of seeding on yield and quality of malting barley. Agronomy Journal 62:216-219.

9
PEARL MILLET

L. N. Skold

The word millet (or miscellaneous cereal) is used as a collective term to embrace a group of annual grasses, including 10 genera and 12 to 14 species, all of which are grown primarily for their edible grain. Worldwide, millets provide the staple food for at least 400 million people, and have the great merit of growing in various harsh, unfavorable environments where other grain crops will not grow or are less productive. Since one or more of the various millets is grown in varying amounts in more than 60 countries, often in scattered localities, and since they do not enter into international trade but are consumed locally, total world production can only be estimated. It seems safe to assume, however, that they are grown on about 70 million ha, producing at least 45 million mt of grain annually, or about 4% of the total grain production.

Of this group, pearl millet, Pennisetum americanum, is the most important. Other common names for this crop are cattail millet, bulrush millet, bajara, kambu and many others.

In addition to being the most important, pearl millet probably has the greatest potential for further development. It is more tolerant of heat than sorghum or maize and is generally considered to be more efficient in utilizing soil moisture than either crop. Being a vigorous plant with large stems, leaves and heads it has an exceptional yielding capacity in favorable environments. It is adapted to light textured, well-drained soils and is grown successfully where rainfall is inadequate for sorghum production. Production as a cereal crop is centered in the drier areas of west and south India, and south of the Sahara Desert in Africa. As a pasture and forage crop it is grown in the southeastern USA and in Australia and South Africa.

Possibly 85% of pearl millet grain is used as human food, the balance being used for seed or animal feed.

[1]Professor of Plant and Soil Science, University of Tennessee, Knoxville, Tennessee 37901, USA.

The grain is prepared for consumption in many ways. Using hand grinding stones it may be ground into a meal to be used for baking a flat, unleavened bread, or to be fried, or cooked as a soup or gruel. Sometimes it is cracked and cooked much as rice would be. Pearl millet is among the most nutritious of cereals. Protein content commonly ranges from 9 to 15% and is of high biological value having a good distribution of amino acids except for a deficiency in lysine. The fat content is higher than in most other grains and the nutrient mineral content is good. Pearl millet is a particularly good source of Fe and P. If populations whose staple food is pearl millet were to change their food habits to consume more of the cereals preferred by the western world, their nutritional status would almost certainly suffer.

Possibly 5% of the total area planted to pearl millet is utilized as forage, either for grazing or a green chop feed, the largest area being in the southeastern USA. As a green forage to be cut and fed to livestock it is nutritionally equal to other forages and has a very low hydrocyanic acid content. The dry stover with the grain removed has a low feeding value for livestock and in the traditional production areas is often used as a thatch or fuel.

ECOLOGICAL ADAPTATION

Pearl millet production is concentrated in regions from the equator to 35° north and south latitude, and at lower elevations up to 1400 m. Bright sunny weather with occasional showers is most favorable for satisfactory yields. Optimum temperatures appear to be a few degrees higher than for sorghum. The crop is grown in areas with a rainfall of from 125 to 900 mm during the growing season, and pearl millet grows better than sorghum under dry conditions because the life period of the crop is often shorter. Pearl millet is considered to be more drought resistant than sorghum. However, successful production in areas of scant rainfall is probably as dependent upon its drought-evading qualities and heat tolerance as upon drought resistance as such. There is a remarkable amount of genetic diversity in pearl millet and certain ecotypes can mature a crop in as little as 60 days, taking advantage of the small amount of available soil moisture. Furthermore the crop is deep rooted and exploits soil moisture to depths of 150 cm. Apparently pearl millet is not able to go into a dormant condition during periods of moisture stress as can sorghum. Under more favorable moisture conditions and good management pearl millet has the potential for grain yields up to and probably in excess of 6000 kg/ha.

Pearl millet is best adapted to light textured, well-drained soils although it has a wide range of soil adaptation. Where sorghum and pearl millet are included in cropping systems, pearl millet often occupies the sandier soils, and sorghum is grown on soils with finer texture. Heavy rains may cause much damage to the crop because it is intolerant of soil water-logging, or excess soil moisture.

Pearl millet appears to be tolerant of a wide range of soil pH. In general, yield is not affected within a pH range of 4.5 to 6.8, i.e. strongly to mildly acid. In regions where the crop is most commonly grown, the use of soil amendments to adjust soil pH is not common, and critical information as to salt tolerance is not available. However, since it is grown, for example, on the high sodic soils of Haryana, India, the crop must have considerable tolerance to salt. Frequently the crop is grown on soils of low fertility, producing modest yields even under these conditions. However, it responds well to additions of organic or inorganic fertilizers.

Pearl millet is grown in a wide variety of farming systems. It may be the only main crop year after year, grown in regularly planned rotations with legumes or various cash crops, or grown in a mixed crop situation being interplanted with various legumes, sesame, sorghum, or other crops. A common rotation is groundnuts alternated with pearl millet, in which case the millet crop is the one receiving such fertilizer as may be applied. Residual N fixed by the legume crop benefits the following pearl millet, either in a rotation or in a mixed cropping.

Seedbed preparation methods likewise vary widely, from thorough preparation by repeated plowing, harrowing and smoothing to virtually no land preparation at all. Seed may be sown broadcast, drilled in rows, or hill planted. In any case, thinning the crop to the desired stand compatible with expected rainfall is an important aspect of culture. Weed control is usually accomplished by mechanical means and is critical to crop growth. In areas where monsoon rains occur, weed control may be difficult and often limits grain yields. Plant populations of pearl millet vary widely depending largely on soil fertility, rainfall and potential for mixed cropping. Time of sowing is largely dependent upon the rainfall pattern of the locality unless it is grown under irrigation. Planting is usually timed to coincide with the onset of rain, and the life period of the variety sown is dependent on the length of time during which adequate soil moisture conditions prevail. Critical growth stages for soil moisture are the seedling stage and the heading and flowering period.

The recent development of improved hybrids with much greater yield potential and the development of more sophisticated systems of farming may result in an expanded role for pearl millet as a main season crop and for various multiple cropping systems. Traditional cropping methods are currently experiencing changes which may be even more profound in the future.

NUTRIENT REQUIREMENTS

With the advent of superior hybrids and composites that are more responsive to management inputs, with shorter plants and stronger stems, and with tolerance to higher plant populations, the yield potential of pearl millet is beginning to be realized. In numerous tests throughout India under favorable climatic conditions, various combinations of plant spacing and rates of N fertilization have been evaluated. Rates of 120 to 180 kg N/ha with 40 x 15 cm plant spacing have often given yields exceeding 5000 kg/ha.

In India, pearl millet has responded most to applications of N and somewhat less to P. In general, applications of N and P or N, P and K are superior to applications of N alone.

As indicated previously, accurate estimates of pearl millet yields are not available. It is probable, however, that yields of 500 to 600 kg grain/ha are commonly achieved. Thus the yield potential under the best management practices may be about ten times that of the present average yield. And some of these outstanding yields have been under rainfed conditions, as well as under irrigation. It is probable that pearl millet is as responsive to management and fertilizer inputs as maize and sorghum, and its potential productivity is probably at least as high.

Until comparatively recently, organic fertilizers (animal manures) were the only kind used on pearl millet, and inorganic fertilizers were mostly unavailable in pearl millet areas. The value of organic matter in the form of manure, compost, or cover crops incorporated into the soil prior to planting has been demonstrated. Additions of approximately 10 mt/ha of manure or compost often give 50 to 90% yield increases over unfertilized controls in both India and Africa. Animal manures may also contribute to the supply of essential minor mineral nutrient elements.

Fertilizer response depends on a number of variables; weather, soil moisture, pests and diseases, variety, previous cropping, management practices, and inherent soil fertility can profoundly affect crop response to applied fertilizer. In parts of Africa bird damage can be devastating. Good information on pearl millet nutrition is difficult to obtain, and many research results are inconclusive or misleading. However, over time, a reasonably good picture of fertilizer response is emerging, and there is a clear pattern of positive and economic response to applied fertilizer. The exceptions are remote areas where transportation costs of fertilizer are prohibitive, or where lack of rainfall or factors such as the soil physical properties limit land productive potential.

Soil Testing

While soil testing is commonly practiced in some countries where pearl millet is produced as a food crop, the experiments necessary to correlate accurately soil tests results with fertilizer response generally have not been conducted, although considerable progress is being made.

The need for minor elements in pearl millet production is not generally recognized and there is little evidence to indicate--except in a few isolated cases--that such deficiencies occur. It may prove in the future, as it has elsewhere, that as productivity increases, minor element deficiencies will appear.

Visual Symptoms

Not much information is available on deficiency symptoms in pearl millet, but there is little reason to believe that they are very different from those in maize and sorghum. Please refer to the chapters on those crops for deficiency symptoms that may be seen in pearl millet.

In areas of lower rainfall and higher soil pH, Zn deficiency is not uncommon. The appearance of yellow streaks between the veins of the lower leaves, often accompanied by a reddening of leaf sheaths and veins is a good indication of Zn deficiency. Chlorosis may become severe and young leaves may be almost white in color.

10
SOYBEANS

Stanley A. Barber and W. B. Hallmark

Soybeans are grown throughout the world as a source of oil and of protein for human and animal consumption. They constitute a major part of the vegetable oil and protein used. Oil content of soybean grain varies from 15 to 22 percent, while protein varies between 30 and 44 percent. Soybeans grown in the USA have an average oil content of 20% and an average protein content of approximately 40%.

The area of world soybean production has been increasing faster than for any other major crop. Production by year for the period 1970 to 1979 has been dramatic; the area has been increasing at a rate of 2.5 million ha/year. In addition to this increase, yield per hectare has also been increasing even though new lands used for soybean production may not be as high yielding as those previously in production. Average world soybean yield increased 0.04 mt/ha/year in the 1970's.

Soybeans are grown mainly in North America, Asia and South America. The 1979 soybean area and average yield in major producing countries are given in Table 10.1. Over half the crop area is in the USA, where rapid expansion has continued in recent years just as it has in other countries.

CLIMATIC ADAPTATION

A wide diversity of soybean genotypes is available with maturities ranging from 85 to 180 days. Both determinate and indeterminate types are used.

Soybean is a photoperiod sensitive crop and transition from vegetative to flowering stages is in direct response to day length. The key to the flowering mechanism is the length of the dark period. Most varieties flower soon after the day length begins to shorten. Since day

---

[1]Professor of Agronomy, Purdue University, West Lafayette, Indiana 47907 and Professor of Agronomy, Louisiana State University, Baton Rouge, Louisiana 70893.

length varies with latitude, a variety is only suitable for growth in a band no wider than 160 to 240 kilometers north and south. In the USA there are 10 maturity classes of soybeans that extend across the various latitudes between Florida and the Canadian border. Hence it is necessary to have varieties that are adapted to the latitude where they are to be grown.

TABLE 10.1
Soybean area and yield in 1979

Country	Area (1000 hectares)	Average Yield metric tons/ha
USA	28,543	2.16
China	9,200	1.14
Brazil	8,500	1.73
Argentina	1,850	2.11
USSR	840	0.71
Indonesia	710	0.80
Paraguay	433	1.50
Mexico	400	1.50
Canada	283	2.37
Korea	280	0.92
Romania	270	1.39
Other	1,681	
Total	52,990	1.82

SOIL ADAPTATION

Soybeans are adapted to a wide variety of soils. They grow as well if not better than maize on heavy textured soils where poor drainage often occurs in the spring. Soil pH of 6.0 or higher (mildly alkaline) is preferred; however, on some soils manganese (Mn) deficiency occurs with pH higher than 6.2. Iron deficiency may occur on soils with pH above 7. To accomodate root growth the subsoil should be high enough in pH to reduce the probability of Al and Mn toxicity. Aluminum and Mn toxicities occur on some soils at pH's below 5.2 (significantly acid).

While soybeans are tap-rooted, the tap root growth is usually inhibited by high soil bulk densities (compact soils) and low fertility subsoils, so that many roots branch out from the tap root. The resulting root system is more like a fibrous rooted crop. A large proportion of the root system is usually in the top 20 cm of soil. Soybeans differ from maize in this respect since maize usually has a greater proportion of its root system in the subsoil. Nevertheless, where soil conditions are suitable some soybean roots extend deep into the subsoil.

NUTRIENT REQUIREMENTS

Nitrogen (N): A legume, soybean should be inoculated with appropriate rhizobia bacteria, or grown in soil that contains the right rhizobia from previous soybean crops. The soil should also have pH and fertility levels conductive to growth of the rhizobia in the root nodules. Usually this requires a soil pH above 6.0 (mildly acid) and a medium level of available P and K. A higher level of molybdenum is required for N fixation than for growth of the soybean plant.

Calcium (Ca) and Magnesium (Mg): Soybeans are higher in Ca and Mg than cereal grain crops such as maize, wheat and barley. With a soil pH above 6.0 the levels of Ca and Mg are usually adequate unless the soil is extremely low in cation exchange capacity. The ratio of Ca to Mg is not critical so long as there is more Ca than Mg and the exchangeable Mg level is 100 kg/ha or more.

Phosphorus (P): Soybeans respond to phosphate fertilization similarly to maize. The soybean has a large seed and the quantity of P in the seed used per ha is greater than for maize; consequently, soybeans do not show the early growth response to P that is characteristic of maize and wheat. However, the yield is determined largely by P taken up from the soil during the plant growth cycle, and soybeans show the same relative response as maize to applied P fertilizers. Soybean plants older than 50 days increase in P concentration during growth and it is important that P is distributed through the root zone so the plant can get the required amount during the latter part of the growth cycle when the grain is being formed. In contrast, maize plants decrease in P concentration with age. Soybeans respond to P plowed under and mixed with the soil. Frequently P is sprayed on the soybean foliage. Banded P fertilizer keeps the soil near the band acidic in condition and minimizes oxidation of Mn to unavailable forms. Alternatively, reducing soil pH by adding elemental sulfur will minimize Mn deficiency.

Iron (Fe): Iron deficiency occurs on alkaline soils with pH above 7.0 (neutral). Spraying the foliage with Fe salts is the suggested corrective procedure. Adding ferrous iron to the soil does little to increase Fe availability because it rapidly changes to an unavailable form.

Molybdenum (Mo): Molybdenum deficiency occurs in certain soils when the pH is low (acid), usually below pH 5.5. It can be corrected by seed treatment with Mo because the amount needed is only 10 to 20 g of sodium molydate/ha. Molybdenum is commonly applied in southern

USA soils where the pH is below 6.2. Southern USA soils are more consistently low in Mo than those in areas such as midwest USA, north of the Ohio and Missouri rivers, where only a few soils give soybean yield responses to added Mo.

NUTRIENT DEFICIENCY SYMPTOMS

Nitrogen (N): Soybean growth will be stunted and leaves a very pale green. Nitrogen deficiency occurs because the soybean roots are not nodulated or nodules are not effective because of poor soil fertility or low levels of Mo.

Phosphorus (P): Soybean growth is stunted and stems are reddened under severe deficiency. Leaves are tilted upward as compared to leaves of soybeans grown with ample P.

Potassium (K): Deficiency in early growth stages shows up as irregular mottling around the edges of leaves. These chlorotic areas increase as deficiency becomes more severe, then they merge so that chlorosis occurs around the edges of the leaf. As deficiency becomes more severe, chlorosis progresses toward the center of the leaf. In early growth, necrosis may be on lower leaves but later in the season it may be on leaves in the upper parts of the plant.

Calcium (Ca): On acid soils, Ca deficiency and Al toxicity may occur. Growth is stunted and yellow. Nitrogen fixation will be low because of poor nodulation and little N fixation.

Magnesium (Mg): In early stages of deficiency the areas between the veins become yellow. These areas later turn deep yellow and rusty specks and necrotic blotches may appear between the veins and around the edges of the leaves. In later stages, Mg deficiency gives the appearance of early maturity. Magnesium deficiency may occur on acid soils, especially where high rates of K have been applied. Excess K depresses Mg and Ca uptake and intensifies their deficiency symptoms.

Manganese (Mn): Manganese deficiency is the most common micronutrient deficiency of soybeans. It frequently occurs early in growth in low areas of the field. Later in the season, under some situations new growth may be normal as roots grow into soil that has a higher Mn supply, or climatic conditions change so that available soil Mn increases. Manganese deficiency is associated with sandy and/or high organic soils with a soil pH above 6.2, and also occurs in heavy textured soils. It frequently is only present in low spots or depressions in the field.

Leaves become chlorotic in the interveinal areas while the veins remain green. Symptoms differ from Fe where the veins also become chlorotic. Whole leaves, veins excepted, become pale green and then pale yellow. Brown spots and necrotic areas develop on lower leaves as the deficiency becomes more severe. The deficiency occurs on the new leaves, however, when later growth is normal the chlorotic leaves are no longer at the top of the plant.

Iron (Fe): Iron deficiency of soybeans occurs on some soils when the pH is high. Frequently it is on soils which contain considerable quantities of free lime. With Fe deficiency, the whole leaf including the veins turns yellow. Interveinal areas turn chlorotic first then the veins become chlorotic and finally, under severe Fe deficiency, the leaves turn almost white.

Zinc (Zn): Zinc deficiency of soybeans is not common. The leaves become chlorotic, then rusty brown in color. The veins remain green. The chlorosis is uniform over the leaf and not concentrated initially on the edges as occurs with deficiencies such as K. Zinc deficiency often occurs where plants are growing mainly on the subsoil.

Molybdenum (Mo): Molybdenum deficiency symptoms are rare. more Mo is needed for n fixation than for plant growth. Hence,the first symptoms which occur are those of N deficiency unless N supplies in the soil from organic matter mineralization and/or previous N fertilization are adequate to supply N to the plant.

TISSUE ANALYSIS

The adequacy of the nutrition of the soybean plant can be appraised from chemical analysis of the upper leaves. Table 10.2 gives the composition range showing adequate nutrition.

FERTILIZING SOYBEAN

Phosphorus (P): Results from a 25 year experiment at Lafayette, Indiana, (Table 10.3) illustrate the response of soybeans to P and show how the reponse is similar to that for maize. In a rotation of maize and soybeans, P is applied for the preceding maize crop, since soybeans respond to the P level in the soil rather than directly to P fertilizer application. On very deficient soils such as with lateritic soils where P fixation is high, placement of P may be needed to increase efficiency of use. However, in almost all situations of soybean production, P is broadcast and worked into the soil. Row fertilizer placed near the seed can injure germination, hence it is seldom used. After the P level in the soil has reached a medium level, 25 kg P/ha/year is required to maintain an adequate level of P in the soil.

TABLE 10.2
Soybean nutrient composition of adequately fertilized soybeans

Nutrient	Composition of upper leaves % (on dry leaf basis)
N (Nitrogen)	4.26 - 5.5
P (Phosphorus)	0.26 - 0.50
K (Potassium)	1.71 - 2.50
Ca (Calcium)	0.36 - 2.00
Mg (Magnesium)	0.26 - 1.00
	ppm
Mn (Manganese)	20 - 100
Fe (Iron)	51 - 350
B (Boron)	21 - 55
Cu (Copper)	10 - 30
Zn (Zinc)	21 - 50
Mo (Molybdenum)	1 - 5

Potassium (K). Soybeans require large amounts of K. Exchangeable K levels of 200 kg/ha or more in the surface 15 cm are required for maximum soybean yields. Potassium fertilizer should not be banded with the seed because of danger of salt injury to the seedlings. Broadcast applications worked deeply into the soil are preferable. Annual applications of 80 kg K/ha are needed to maintain yields after a medium or higher level of available K in the soil has been reached. The level of K required should be determined by soil testing.

MICRONUTRIENTS

Soybeans are susceptible to deficiencies of Mn, Fe, Mo, and Zn. Manganese deficiencies occur in soils low in total Mn when the soil pH is raised above 6.3 (mildly acid). The deficiency can be corrected by banding divalent Mn with P fertilizer or by applying a foliar solution.

AVERAGE YIELDS VERSUS POTENTIAL YIELDS

The average yield of soybeans in the world has been increasing about 45 kg/ha per year over the past 10 years. This increase is due to a combination of improvements which includes adapted variety, improved weed control, and suitable cultural practices, as well as increased rates of fertilization. The continued increase in yield will depend on adoption of new practices and development of new varieties and increased fertilization, as well as the inherent productive capacity of new land areas. As production is extended into areas less well adapted, yields may not be as high as in better adapted areas.

TABLE 10.3
Response of soybeans to an average annual phosphate application over 25 years in a maize-maize-soybean-wheat rotation (yields are average for 25 years).

Phosphate added kg/P/ha/yr	Soybean yield average kg/ha	Relative yield	Maize yield kg/ha	Relative yield	Soil test Bray P[1] at end of 25 years ug/g
0	2740	.85	7560	.88	14
5	2800	.90	7850	.92	18
11	3130	.98	8460	.98	30
22	3180	.99	8640	1.00	43
44	3180	.99	8610	1.00	89
49	3200	1.00	8620	1.00	--
54	3210	1.00	8500	.99	--

FERTILIZER PLACEMENT

Soybeans do not usually respond to P and K fertilizer banded by the row at planting. The nutrients supplied in the seed are often enough for early seedling growth. Soybeans respond to fertilizer distributed throughout the soil. Some believe they respond as well to residual fertilizer applied to a previous crop as to fertilizers applied directly to soybean. Building up soil P and K to medium or higher levels is a satisfactory method of guaranteeing proper nutrition for soybeans.

Soybean continues taking up nutrients from the soil until grain maturity. The average plant P concentration increases gradually from 50 days to maturity. There is a belief that soybeans may not be able to get enough nutrients from the soil late in the season to maximize yield yield, so low concentration foliar sprays have been used with uncertain results.

CROP ROTATION

Soybeans are not usually grown continuously year after year since parasites (cyst nematode), disease, and/or insect problems tend to reduce yields after two or more years of continuous soybeans. In the midwest USA, soybeans are frequently alternated with maize. this rotation is beneficial since maize yields are higher when grown in a rotation with soybeans than when grown continuously. Corn root worm can also be controlled without the use of pesticide.

In areas where the cropping season is long enough, soybeans are planted after wheat harvest. The practice is called double cropping because both wheat and soybeans are grown in one year. The wheat crop is winter wheat planted the previous fall. In southern Brazil, soybeans are alternated with wheat. The wheat is grown in the winter and soybeans are grown during the summer.

11
DRY BEANS

Roger F. Sandsted

The common dry bean Phaseolus vulgaris is an annual which originated in Mexico and/or the Andean region of South America. There are other species which are included in dry bean production statistics. Some of these are the Lima Phaseolus lunatus, scarlet runner Phaseolus coccineus, and the Tepary bean Phaseolus Gray var. latifolius Freem. acutifolius. Production practices and uses of each are similar.

Table 11.1 shows dry bean areas of production, production and yields, by selected regions and it also shows how this crop has spread throughout the world since its discovery in the Americas by Columbus.

In those countries where meat production is low and dry beans are grown and consumed they are an important source of protein and energy for the human population. Beans contain 22-23% protein but have a low content of some essential amino acids, particularly methionine. The human requirements for protein generally are nearly satisfied if beans are consumed with maize or other cereal grains which are deficient in lysine and tryptophan, amino acids which beans contain. The importance of this combination was discovered by the inhabitants of the countries where beans and maize are grown and may explain why intercropping of these two became a standard practice long ago.

Beans provide significant amounts of vitamin E, folic acids, small amounts of thiamine, riboflavin and niacin for the human diet. They are a good source of calcium and iron, and to a lesser extent phosphorus and potassium.

---

[1]Professor, Department of Vegetable Crops, College of Agriculture, Cornell University, Ithaca, New York 14853, USAID

ECOLOGICAL ADAPTATION

The bean is a warm season crop, but is not well adapted to the humid tropics. It yields poorly at high temperatures particularly when accompanied by high humidity with resultant attacks by bacterial and fungal diseases. Days to flowering and maturity are extended by high temperature, even in the absence of disease. High temperature may cause reduced pod set. Most bean production is in areas with a mean temperature of 18 to 25°C. On either side of this range, production declines sharply and practically ceases below 13°C. The optimum temperature for development of beans, that giving earliest flowering and maturity, is about 23 to 24°C. At temperatures above or below this beans will take longer to flowering and maturity. There are differences in variety response to both low and high temperatures.

Varieties which require 80 days from planting to maturity at 18 to 25°C may require 150 days at 13 to 15°C. Increasingly later flowering and maturity with lowering of temperature is called the "general temperature response." Later flowering and maturity with raising of temperature, which occurs at above-optimum temperatures, is called the "high temperature response."

Some cultivars are highly sensitive to both day length and temperature, while others are insensitive to both, and still others are sensitive to daylength or temperature but not to both. No variety is totally insensitive to either, especially temperature, and each cultivar has a specific genetically controlled level of small to intermediate, high, or very high sensitivity. The importance of this phenomenon to a farmer is that the maturity of a variety may be greatly changed if it is moved long distances north or south, or simply moved to where growing temperatures are different because of higher or lower altitude.

Length of the desirable growing period of a cultivar is determined by the number of days between killing frosts or by the duration of available water from rain or irrigation. When the climate is favorable, more than one crop may be grown per year.

Beans are grown on several different kinds of soil. The best soils are those of intermediate texture (loams, silt loams) which are well or moderately well drained yet have good moisture holding capacity. The fine textured soils (clays, clay loams) have a high moisture holding capacity and high mineral content but often are somewhat poorly drained. Beans tolerate excess soil water for only a very short time. Crusts which form on the clay or high silty clay surfaces interfere with seedling emergence. Clays require the highest input of energy for farming whether it be from machines, animals or people. The coarse textured soils (sand and gravels) often are so well drained they do not hold sufficient moisture to support the plants during periods of deficient moisture. Coarse textured soils tend to be low in available mineral or organic matter content. Of the mineral soils, they require the least energy to farm and can be very productive with additional inputs of fertilizers and irrigation water. Beans often grow well on organic soils (more than 20% organic matter) if the soils are properly drained.

When grown as a single crop, beans fit into several rotational systems, following maize, other cereal grains, and legume or grass sods, or other cash crops. The rather weakly rooted beans benefit from soil-improving crops, particularly forage legume sods. Not only do such crops help maintain soil organic matter and supply N but they also improve soil structure which allows better root growth and higher yields. If beans follow grass sods (without legumes) or other high residue crops, additional N fertilizer may be necessary. Because of soil borne root diseases or leaf diseases which may be present in previous crop plant residues, it is recommended that beans appear in the rotation once every 3 years or more.

In most instances where beans are grown in some form of multiple cropping, that system evolved from many years of farmer experience. New technological inputs (fertilizers, tillage, crop protection) into the system may improve bean yields but the basic system itself will endure long into the future. Considerable yield improvements have been obtained with only slight modifications in plant spacing and time of application and placement of fertilizer. Some yield increases have been obtained with only moderate or no extra costs.

Because of various plant growth habits, bean plant types are discussed here:

Type 1: Strictly a bush, there are no runners. Often planted in pure culture or between rows of maize, potatoes or other row crops. Plants do not grow into spaces left by missing hills so do not compensate greatly for low plant populations. They are easily cultivated and harvested.

Type 2: Bush type growth habit but they do produce very short runners. Type 2 plants compensate somewhat more for missing hills and are more productive than Type 1. Plant breeders are developing erect, strong stemmed, tall growing cultivars of Type 2 because of their high yield potential and their ability to hold pods high off the ground. They are easy to cultivate and harvest.

Type 3: Plants produce long runners, which do not climb but grow along the ground. They often are late in maturity, compensate well for missing hills and are high yielding. Disadvantages of Type 3 are excessive foliage particularly if overfertilized with N, pods are on or near soil surface and subject to rots and molds, and they are difficult to cultivate and harvest.

Type 4: Plants have long, climbing runners. Type 4 cultivars can be grown in pure stands on poles, but most often they are intercropped with maize where they climb the stalks. Both crops can be grown at the same time, or the beans may be planted after the maize is mature and use the maize stalks for poles. Type 4 cultivars are high yielding but require extended harvesting because pods mature over an extended period of time.

NUTRIENT REQUIREMENTS

Ripe bean yields of 3000 to 3500 kg/ha are uncommon but do occur on good soils with weed control, adequate inputs of fertilizer, chemical protection from diseases and insects, and adequate soil moisture. These yields are approximately 6 to 7 times that of the world average (Table 11.1) and several times that in some countries. High yields can be obtained occasionally any place where beans are grown but the most consistent high yields are found in those places with a dry climate, good soils with sufficient fertilizer, and irrigation.

In most developed countries the mineral nutrient requirements for beans are adequate or available. In developing countries the mineral requirements are not being met because of lack of natural resources, fertilizer supplies, transportation to the more remote farming regions, and poor financial status of the small farmer.

Of the soil nutrients required by beans N, P, K and Ca are required in the highest amounts, followed by Mg and S. Boron, Cu, Fe, Mn, Mo, and Zn are required in very small amounts but are essential.

SOIL TESTING

The most frequent soil tests made on bean fields are for acidity (pH), organic matter (O.M.), P and K. Tests for micronutrients generally are made only when trying to diagnose for a nutritional problem other than N, P, or K.

Soil reaction (pH). Beans often grow well within a pH range of 5.5 to 7.5. The optimum range is between 6.5 and 7.0. Within this range the required mineral nutrients are available at the optimum level, being neither unavailable for plant growth nor in amounts toxic to plants.

Soil organic matter (O.M.). No definite soil O.M. content has been established as being best for beans. In general an O.M. content of 2% is adequate. If the content of a mineral soil exceeds 4% look for poor water holding capacity.

TABLE 11.1
Estimated world dry bean production (FAO, 1979)

Selected Regions	Area (1000 ha) 1969-71	Area (1000 ha) 1979	Production (1000 met. tons) 1969-71	Production (1000 met. tons) 1979	Yield (kg/ha) 1969-71	Yield (kg/ha) 1979
World	23,692	25,486	12,377	14,781	522	580
Africa	1,897	2,157	1,108	1,259	584	584
N. America[1]	2,379	2,197	1,745	2,062	1,163	1,369
C. America	470	521	300	316	638	639
S. America	4,172	4,924	2,682	2,814	643	571
Asia	12,296	14,072	5,619	7,545	457	536
Europe	2,438	1,559	846	699	347	449
Oceania	5	7	1	5	322	765
USSR	35	50	75	80F	2,132	1,600

F = FAO Estimate.
[1]Includes Mexico.

Aluminum (Al) toxicity. Aluminum is a natural component of soils, but is toxic to plants in excess amounts. The more acid (low pH) the soil, the more Al is available to plants and the greater the problem. Liming the soil to a pH of 5.2 or higher decreases the availability of Al and is the method of control. Manganese (Mn) toxicity can also occur in some acid soils. Liming is the recommended remedy.

Boron (B) deficiency. Deficiency can occur in acid, coarse textured soils with low organic matter, high aluminum or iron hydroxides, or in alkaline or over-limed (pH over 7) alluvial soils. Plant symptoms: (Excellent photographs of mineral deficiency symptoms of beans can be found in Schwartz and Galvez, 1980). The symptoms are crinkled leaves, curled down with necrotic spots, stems thick. Leaf and soil analysis: Critical level in leaves is 20-25 ppm, in soils 0.65 ppm. Remedies: Apply manure and/or grow cover crops for organic matter (O.M). Broadcast and incorporate 1 to 2 kg B/ha in acid soils. A foliar application of a 1% solution of Borax or Solubor will help if plants are not too severely stunted.

Boron (B) toxicity. Occurrence: Over-application of B fertilizer to previous or present crop, or banded too close to the seed at planting. Leaf or soil analysis: Critical level in leaves is 300 ppm. Toxicity level in soils is 5 ppm. Remedy: Use B cautiously and accurately. Some cultivars are more tolerant to excess B than others.

Calcium (Ca) deficiency. Occurrence: In a few acid soils. Plant symptoms: Dark green, crinkled leaves, curled downward with yellowish margins. Plant internodes short, root growth reduced. Leaf analysis: Critical minimum level in developed leaves is 1.44%, optimum or normal level 2%. Remedies: Before planting incorporate (deeply) calcitic or dolomitic lime. Low rates (500 kg/ha) may reduce Ca deficiency but higher rates are needed to reduce soluble Al or Mn toxicities. Caution: do not over-lime.

Copper (Cu) deficiency. Occurrence: Common only in few organic soils. Plant symptoms: Stunted, internodes short, leaves blue green. Leaf analysis: Normal Cu content in upper leaves is 15-25 ppm. Remedies: Before planting incorporate 5-10 kg/ha of copper sulfate.

Iron (Fe) deficiency. Occurrence: In some alkaline or overlimed soils. Plant symptoms: Upper leaves light yellow to white, green veins. Leaf analysis: Normal content of leaves is 100 to 800 ppm. Remedies: Band iron chelates (special organic compounds) with fertilizer at planting, or make foliar applications to the leaves. Follow label instructions.

Magnesium (Mg) deficiency. Occurrence: In some
sandy or acid soils or volcanic ash soils high in Ca or
K. Plant symptoms: Older lower leaves yellow at outer
surfaces. Upper leaves may be more yellow with green
veins. Leaf analysis: The critical minimum level is
0.2% to 0.3%; normal content is 0.35% to 1.3%. Remedies:
Band those fertilizers containing 2% magnesium oxide or
magnesium sulfate, or apply a foliar spray containing 1%
magnesium sulfate. Repeated foliar applications may be
necessary at 1 to 2 week intervals.

Molybdenum (Mo) deficiency. Occurrence: Common
only in some acid soils. A deficiency may be corrected
with application rates as low as 30-60 g/ha of sodium
molybdate. Caution; excessive Mo is toxic to beans.

Nitrogen (N) deficiency. Occurrence: Common in
soils with low organic matter (O.M.), or in acid soils
with toxic levels of Al or Mn, or deficient levels of Ca
or Mg. All of the above problems limit microbial breakdown of soil organic matter, or inhibit N fixation by
Rhizobia (root nodule) bacteria. Plant symptoms: Small
plants with yellowing of the lower leaves progressing to
upper leaves. Leaf analysis: Deficiency symptoms will
appear if the N content is less than 3% at time of initial flowering. The normal content is about 5% N. Soil
tests are not reliable for predicting a N deficiency or
rates of N fertilizer application. Remedies: Lime acid
soils. Apply animal manure or grow N-fixing crops (legumes) in the rotation, and/or apply N fertilizers.
Caution: Avoid placing N fertilizer in direct contact
with seed.
Some cultivars of beans and some Rhizobia fix more N
than others. Researchers are attempting to find the combinations that will fix the most N. To date only limited
success applicable to a farmer's situation has been
achieved. However, beans planted in fields with little
or no bean cropping history should be inoculated with
Rhizobia of the best strain available.

Phosphorus (P) deficiency. Occurrence: Of the soil
nutrient deficiencies, P is the most common. P is less
available to plant roots in acid soils. Plant symptoms:
Stunted with few branches, lower leaves yellow, and becoming necrotic, may fall off. Upper leaves dark green
and small. Flowering is reduced. Leaf and soil analysis: Deficient leaves contain less than 0.2% P. Upper
fully expanded leaves at 10% flowering should contain 0.2
to 0.4% P. Critical levels in soil vary according to the
kind of soil and the soil test method, but range from 8
to 15 ppm. Remedies: Apply lime to acid soils. Band
superphosphate fertilizers. Broadcast and incorporate

rock phosphates or basic slag ahead of planting. Use superphosphate fertilizers in S-deficient soils. Researchers are attempting to develop cultivars which are tolerant to low soil P availability.

Potassium (K) deficiency. Occurrence: In soils with low clay content and low K-supplying minerals. Plant symptoms: Outer margins of lower and older leaves are yellow and necrotic if deficiency is severe. The symptoms progress to the upper and younger leaves. Leaf analysis: Normal leaves contain 2% K. Deficiency symptoms show if the K content is less than 2% at initial flowering. Remedies: At planting time band 50 to 80 kg K/ha. Adjust rates in accordance with soil tests. Caution: Avoid fertilizer contact with seed.

Sulfur (S) deficiency. Because S is a component of air pollutants its deficiency is seldom found except in regions distant from industrial plants. Plant symptoms: Top growth symptoms may be confused with N deficiency. Root growth is normal. Leaf analysis: The critical minimum level in leaves is 0.2% to 0.25%. Remedies: Broadcast and incorporate 10 to 20 kg/ha of elemental S before planting. Also, gypsum may be effective but check its use with local agricultural specialists. Use S-bearing fertilizers as ammonium sulfate, simple superphosphate, potassium sulfate, etc.

Zinc (Zn) deficiency. Occurrence: In highly alkaline or overlimed soils, and often found in high lime subsoils after topsoil erosion. Plant symptoms: The leaf interveinal tissue is yellowish between green veins. Necrotic areas appear if deficiency is severe. If soil deficiency is slight the first leaves may show deficiency symptoms but the newer leaves do not as the roots grow deeper into the soil. Pod set may be reduced. Leaf and soil analysis: The critical minimum level of Zn in leaves is 15 to 20 ppm. The normal level is 42 to 50 ppm. (Levels of 120 to 140 ppm are toxic to the plants.) Remedies: Broadcast and incorporate 5 to 10 kg/ha of zinc sulfate or band 6 to 7 kg/ha of zinc sulfate well mixed into the fertilizer. Special Zn-containing fertilizers, if available, will help. Some cultivars are more tolerant of low available Zn in soils than others.

CULTIVAR DIFFERENCES

In the past it has been noted that some bean cultivars are more susceptible or tolerant of soil mineral toxicities or deficiencies than others. It has been relatively recent that bean breeders have begun to take advantage of these differences. Prospects seem good for the development or selection of commercial cultivars which will be tolerant of excessive soil Al, and low B, N and P. There are no commercial hybrids.

FERTILIZING BEANS

Where available, animal manure can be of value as a bean fertilizer. Its principal mineral nutrients are N, P_2O_5 and K_2O. The average composition in cattle, hog and horse manure is approximately 5 kg N, 3 kg P_2O_5 and 5 kg K_2O per ton. Note the low value of P_2O_5. The N and P_2O_5 in chicken manure, and the N and K_2O in sheep manure are 2 to 3 times the values shown above. The fertilizer values of manure can be improved by mixing superphosphate with it as it is collected or by supplemental applications in the field. Approximately one-half the N, one-fifth the P_2O_5 and one-half the K_2O are available to the first crop. Thus for each ton of average manure applied per hectare 2.5 kg of N, 0.6 kg P_2O_5, and 2.5 kg of K_2O are available to the beans, and the residual nutrients are available to crops which follow. These figures if multiplied by the number of tons of manure applied per hectare, can be subtracted from mineral fertilizer recommended for the bean crop. Caution: Too much chicken or sheep manure will cause excessive N fertilization and result in excessive plant growth and delayed crop maturity.

Manure also helps maintain soil organic matter and humus content, improves soil structure and soil tillage, and supplies or promotes growth of beneficial soil microorganisms. When applied to acid soils it helps prevent or lessen any toxic effects of Al or Mn.

Green manure or cover crops provide somewhat the same beneficial effects as animal manure, but to a lesser extent. They will help maintain organic matter but not apply much N unless the cover crop is a legume.

The most efficient use of fertilizer is to band it to the side and as deep or deeper than the seed. If broadcast before or after plowing, the fertilizer requirement per ha may be 2, 3 or more times that to obtain the same yield from banding. Some types of fertilizers should be broadcast; for example, rock phosphate and basic slag.

There are two extremely important points to remember: (i) N and K_2O fertilizers contain salts which will injure or kill seed or seedlings on direct contact. Keep the band 5 to 10 cm to the side of the seed. (ii) If the banded N and K_2O fertilizers added together exceed 80-100 kg/ha, broadcast the excess before or after plowing.

When grown on favorable soils, beans do not require high inputs of fertilizer. In general, rates of 20 to 45 kg N, 30 to 90 kg P_2O_5, and 30 to 90 kg K_2O per ha are sufficient for good to optimum yields. Adjust the rates according to soil tests, amount of manure applied, and previous cropping history.

LIFE CYCLE OF BEANS

The life cycle of a bean plant is relatively short. This means that the mineral nutrient supply for optimum yields must be available early in the plant's life. Even if soil tests show that the soil nutrient supply is high, yields may be improved with small amounts (10 to 40 kg/ha) of N, P_2O_5 and K_2O banded at planting. These nutrients are quickly available to the plants before the roots can grow widely to absorb nutrients from the soil.

With the exception of N, side dressing of fertilizer after seedling emergence is of little value for beans. P_2O_5 and K_2O remain mostly on the soil surface and are unavailable to roots. If N is needed it should be applied by the time the plants have 2 to 3 trifoliate leaves. Late maturing, climbing beans may respond to N side-dressed at the initiation of flowering. Late application of N often results in excessive plant growth without an increase in yields.

The plant response to micronutrients generally is best when banded with the other fertilizers. If the micronutrient deficiency is severe, plant growth may be so limited that there is insufficient leaf area to absorb a foliar spray.

Healthy, well-nourished plants are better able to compete with weeds and to survive insects and diseases. Over-fertilization, particularly with N, can cause excessive plant growth and a decrease in aeration and drying of plants after rains or nights of heavy dew. This situation results in increased attacks of leaf diseases, reduction in yields and a decrease in the quality of the crop.

MEASUREMENT OF BENEFITS

Because of the residues from fertilizers applied to other crops, beans often respond only modestly to fertilizers in developed countries such as the USA and Canada. In these countries soil tests are particularly valuable to determine how much fertilizer <u>not to use</u> as well as to determine how much to apply.

According to FAO reports of trials in developing countries substantial yield increases often result from the application of fertilizer particularly P and K. Yield increases often are 1000 to 1500 kg or more per ha with a value to cost ratio of 4 to 5 or greater.

SUPPLEMENTAL READING

KELLY, W. C. 1981. Handbook of Soil Testing for Vegetables. Veg. Crops Dept., Cornell Univ. mimeo. 30 pp.

SCHWARTZ, H. F. and G. E. GALVEZ (editors). 1980. Bean Production Problems: Diseases, Insects, Soil and Climatic Constraints of Phaseolus vulgaris L. CIAT, Cali, Colombia. 424 pp.

VITOSH, M. L., D. R. CHRISTENSON and B. D. KNEZEK. 1978. Plant Nutrient Requirements. In: Dry Bean Production-Principles and Practices. Mich. State Ext. Bull. E-1251. pp. 94-111.

12
DRY PEAS

F. J. Muehlbauer and R. J. Summerfield

Dry peas (<u>Pisum</u> <u>sativum</u>) are a cool season annual crop adapted to temperate regions of the world and to the winter period of the tropics and subtropics. In the tropics, dry peas seldom yield well below 1200 m elevation and often do best at 1800 m or higher. In the temperate regions where nearly the entire world crop is grown, they are sown in rotation with cereals in areas where sufficient moisture is available for annual cropping. The winter dry pea (<u>P</u>. <u>sativum</u> ssp. <u>arvense</u>) is also sown in rotation with cereals but is fall planted, overwinters and produces a dry seed crop the following summer.

Much of the world's dry pea crop of over 13 million hectares (M ha) is grown in China (7.5 M ha) and the USSR (4 M ha). India, Ethiopia, the USA and more recently Canada and Brazil are large producers.

Dry peas are used in soup making, canning, and in products that include flours, powders, and confections. Canned dry peas, produced by soaking the peas in water and adding flavorings prior to canning, are consumed as a vegetable similar to fresh green peas except they contain a considerably higher starch content. Pea flour, when combined with wheat flour for baking, produces bread of improved nutritional quality.

The composition of dry peas is similar to other food grain legumes. They contain about 10% water, 25% protein, 60% carbohydrate, 1.0% fat, 3-3.5% minerals but only trace amounts of oils. They are a high protein food, but are deficient in the sulfur containing amino acids, methionine and cystine. Because of their relatively high content of lysine, dry peas are often used to supplement cereal protein in human diets.

---

[1]Research Geneticist, Legume Breeding and Production Research Unit, USDA-ARS, Pullman, Washington 99614, USA; and Plant Physiologist, Plant Environment Laboratory, University of Reading, Reading, England.

As a feed grain, dry peas are an excellent protein supplement because of their high protein content and high lysine content, which makes them quite suitable for replacement of the more traditional forms of protein supplements (e.g. soybeans) that may be expensive or in short supply. Rations that contain up to 25% dry peas have been used successfully in livestock and poultry feeding. In Canada, pea flour protein has been concentrated by an air classification system into a high protein fraction (50%+ protein) and a starch fraction; however, for economic reasons the process is not widely used. The high protein fraction can be used for supplementation in cereal based rations, whereas the starch fraction can be used in various industrial processes (e.g. glue making, alcohol production).

ECOLOGICAL ADAPTATION

Dry peas are best adapted to cool climates with moderate rainfall (approximately 50-55 cm seems best). In temperate regions dry peas are sown early in the spring, after danger of killing frosts has passed, or when fields can be prepared for sowing. The crop can withstand light and moderate frosts without injury; however, in cases of frost injury, the plants generally recover by initiating branches from lateral buds. Most dry pea cultivars intended for spring sowing have no vernalization requirement; however, winter peas flower sooner after a brief cold period. The crop will produce acceptable yields on limited rainfall during the growing season provided sufficient moisture is stored in the soil profile. In tropical and subtropical zones the crop is grown during the winter season or sown at the onset of the rainy season.

Dry peas perform well on soils moderately well supplied with Ca that have a neutral or slightly acid pH. Peas are not considered a deep rooted crop, and they tend to be intolerant of shallow or poorly drained soils, possibly because of increased root disease. Peas have a tap root system with numerous lateral roots and are known to root to depths of 1.5 m or more. Peas are not well suited to highly leached soils normally found in high rainfall areas of the tropics and subtropics because of low soil pH and high temperatures.

Dry peas are often grown in rotations with cereals where they offer a number of advantages that include better control of cereal diseases and opportunities for control of grassy weeds, particularly wild oats (Avena fatua). Their ability to fix N in symbiosis with rhizoba is an added benefit. It is important in cereal producing regions that cropping sequences are developed that include legumes to supply at least part of the N needs of the cereal-legume rotation. In well-planned rotational systems legumes are known to increase the N and P status of the soil when compared with cereal or fallow rotational systems. The improved P status is thought to be the result of solubilization of P by soil micro-organisms, a process promoted by legumes in the rotation.

MINERAL NUTRIENT REQUIREMENTS

Large variations in dry pea yields exist worldwide that may be due to inadequate fertility, moisture supply, or growing season. According to recent FAO reports, yields average 1120 and 1500 kg/ha in China and the USSR, respectively. In India and Ethiopia, yields average between 600 and 700 kg/ha, whereas yields in the USA and Canada average 1600 kg/ha.

Most of the world's dry peas are grown on marginal lands with little or no fertilizer. Usually no fertilizer N is applied to the crop. In addition to N supplied by symbiotic N-fixation, remaining requirements are met from either residual fertilizer applied to previous crops or from the breakdown of soil organic matter. In most cases virtually no response has been obtained from large N applications primarily because of increased weed growth and weed competition, reduced N fixation, and delayed crop maturity. In general, high soil N levels reduce nodulation and N fixation activity of Rhizobia.

Peas respond to P fertilization where soils are deficient in P. In most cases broadcast applications and incorporation during tillage are used prior to planting. However, where equipment is available, excellent responses to small amounts of P are obtained by placing the fertilizer in bands beneath the seeds, thereby obtaining good uptake by the pea plants while avoiding the creation of a fertilizer salt concentration in the immediate vicinity of salt-sensitive pea seedlings.

Larger yields in the Western Hemisphere (e.g. the Palouse region of northwestern USA, certain provinces of central Canada, Colombia, Argentina, and Brazil) can be attributed to improved culture techniques, higher soil fertility, and usually adequate soil moisture.

DIAGNOSIS OF NUTRITIONAL DISORDERS

Nutritional disorders in dry peas can be diagnosed by soil testing, visual plant symptoms, or by plant tissue analysis. Unfortunately, the deficiency symptoms are not distinct and can often be confused with other disorders, e.g., pathological or physiological problems. Visual symptoms of N deficiency are rare because sufficient N is usually supplied by N fixation. Deficiencies of P and K are difficult to diagnose because deficiencies of those elements usually result in reduced growth and yield rather than observable symptoms brought about by translocation of these elements from older leaves to younger leaves. The large cotyledons of pea seeds contain adequate quantities of most micronutrients (at least adequate for a single cropping season) and where deficiencies of micronutrients have been detected, the seeds and the crop were produced on nutrient deficient soils (e.g. highly leached tropical soils).

Soil Testing

Critical levels of pH and mineral nutrients in soils for acceptable yields of dry peas are shown in Table 12.1. The values shown are not absolute and can vary depending on soil type and soil moisture conditions and the method of soil extraction. N-fixation in root nodules of dry peas is sensitive to extremes of pH (Table 12.1). Below pH 5.5 effective nodulation is restricted and results in reduced N-fixation. High soil salinity also reduces nodulation and N-fixation. In general, where P and K are adequate, dry peas do not respond to added N and often smaller yields are obtained with N applications.

Plant Tissue Testing

Critical concentrations for deficiencies in pea tissue have been established for most nutrients and are summarized in Table 12.2. The plant tissue most often sampled is the third leaf down from the top of the plant when the plants are in the 4th to 8th node stage of growth counting the cotyledonary attachment node as zero. Leaves sampled at the stage are near full expansion.

Visual Deficiency Symptoms

Nutritional deficiency symptoms have been observed for most elements either in the field or in nutrient solution culture. In general, foliar symptoms of nutrient deficiencies are not always clear unless the deficiencies are severe. Certain deficiencies can affect seed quality, pod formation, N-fixation and overall plant growth and yields. Symptoms of nutrient deficiencies or toxicities in peas are briefly summarized for the major elements (N, P, K, Ca, Mg, S) and the minor elements (B, Fe, Mn, Zn, Mo, Cu).

Nitrogen (N). In the rare occurrences of N deficiency, plants exhibit a uniform yellow appearance and reduced overall growth. N deficiency may occur in cool wet springs when symbiotic N-fixation in the root nodules is likely to be slow in starting.

Phosphorus (P). Phosphorus deficiency in dry peas is difficult to diagnose; only very slight yellowing of the lower leaves may be present. Under field conditions factors such as soil type, weather, and volume of soil occupied by roots determine the amount of P the plants will obtain and to a large degree determine deficiencies.

TABLE 12.1
Critical soil test levels for pH and mineral nutrients for dry peas

pH or Mineral Nutrient	Critical Level	Procedure[1]
pH (soil reaction)	5.5	Soil paste with distilled H_2O. Let set 30 minutes before reading
N as NO_3 (nitrogen)	42 kg/ha	Water extraction - shaking time 30 minutes
P (phosphorus)	20 ppm	Extraction with 0.5 M $NaHCO_3$, pH 4.8, 1:5 soil-extractant, shaking time 30 minutes
P	4 ppm	Extraction with 10% NaOac, pH 4.8, 1:5 soil-extractant, shaking time 30 minutes
K (potassium)	75 ppm	
Ca (calcium)	10 meq/100 g soil	
Mg (magnesium)	2 meq/100 g soil	
S (sulfur)	10 ppm	Water extraction - (Ba ppT - turbidimetric) shaking time 30 minutes
B (boron)	0.5 ppm	1:2.5 soil-water ratio, filter curcumin color development
Mo (molybdenum)	0.09 ppm	Estimate only
Mn (manganese)	67 ppm	Almost impossible to test for because of different forms. No satisfactory correlation has been established between soil test for Mn and need for fertilizer. Plant analysis should be better for deter-mining deficiencies.
Fe (iron)	97 ppm	Estimate only
Zn (zinc)	0.8 ppm	Estimate only
Cu (copper)	1-2 ppm	Estimate only

[1]Procedure used by Washington State University soil testing laboratory.

TABLE 12.2
Critical nutrient concentrations for deficiencies in pea plant leaf tissue

Element	Plant[1] Tissue	Concentration	
		Normal	Deficient
N (nitrogen)	leaf	4.4 - 4.6%	2.4%
P (phosphorus)	leaf	0.4%	0.1%
K (potassium)	leaf	2.3 - 2.6%	0.3 - 1.5%
Ca (calcium)	leaf	1.0 - 2.8%	0.7%
Mg (magnesium)	leaf	0.3 - 0.5%	0.1%
S (sulfur)	leaf	0.6%	0.3%
Fe (iron)	leaf	110 - 117 ppm	(<50 ppm - estimated)
Mg (manganese)	leaf	75 ppm	(<15 ppm - estimated)
B (boron)	leaf	20 ppm	(<5 ppm - estimated)
Mo (molybdenum)	leaf	0.05 ppm	--
Zn (zinc)	leaf	33.0 ppm	20 ppm

[1]Third leaf down from the top of the plant when the plants are in the 4th - 8th node stage of growth counting the cotyledonary attachment node as zero.

Potassium (K). Plants show retarded growth, yellowing of the lower leaves and poor pod filling.

Calcium (Ca). Deficient plants have reduced root and shoot growth and some chlorosis of the leaves near the top of the plant.

Magnesium (Mg). Deficiency causes reduced vine growth, browning of the leaf tips, reduced pod formation, and developmental failure of seeds. Where increased pea yields were noted from Mg applications, there was a concurrent increase in both Mg and P uptake.

Sulfur (S). Deficient plants become chlorotic within 20 days of germination. New leaves are first affected, but gradually the entire plant becomes uniformly chlorotic (yellowish).

Boron (B). Sprouts of deficient plants are somewhat pale, stunted and lack the recurved plumular bud typical of normal sprouts, and the buds fail to develop. Multiple branching and thickened stunted shoots may also be present. Seeds from a pea crop grown on land that was deficient in B will grow normally if planted in soils with adequate B, but will not grow normally in B deficient soils.

Iron (Fe). Deficient plants become chlorotic when they reach the 5th-6th node stage of growth, affecting mostly the young leaves and growing tips. All subsequent leaves are chlorotic and some may appear bleached. Flowering and pod formation are reduced.

Manganese (Mn). Deficiency is recognized by interveinal yellowing of leaves, and leads to formation of a brown spot in the center of the pea seed, commonly termed "marsh spots" which makes the peas unsuitable for food or seed. Severely affected plants have very yellow leaves and die back from the growing tips, flowers abort, and pod formation is decreased. This disorder should not be confused with hollow heart, an abnormality that is primarily related to high temperature during seed formation. Mn deficiencies tend to occur irregularly in fields because of soil variability.

Zinc (Zn). Deficiency is recognized by yellowing of the older leaves and interveinal chlorosis of the younger leaves. Flowering also is reduced.

Molybdenum (Mo). Molybdenum is essential for active symbiotic N-fixation by Rhizobium bacteria in the root nodules. Symptoms of deficiency appear as reduced growth and general yellowing of the plants, symptomatic of N deficiency.

Copper (Cu). Deficiency is recognized by wilting of the terminal stem tips, flower abortion and reduced pod formation.

Toxicity symptoms

Potassium (K). Excess K can induce Mg deficiency when exchangeable soil Mg levels are low. Excess K can also lead to early hardening of pea seeds.

Calcium (Ca). Excess Ca leads to formation of excess intercellular calcium pectate in the seed coats and cotyledons that results in long cooking times and lack of tenderness. The hardness or toughness can be overcome by the addition of common salt (NaCl) during cooking.

IMPROVED CULTIVARS

There are a great number of regional types, cultivars and strains of dry peas that have been bred for the temperate regions of Northern Europe, North America, and the USSR. Many of these cultivars were bred originally for use as canning peas and later found to be of good quality as dry peas. Very recently, serious attempts have begun to improve yield and quality of dry peas for reconstitution and for splitting. While no international research center has a mandate for improving yields and quality of dry peas, national programs in the USA, United Kingdom, India, USSR, Poland, and countries of Eastern Europe and South America have made strides in improving stability of production through emphasis on disease resistance and altered plant canopy designs. Most noteworthy of these attempts is the "semi-leafless" type developed through the use of the af gene (heritable trait) isolated in 1965 by Goldenburg in Argentina. The af gene in combination with the st gene gives a unique "leafless" character to peas. Such types have shown potential for reducing the incidence of foliar disease, improving drought tolerance (less leaf area and thus reduced transpiraton), hastened maturity (the upright canopies with reduced foliage dry more rapidly at maturity). Improved seed quality is a benefit from these types and is brought about by hastened maturity and escape from pathogens that often lower seed quality.

Future cultivar development will likely be in the direction of multiple disease resistance with improved seed quality (color, size and uniformity) with the goal of stabilizing production and expanding acceptance of the crop. Dramatic yield increases in the short-term from improved cultivars seem unlikely.

NUTRIENT REQUIREMENTS FOR PRODUCTION

A dry pea crop of 1000 kg/ha will contain 43 kg N, 4.2 kg P, 9.2 kg K, 0.6 kg Ca, 1.2 kg Mg, and 0.8 kg S (Table 12.3). The amounts of these nutrients contained in dry peas are very similar to that contained in lentils and chickpeas, but considerably lower than in soybeans.

TABLE 12.3
Estimates of nutrient removed (kg per hectare) in dry pea seeds for crops producing an economic yield of 1000 kg per hectare. Data for soybeans, lentils and chickpeas are included for comparison

Crop	Nutrients removed (kg/ha)					
	N	P	K	Ca	Mg	S
Dry peas	43	4.2	9.2	0.6	1.2	0.8
Lentil	43	5.0	11.7	0.7	1.2	2.0
Soybean	71	6.1	20.7	3.0	3.0	1.7
Chickpea	41	4.8	10.7	0.8	1.4	0.7

Dry peas are sensitive to both low pH (5.5--strongly acid) and high pH (9.5--strongly alkaline), possibly because inoculation of pea roots by Rhizobia is inhibited at those extremes of pH and thus nodule formation and symbiotic N fixation are reduced. Correction of low pH by liming brings about improved growth, reduced root disease and improved yields. However, liming of acid soils may cause Mn deficiencies (marsh spots) because Mn is less available at high pH.

Nitrogen removed by a dry pea crop is the result of uptake of soil N, combined with fixed N from the root nodules. High residual N in the soil inhibits N-fixation and reduces the amount accumulated from that source. Nitrogen fertilizers are usually not applied to dry pea crops because such applications promote unneeded and excessive vine growth and promote weed competition with little if any effect on seed yields. Nitrogen is not usually needed because of symbiotic N-fixation in the root nodules. However, the specific Rhizobia (root nodule bacteria) for peas must be present in the soil, or applied to the seeds or soil at the time of planting.

In the Palouse region of the northwestern USA "starter" fertilizer solutions consisting of 13 kg N, 36 kg P_2O_5, and 8 kg S applied at the rate of 120 kg per hectare have been used with limited success. Starter fertilizers have been successful when applied to early plantings of dry peas in cold wet soils where the small amount of N stimulates early N-fixation by the root nodules. Presumably the starter fertilizer provides a source of N to developing plants that promotes vegetative growth, and supports leaf photosynthesis which is necessary for active N-fixation by the root nodules. The starter fertilizer is applied in bands beneath the seed row at planting time.

Amounts of phosphate fertilizers to use can readily be determined by using a calibrated soil P test. In general, between 44 and 66 kg P_2O_5 per ha are applied if soil tests indicate that available P concentrations are low. Responses to P fertilizer are common on severely eroded soils. Some soils will provide more than enough P derived from mineral decomposition and organic matter mineralization to supply adequate P.

When soil tests show that K is needed, K applications should be made in accordance with the rates determined for the area in question by field trials. Where deficiencies are detected, applications of about 222 kg K_2O per ha have proven beneficial.

Sulfur, where deficiencies are detected, is applied to other crops grown in rotation with dry peas at the rate of about 17-22 kg S/ha.

Fertilizers containing sufficient P, K and S to correct deficiencies are applied early in the spring and worked into the soil by suitable tillage. Where soil tests are unavailable P and K applications of 50 and 25 kg/ha respectively, should generally be adequate for dry peas. Mg should be applied where soil contains less than 5 meq available Mg in 100 g of soil.

Deficiencies of one or more of the trace elements, Mn, Fe, Cu, Zn, B, and Mo (Table 12.2) may be corrected with applications of animal manures that usually contain small quantities of these elements; however, animal manures may not be generally available in the large dry pea-producing areas. The use of organic amendments for soils used for peas can improve production by decreasing disease incidence, promoting root growth, and improving soil moisture status. Severe Fe deficiency can be corrected by foliar applications of ferrous sulfate. Two applications at the rate of 0.9 kg/ha are sufficient to correct Fe deficiency.

To correct Mo deficiency, seed treatment applications at the rate of 35 g/ha in the form of sodium molybdate have been very successful. It is important that the sodium molybdate be uniformly distributed over the seed and applied with a "sticker" to ensure adherence to the seeds. Ammonium molybdate at the rate of 1.1 kg/ha applied broadcast to the soil in combination with gypsum (calcium sulfate) has also been used successfully to correct Mo deficiency.

CROP MATURATION AND HARVEST

Dry peas are usually harvested when seed moisture decreases to less than 14%. Even though N fertilization extends the growth period of the crop, larger yields seldom result. The delayed maturity, however, may cause conflicts with harvest of other crops or result in deterioration of seed quality from foliar disease or adverse weather (in some regions more probable later in the season). Potassium fertilization tends to bring about early hardening of peas.

UTILIZATION OF INCREASED PRODUCTION

Dry peas have numerous uses (whole peas, split peas, canned peas, pea powders, snack items, and protein supplementation of animal feeds) that could absorb additional production. Highest prices are obtained for peas for human consumption, but they must be of good size, color and uniformity with absence of defects and insect infestations. Excess dry peas, beyond those used for human food, can be used effectively as protein supplements in feeds for poultry, hogs and other livestock provided the peas are mixed in proper proportions (usually 25% of the ration) with other feed grains.

Air classification, a system used to separate pea flour into a protein rich fraction and a starch fraction, offers potential for the continued development of dry peas as a crop for protein supplementation of animal feeds (a likely use of the protein rich fraction) in regions of the world unsuited to production of the more traditional protein crops (e.g. soybeans). The starch fraction has potential for alcohol production, glue making and other industrial uses. The suitability of dry peas as a protein crop and its ability to produce on lands unsuited to soybeans may ultimately mean that dry peas will be more widely grown in the future.

The potential use of dry pea flour in bread making and certain snack items is very feasible and results in products of improved nutritional quality when compared to products made from wheat flour alone. Loaf volume of bread is not depressed by additions of up to 10% dry pea flour.

SUPPLEMENTARY READING

BISHOP, R.F., C.R. MacEACHERN, J.S. LEEFE, and H.B. CANNON. 1968. Effect of nitrogen, phosphorus and potassium on yields and nutrient levels in the leaves of processing peas. Can. J. Plant Sci. 48:255-233.

MacLEAN, K.S. and D.L. BYERS. 1968. Nutrient content of field grown peas. Can. J. Plant Sci. 48:155-160.

SUTCLIFFE, J.F. and J.S. PATE, eds. 1977. The Physiology of the Garden Pea. Academic Press. New York.

13
MUNGBEAN

C. S. Ahn and S. Shanmugasundaram

Mungbean, _Vigna radiata_, and the other oriental beans, black gram (_V. mungo_), rice bean (_V. umbellata_), adzuki bean (_V. angularis_) and moth bean (_V. aconitifolia_) belong to the subgenus _Ceratotropis_ of the genus _Vigna_.

Mungbean is believed to be native to the India-Burma region of Asia but its wild form is unknown. _V. radiata_ var. _sublobata_ race I which grows wild in the Western Ghats of India has been suggested as the closest relative of mungbean and its possible progenitor.

IMPORTANCE OF MUNGBEAN IN WORLD AGRICULTURE

Mungbean has been widely cultivated in India and the neighboring countries from ancient times and was introduced early into other Asian countries and Africa. Recently, it has been introduced into the USA, Australia and other parts of the world. Present world production is estimated to be 1.4 million mt harvested from 3.5 million hectares, of which more than 60% is in India (Table 13.1). The other major mungbean producing countries are Thailand, Indonesia, Pakistan and the Philippines.

One hundred grams of dry beans contain 10.7% water, 340 calories, 24 g protein, 1.3 g fat, 60 g carbohydrates of which 1.19 g is crude fiber, 118 mg calcium, 340 mg phosphorus, 7.7 mg iron, 6.0 mg sodium 1,027 mg potassium, 79.3 I. U. Vitamin A, 0.38 mg thiamine, 0.21 mg riboflavin and 2.6 mg niacin. With sprouting there is an increase in protein, thiamine, riboflavin, niacin and ascorbic acid on a dry weight basis.

[1]Mungbean Breeder, and Legume Crops Program Leader, respectively, Asian Vegetable Research and Development Center (AVRDC), Shanhua, Taiwan, Republic of China.

TABLE 13.1
Mungbean production in selected countries

Country	Year	Area (ha)	Yield (kg/ha)	Production (mt)
Bangladesh	76/77	15,607	611	9,540
Burma	75	34,994	280	9,798
India	78	2,225,000	338	854,000
Indonesia	78	193,000	575	111,000
Iran	77	27,500	550	15,129
Korea, Republic of	78	6,000	900	5,400
Malaysia, West	73	70	450	32
Pakistan	78/79	66,000	455	30,000
Philippines	78	45,120	580	26,177
Sri Lanka	79	12,188	794	9,700
Taiwan	79	4,691	764	3,583
Thailand*	80	525,300	606	318,400
Total		3,455,470	403	1,392,759

* Includes black gram

Due to its easy digestibility and low production of flatulence, mungbean is an excellent supplementary protein source in cereal-based diets. The whole or split seeds are cooked, fermented or milled to make "dahl", soup, curries, porridges, noodles, breads, cakes, cookies and other culinary products. Mungbean sprouts have long been used as a fresh vegetable in China and other Asian countries. In recent years, the sprouts have increasingly become popular in the USA and other Western countries. The crop residues or whole plants are used as animal feed or incorporated into the soil for soil improvement purposes. Mungbean is an ideal component in multiple cropping systems due to its short growth duration (60-100 days), flexibility in terms of adaptation to different seasons, and its ability to fix atmospheric nitrogen through a symbiotic relationship with rhizobium.

ECOLOGICAL ADAPTATION

Mungbean is classified as a short day plant but there are varietal differences in response to photoperiod. Flowering of mungbean is also delayed by low temperature. Genotypes with low photoperiod sensitivity are desired in the tropics and subtropics where mungbean is grown as a short duration crop in multiple cropping systems or at higher latitudes with short growing seasons.
Mungbean is a warm season crop and therefore well adapted to the tropics. In the subtropical and temperate zones, its adaptation is limited to warm seasons. It is very sensitive to frost damage. The optimum temperatures for plant growth range from 28 to 30° C. Mean tempera-

tures of 20 to 22° C may be the minimum for productive
growth. At 18° C plant growth is stunted. Yields are
positively correlated with the mean temperatures during
vegetative and reproductive growth. Warm temperatures
are essential for good and rapid germination. Optimum
temperature for germination is 29 to 31° C. Germination
rate declines slowly below 25° C and drops off sharply
below 14° C. Flower shredding is high and yields are
poor at temperatures above 33° C, especially when moisture is deficient.

Mungbean is generally unsuitable for the wet tropics
where the annual precipitation exceeds 1,000 mm. Seedling emergence is inhibited by both excessive moisture
and drought. Mungbean is susceptible to waterlogging.
High humidity during the growing season increases the incidence of foliar diseases. Heavy rains at maturity result in seed damage by molds or seeds sprouting in the
pod. Mungbean is commonly assumed to be tolerant to
drought stress, however, reduction in photosynthetic rate
when the leaf water potential is below -2 bars indicates
its extreme sensitivity to drought, especially before and
during flowering.

Mungbean is grown in Southeast Asia as a catch crop,
relay crop, intercrop or mixed crop. It is generally
found in cropping systems based on cereals. Mungbean can
also be intercropped with sugarcane, maize, sorghum, millet, cotton, jute, cassava, pigeon pea, castor bean, sesame or sunflower depending on the season and location.
In India, Pakistan and Nepal it is also mixed-cropped
with rice, wheat, corn or sorghum. In temperate regions,
mungbean is planted in May to July after wheat or barley
and harvested in September or October.

MINERAL NUTRIENT REQUIREMENTS

Mungbean is grown on a wide range of soil types,
however, for good production, a loamy or sandy loam soil
with good drainage is desirable. Heavy soils and waterlogged soils are unsatisfactory. Optimum soil pH for
plant growth and symbiotic N fixation ranges from 5.5 to
6.5. Traditionally, mungbean has been grown without commercial fertilizers. Generally, fertilization will not
be beneficial if mungbean is grown with inadequate soil
moisture or poor weed and pest control.

Fertilizer recommendations vary depending on locations and seasons. Soil and plant analyses are necessary
for determining adequate fertilization to avoid nutrient
deficiencies. To produce 1.5 to 1.6 t/ha of mungbean the
plants should absorb about 98 kg N, 25 kg P_2O_5 and 85 kg
K_2O/ha.

For nodule formation and symbiotic N fixation, rhizobia belonging to the cowpea cross-inoculation group and
a suitable soil environment for the bacteria to function
must be present. Mungbean grown in soils lacking effect-

ive indigenous rhizobia may benefit from inoculation. Yield increases of 10 to 37% by inoculation have been reported. At the Asian Vegetable Research and Development Center (AVRDC), N fixation activity is highest at the fifth and ninth week after planting in fall and spring plantings respectively, and then declines sharply, indicating that supplemental N application after flowering may be needed to meet the heavy N demand during pod and seed development. Indiscriminate N application may interfere with nodulation and nitrogen fixation in legumes. Under favorable conditions mungbeans are known to fix about 50 to 100 kg N/ha. Mineral nutrients necessary for satisfactory functioning of nodule bacteria include P, S, K, and traces of Mo, B, Cu, Zn, cobalt, and Mn.

Many research reports show that a small dose of N, about 10 to 30 kg/ha applied at early stages, is beneficial in mungbean. This may be due to slow plant growth at early stages when N fixation is barely detectable for the first 3 weeks after planting. At AVRDC a basal application of 15 kg N/ha and a side dressing of 15 kg N/ha at flowering stage are recommended.

Mungbean responds to P fertilization in a variety of soil types and agro-climatic conditions, and rates of 40 to 100 kg P_2O_5/ha are often used. In solution culture the critical concentration of P_2O_5 at which mungbean showed deficiency symptoms was 0.1 to 0.5 ppm.

There is only a limited number of reports on K fertilization in mungbean. Application of K has been recommended in the Philippines, Thailand, Indonesia, Taiwan and at AVRDC, ranging from 15 to 100 kg K_2O/ha. In solution culture, the critical concentration of K_2O at which mungbean showed deficiency symptoms was 2.4 ppm.

Reports on the effects of other mineral nutrients on mungbean are limited. In solution culture, the critical concentration of Mg at which mungbean showed deficiency symptom was 0.8 to 1.6 ppm. Zinc content in mungbean (150 to 300 ppm) is higher than in other crops (50 to 100 ppm).

In India, basal fertilizers placed in bands below the seeds gave better response than broadcast applications. Liming has been recommended in the Philippines on soils with pH below 6.5.

MINERAL NUTRIENT DEFICIENCIES

The general nutrient deficiency symptoms in legumes are described below; most of them are probably similar and applicable to mungbean.

Nitrogen (N). Legumes with proper nodulation generally do not respond to N fertilizers. For small seeded and short season legumes, however, a small amount of N as a starter is beneficial due to the time required for nodules to become active. The leaf color of the N deficient plants turns pale green with a yellowish tinge and finally yellow. The deficiency usually appears first on the lower leaves, but spreads quickly to the upper parts. Eventually, the deficient plants lose their foliage, the lower leaves dropping first.

Phosphorus (P). Legumes as a group have a relatively high P requirement. Response to P fertilization is often marked during the early growth period when plants have limited root systems. The chief symptoms of P deficiency are retarded growth and spindly plants with small leaflets. Leaf color changes to dark or bluish green and maturity is delayed.

Potassium (K). Compared to other mineral nutrients, K is removed in relatively large quantities by legumes. In broad-leaved legumes, such as soybean, the deficiency symptoms are yellowing of the leaf tips and margins of leaflets. While the centers and bases remain green, necrosis in the chlorotic area follows, giving the leaflets a ragged appearance. Maturity is delayed.

Calcium (Ca). Ca deficiency symptoms are rarely observed in the field. Delayed emergence of primary leaves which are cup-shaped is a Ca deficiency symptom in soybean in sand culture. Tips of primary leaves become necrotic and narrow chlorotic bands develop around leaf margins. Tissues between the veins tend to ridge. Terminal buds deteriorate and petioles break down. Primary leaves become soft and drop off.

Magnesium (Mg). Mg is a constituent of chlorophyll and tends to improve utilization and mobility of P. Mg deficiency hinders N fixation. At early stages of Mg deficiency in soybean, the interveinal areas of leaves become pale green and then turn deep yellow except at the base of the leaves. Lower leaves are likely to be affected first. Rusty specks and necrotic blotches may appear between veins around edges of the leaflets.

Sulfur (S). S is a constituent of a number of plant compounds and is particularly important in legumes. With S deficiency the young leaves including veins turn pale green to yellow. In later stages the older leaves turn yellow.

Boron (B). Although the amount needed is small, deficiency may occur in soils that are very low in available B. With B deficiency the upper internodes of the stem are shortened, giving the plants a rosette appearance. Upper leaves near the growing points turn yellow and sometimes red. Symptoms are most severe at the leaf tips while the leaf bases remain green.

Cobalt (Co). Co is an essential element for nodule bacteria on legumes. Deficiency symptoms have been observed in nutrient culture experiments but have not been reported in the field. With Co deficiency, interveinal mottling occurs first on older leaves and later on the newer leaves, while the veins remain light green. As deficiency continues the leaf tip and margin become necrotic with the leaf curling inward.

Copper (Cu). Cu deficiency is mainly found on peat or muck soils. Symptoms occur first in the youngest tissue. The plants are poor in growth and the color changes to grayish green, blue green or olive green. Leaves die and internodes are shortened.

Iron (Fe). Fe deficiency occurs more frequently on high lime soils because of low availability of Fe. Deficiency symptoms start from the newer leaves. In soybean, the areas between veins turn yellowish white, and as deficiency advances, the veins also turn yellow and finally whole leaves become almost white. Brown necrotic spots may occur near leaf margins.

Manganese (Mn). Mn deficiency usually occurs in soils low in total Mn with pH above 6.2. In soybeans with Mn deficiency, the interveinal areas of the leaves become chlorotic first while the veins remain green. Whole leaves except the veins turn pale green and then pale yellow. Brown spots and necrotic areas develop as the deficiency becomes more severe.

Molybdenum (Mo). Mo deficiency may occur in soils of low pH. Molybdenum is needed for N fixation by nodule bacteria and N assimilation by plants. Mo deficiency symptoms are similar to those of N deficiency. In soybeans grown on Mo-deficient nutrient solutions, leaves become pale green and necrotic areas develop in the interveinal areas adjacent to midribs and along the margins of the leaflets. Occurrence of cup-shaped leaves is a typical symptom of Mo deficiency.

Zinc (Zn). Zn deficiency occurs under a wide variety of soil conditions. In soybeans and beans, Zn deficient plants fail to develop to natural size. The interveinal areas of leaves become yellow and chlorotic. The chlorotic tissues may turn brown and grey and die prematurely. The deficiency symptoms appear first on upper active young leaves.

VARIETAL IMPROVEMENT

Breeding in mungbean has received far less attention than in other important food crops, especially outside India. Varietal improvement of mungbean in India started in the early part of this century. In recent times active breeding programs have been developed in the Philippines, Indonesia, Thailand, the USA, Australia, and other countries. AVRDC, as an international agricultural research center, initiated a mungbean research program in 1972 and now maintains 5,108 accessions of germplasm. Common mungbean breeding objectives are high yield, early and uniform maturity, resistance to major diseases and insects, tolerance to extreme temperatures and moisture stresses, good seed quality and better responsiveness to management inputs. New varieties developed in different national programs and at AVRDC exhibit a higher yield potential under optimum fertilization.

The International Mungbean Nursery (IMN) was initiated by the University of Missouri in 1971 to test diverse mungbean germplasm internationally and has been coordinated since 1976 by AVRDC. The results of the IMN show that the highest yields of mungbean range from 1.8 to 3.1 t/ha under appropriate management conditions in most countries. In Asia, mungbean is grown as a subsistence crop in marginal lands under rainfed conditions with little or no inputs. Under these conditions the average yields are only 280 to 900 kg/ha.

MUNGBEAN PRODUCTION

In tropical and subtropical areas where mungbean is grown as a short season catch crop, it can be planted almost year-round and it is common practice to fertilize the principal crops and grow mungbean mostly on residual fertilizer. At higher latitudes mungbean is planted after winter cereals as soon as temperatures rise to around 25° C, to ensure its harvest before the first frost.

The broadcast method of planting widely practiced in Asia requires a higher seed rate than row planting. Sowing in rows facilitates accurate spacing of plants, weeding, cultivation, spraying and harvesting. Row spacing may vary from 25 to 75 cm depending upon season, soil fertility, plant growth habit, and levels of management. The recommended inter-row distances are 25 to 30 cm in India, 50 cm in Thailand and at AVRDC, and 50 to 75 cm in the Philippines and the USA. Seeds may either be drilled or dibbled at 10 to 30 cm within rows. Populations of 300,000 plants/ha in the wet season and 400,000 plants/ha in the dry season may be optimal.

Mungbean has an indeterminate growth habit; flowering and pod maturation on the same plant occur over a period of several weeks. Ripe pods remaining on the plant for some time may shatter or the seeds may mold or sprout in the pods under rainy or humid conditions. In tropical countries, mungbean is harvested three to five times by hand picking which is laborious and expensive. Harvesting utilizes 25 to 30% of the total production cost and

40 to 50% of the total labor cost. In varieties with synchronous maturity, most of the pods can be harvested in one picking. Such varieties are being developed.

Mungbean is generally perceived as a low income crop mainly due to low yields. From a comparative economic study of mungbean production (Calkins, 1978), the production cost and net return per hectare were US$37.70 and $34.10 in Thailand, $181.60 and $43.10 in the Philippines, and $453.70 and $100.90 in Taiwan. The net return was highest in Taiwan where the yield (0.8 t/ha) was highest due to higher management inputs. The returns for land, labor and management in Oklahoma were doubled in combined cropping of wheat and mungbean compared to single wheat cropping.

Mungbean is grown mainly for domestic consumption in most of the Asian countries. Thailand, Burma, Australia, Indonesia, Kenya and Peru are known to export mungbeans. Japan, the USA, Taiwan, Singapore, Malaysia and Hongkong are major importers.

SUPPLEMENTAL READING

Asian Vegetable Research and Development Center. 1982. Progress Report for 1981. AVRDC, Taiwan, R.O.C.

CALKINS, P.H. 1978. Economics of mungbean production and trade in Asia. In: Proceedings of 1st International Mungbean Symposium. AVRDC, Taiwan, R.O.C. pp. 54-63.

LAWN, R.J. and J.S. RUSSELL. 1978. Mungbeans: a grain legume for summer rainfall cropping areas of Australia. Journal of Australian Institute of Agricultural Science 44:28-41.

MORTON, J.F., R.E. SMITH and J.M. POEHLMAN. 1982. The Mungbean. University of Puerto Rico, Mayaguez, Puerto Rico.

PARK, H.G. 1978. Suggested Cultural Practices for Mungbean. Report 78-63, AVRDC. Taiwan, R.O.C.

PCARR. 1977. The Philippines Recommendations for Mungo in 1977. Philippines Council for Agriculture and Resources Research.

TUCKER, B.B. and R. MATLOCK. 1969. Fertilizer Use on Mungbeans, Cowpeas, and Guar. Oklahoma State University Extension Facts No. 2224.

14
GROUNDNUT (Peanut)

F. R. Cox and A. Perry

The groundnut or peanut (Arachis hypogaea) is grown extensively throughout the world. According to FAO, the area involved annually is about 19 million ha. Asia is by far the largest producing continent with 11 million ha, and India contributes most of that with over 7 million ha. China, Indonesia and Burma also have extensive areas in groundnut. Africa is the next largest producing continent with about 6 million ha, and Senegal and Sudan are the largest producers, each with about 1 million ha. North and South America each grow about 0.75 million ha, with the USA, Brazil and Argentina being heavily involved. Groundnut is also grown to a lesser extent in Australia and Europe.

The groundnut is an edible oilseed, and most of the world's production is used for oil. Groundnut oil is highly valued as a salad or cooking oil. It is especially good in cooking as it has a high smoking point, does not retain odors appreciably, and thus can be reused many times. When processed for oil, the meal is a valuable by-product. Groundnut kernels are nearly 50% oil and 25% protein. There is considerable direct food usage of groundnuts in the USA, Australia and China, and a growing acceptance for foods of this type in Europe. The edible products include the many roasted and boiled forms, snack products, candies and groundnut or peanut butter (sometimes called peanut paste).

ECOLOGICAL ADAPTATION

The groundnut is native to South America, wild cultivars being found from near the equator to 35 S. It is considered especially sensitive to photoperiod but responds to differences in temperature. Little growth occurs at less than 15° C. The rate of vegetative devel-

---

[1]Professor of Soil Science, and Extension Professor (Peanuts), respectively, North Carolina State University, Raleigh, North Carolina 27695, USA.

opment increases with increasing average temperature to around 25°C, but the optimal temperature for reproductive development is somewhat less. Since cultivars have been selected and bred that vary considerably in the length of growing season required (100 to 175 days), groundnuts are now grown from the tropics to Canada.

Groundnuts are considered somewhat drought tolerant. This may be associated with their rather long period of flowering and fruit set. Dry conditions at that time, however, do inhibit continuous fruit formation and, thus, markedly reduce the potential yield. At either the very early or late growth stages, a short drought may not affect yields appreciably. In fact, drying the soil near maturity may facilitate harvesting the crop. A prolonged drought, especially after flowering, can lower yields drastically.

The crop will grow well vegetatively under a wide range of soil conditions. However, for good pod development and, more importantly, the ability to harvest the pods, well-drained sandy or very friable soils are preferred. In clayey soils, pod losses in harvesting may be extremely high, and the pods that are harvested are difficult to separate from adhering dirt and clods. Also, the crop cannot be grown easily in rocky or stony land except where hand labor is available.

Groundnut is more tolerant of soil acidity than many other legumes. At pH 5.5 to 6.0, many cultivars are able to fix adequate N. Such soils, especially if they are sandy, may be deficient in Ca for pod development. Under more acid conditions, Al toxicity restricts root development. Liming acid soils to a pH above 6.5 may induce a Mn deficiency, and naturally calcareous soils often create an Fe deficiency in the crop. Groundnuts do not tolerate salinity.

Groundnuts should be grown in a rotation to avoid build-up of harmful diseases and insects. Several years of a grass pasture is likely the best for typical row crop culture; however, a 3-year rotation with a grass like maize just ahead of groundnuts is good. Cotton also is a satisfactory rotational crop. Crops to be avoided are other legumes which promote build-up of common diseases, Irish potatoes which require a low soil pH, and tobacco which is heavily fertilized. Selection of rotational crops that facilitate the control of grassy weeds is helpful. It is extremely difficult to grow groundnuts if certain types of grasses, such as bermuda grass (Cynodon dactylon), are present.

YIELD POTENTIAL

The average world yield of groundnut in the shell is about 1,000 kg/ha according to FAO. This average is extremely low, perhaps only 1/10 the potential of the crop under ideal conditions. India, the largest producer, has an 800 kg/ha yield while Israel, one of the smallest producers, averages nearly 4,000 kg/ha. In the USA, yields average about 2,500 kg/ha, but top producers achieve at least twice that figure.

There are many reasons for the wide discrepancy in yields around the world. Perhaps the dominant cause is moisture. In Israel, the entire crop is grown under irrigation. Recently, the national average in the USA was cut in half by a severe drought. Lack of many other good management practices, however, also limited yields. Rotation, as mentioned previously, is one of these. Others include improved adapted varieties, nutrition, and control of the numerous insects, diseases and weeds. It is difficult to say which of these management practices is most limiting in most places, but unless due attention is given to all of them, it is not likely that average yields will change appreciably.

Without doubt there are severe nutritional constraints for groundnut production in many regions. FAO fertilizer trials have shown about a 50% yield increase from fertilization in West Africa and Asia. Apparently in these regions little fertilizer is used at all, perhaps due to the high cost and/or lack of access to the materials. Groundnuts respond well to indirect fertilization and, in the USA, there is little response to applications of P and K if other crops in the rotation have been fertilized.

NUTRIENT DEFICIENCIES

Nutrient deficiencies in groundnut may be assessed by plant symptoms, tissue analysis and soil tests. Unfortunately, in areas where deficiencies exist, there have been few research reports to identify critical nutrient concentrations in plants and soils. Some symptoms have been described, but a few of these are so rare they had to be induced under controlled conditions in the greenhouse. Symptoms and plant and soil critical levels will be presented in this section.

Deficiency Symptoms

Nitrogen (N). Leaf chlorosis, plants turn pale yellow, stunted growth, a reddish coloration of stems may occur.

Phosphorus (P). Dull, dark green appearance, leaves thick and leathery and turn from ovate to elliptical shape, stunted growth, stems may take on a purple pigmentation.

Potassium (K). Slight interveinal chlorosis, followed by a yellowing of the edges of the leaves.

Calcium (Ca). Very rare in leaves and stems, small distorted leaves near branch tips, and terminal buds blacken and fail to continue to develop. Most common symptom in developing fruit is lack of kernel formation, darkened plumule if kernel develops, and reduced seed germination.

Magnesium (Mg). Seldom occurs, but appears as chlorosis on older leaf margins.

Sulfur (S). Leaves formed during the deficient period are light green.

Boron (B). Rarely shown in leaves and stems but leaves would be small, branches stubby, and stems may split. More often shown in kernels as incomplete development of cotyledons (hollow heart) and discoloration.

Manganese (Mn). Interveinal chlorosis and stunted growth.

Iron (Fe). Young leaves are reduced in size and suddenly become chlorotic.

Zinc (Zn). Very rare, but older leaves may show slight chlorosis while young leaves are stunted and small.

Copper (Cu). Rare, with chlorosis of small, distorted leaflets and a scattering of yellowish white spots on leaves.

Molybdenum (Mo). Only an indirect effect shown, that of N deficiency.

Plant Analysis

Plant analysis is an excellent tool to assist in solving problems of poor growth and in monitoring the effectiveness of a fertilizer program. For groundnuts, the most recently matured leaf is usually sampled to obtain, as much as possible, a common physiological age. For certain nutrients, notably N, P and perhaps Zn, concentrations decrease during the season so selection of a common sampling time is important. The stage of growth used most commonly is just prior to flowering. Considering this, a list of the nutrients and estimated critical levels in leaves are given in Table 14.1.

The N percentage in Table 14.1 is an average; it has been indicated that it should be slightly greater for sequentially branched cultivars than the alternately branched ones. The N:S ratio given here is greater than the 15:1 deemed desirable. At 3.5% N or greater the S concentration should be near 0.25% to meet that crite-

rion. Groundnuts are notorious K feeders in fields previously fertilized with K, and little information is
available to establish a critical level. In West Africa,
workers have indicated a relationship between leaf position and the critical K concentration, but generally the
levels cited are lower, nearly 1% lower in their studies. Higher Ca and Fe concentrations may also be required if the soil is high in Mn. The Ca level in the
kernel should also exceed 0.045% or germination will be
impaired.

TABLE 14.1
Critical levels of plant nutrients in groundnut

Nutrient	Leaf Critical Level
N	3.5%
P	0.25%
K	1.5%
Ca	0.5%
Mg	0.25%
S	0.15%
B	20 ppm
Cu	4 ppm
Fe	50 ppm
Mn	20 ppm
Mo	<1 ppm
Zn	20 ppm

Soil Testing

Few soil tests have been properly calibrated for
groundnut production. In regions where this method is
used, soils have been fertilized to the extent that
deficiencies are rare. In other regions, plant analysis
has been relied on heavily. As a result, the only soil
test evaluated in some detail is that for P. Table 14.2
lists a few extractants and the critical level of P for
each.

TABLE 14.2
Suggested critical P soil test values for groundnut,
using different extractants

Extractant	Composition	Critical P (ppm)
Bray 1	(0.025 N HCl + 0.03 N NH_4F	5
Double acid	(0.05 N HCl + 0.025 N H_2SO_4)	7
Olsen	(0.5 N $NaHCO_3$)	9
Bray 2	(0.1 N HCl + 0.03 N NH_4F)	10

There has not been a good relation between soil test K and response to K fertilization. In sandy soils, such as are often used for groundnuts, K applied to the previous crop leaches into the subsoil. Groundnut roots seem to feed readily on this source, and subsoil K levels have never been considered in soil test calibrations.

For certain cultivars the Ca concentration in the fruiting zone must be quite high for proper fruit development. For the Florunner variety this has been estimated to be 0.6 meq/100g extracted with the North Carolina double acid method. For other cultivars, especially among the Virginia type, concentrations at least twice this great may still not be adequate. Other factors such as dry weather during fruiting and the Ca:K ratio may affect this requirement.

Soil tests are useful for establishing an optimum pH for groundnut production. Lime is applied to acid soils to eliminate the toxic effects of Al. For most soils, this is achieved by increasing the pH to 5.5 or above. Liming low Mn soils above pH 6.2, however, may induce a Mn deficiency. The general pH range of 5.5 to 6.0 seems optimum for many conditions.

LIMING AND FERTILIZATION

In amending the soil to improve its potential for plant growth, three factors are considered, the source of material, the rate of application, and the method of application. Soil test information is needed to apply these factors efficiently. Lime may be either calcitic or dolomitic, the latter being preferred if the soil is low in Mg. Lime must be finely ground and thoroughly incorporated to be effective. The rate used should eliminate the toxic effects of Al, usually raising the pH to about 6.

Although groundnut is a legume, N is often used at low rates at planting, especially for varieties with a short growing season such as the Spanish type, but N has not been effective with the Virginia type. Since only about 20 kg/ha is used and the plants are small at that time, banding should be more effective than broadcasting, and there should be no difference among sources. If there is a question of adequate rhizobia in the soil, the crop should be inoculated. Granular forms of inoculants are more effective than seed treatments. On soils with a pH below 6, a seed treatment of Mo at only 15 g/ha is excellent insurance for maximum N fixation.

The best rate and method of application of P depend upon the degree of deficiency and the ability of the soil to fix P. For highly deficient conditions and/or high potential fixation, P should be banded at 40 to 80 kg P_2O_5/ha. Under less severe conditions, 20 to 40 kg P_2O_5/ha broadcast would suffice. Both triple superphosphate and ordinary superphosphate are excellent sources of P, but the latter would be preferred if no other source of S is applied in the rotation.

The usual K source is KCl, but others would be acceptable. The material should be broadcast and soil incorporated, or worked in deeply to reduce the K concentration in the fruiting zone as much as possible. This is especially true for cultivars requiring additional Ca.

If other crops in the rotation are fertilized, it may not be necessary to apply P and K directly to groundnuts. In fact, it is more effective to fertilize other crops liberally and not fertilize groundnuts. The other crops may respond to the additional input, and groundnuts yield just as well and usually grade better with indirect fertilization.

When growing groundnuts that require a high level of Ca in the fruiting zone, a soluble source such as $CaSO_4$ is used. This is banded over the plants, covering the area in which fruits develop at early flowering. Rainfall leaches this into the soil. About 400 kg/ha of gypsum is adequate, but higher rates may result in a slightly better fruit grade and ensure better germination of the seed produced. Gypsum also is a source of S. A shallow incorporation of lime may also provide adequate Ca for some less sensitive cultivars.

Of the micronutrients, B and Mn deficiencies are most common. Very low rates are required and even distribution is facilitated by applying these to the foliage. One-half kg B/ha should be applied before pod set to ensure good fruit quality, and 1 kg Mn/ha should be sprayed on the plants if Mn deficiency symptoms appear. Soluble sources, of course, should be used for this type of application. Iron deficiencies occur on some calcareous soils, but this problem is best handled by growing cultivars that are not sensitive to Fe deficiency.

IMPROVED GERMPLASM

Most groundnut breeding has been devoted to selection and crossing of wild species to adapt production to temperate regions. Yields have increased significantly but there has been a concomitant increase in the level of management required, including nutritional needs. Since a large proportion of the groundnut area in the world is grown under low levels of management, current research is now emphasizing lower levels of management and/or adaptation to stress conditions. Nutrient stress is certainly included in these efforts. Research to adapt cultivars to more acid conditions is in progress. Achieving adequate N fixation is inherent in this problem, so Rhizobium species are also evaluated. From these and other efforts, it appears there may be instances where particular cultivars and rhizobia should be blended in order to maximize yields.

There are several other nutritional aspects in which genetic adaptation either has been achieved or has potential. Iron nutrition has been worked on, and it has been shown that efficient varieties exist that are useful under calcareous soil conditions. The Ca requirement, which is relatively great in some of the current higher yielding varieties, could also be reduced through a breeding program. Little is known of the P and K requirements, but these might also be adjusted downward through plant breeding efforts.

SUGGESTED READING

There are a few references in which nutritional needs as well as many other management practices have been brought together to give a state-of-the-art picture for groundnut production. One of the earliest was The Peanut - The Unpredictable Legume published in 1951 by The National Fertilizer Association, Washington, D. C. This was updated in 1973 as Peanuts - Culture and Uses published by the American Peanut Research and Education Society, Texas A&M University, Yoakum, TX. A further update by that organization may now be available under the title Peanut Science and Technology. Aspects of peanut nutrition are also covered in Fertilizer Guide for the Tropics and Subtropics by J. G. deGeus. The second edition of this book was published in 1973 by Centre D'Etude de L'Azote in Zurich, Switzerland. Another reference is Techniques Agricoles et Productions Tropicales XV L'Arachide by Gillier and Silvestre and published in 1969 by G.-P. Maisonneuve & Larose, Paris.

15
COWPEA

D. Nangju and B. T. Kang

Cowpea (Vigna unguiculata) is known by several names; southern pea and blackeyed pea in USA, beans in Nigeria, niebe in Francophone Africa, cowgram in India and kacang in Indonesia and Malaysia. It is utilized as tender green leaves and dried seeds in Africa, mainly as green pods[2] in Southeast Asia, as green seeds in USA, and to a limited extent as a forage crop and green manure. In other countries, it is utilized as dried seeds or pulse.

Cowpea is the second most important grain legume in the lowland tropics after groundnut (Arachis hypogaea). However, while groundnut is used as a food and industrial crop, cowpea is utilized exclusively as a food and to supplement the protein diets of rural and urban people. The major cowpea producers are Nigeria and other West African countries. Limited amounts of cowpea are produced in East Africa, South Africa, Northeast Brazil, India, Southeast Asia and the USA. Total world crop area is about 5.2 million ha. Yields are generally low, with the world average being about 210 kg/ha. Highest yields are obtained in Egypt (3,100 kg/ha), Japan (1,000 kg/ha) and Sri Lanka (1,000 kg/ha).

The popularity of cowpea as a food legume is attributed to the following factors: (i) short growth duration (70-90 days), fairly drought resistant, fits into various types of rainfed cropping systems; (ii) adapted to a wide variety of well-drained soils; (iii) good seed viability for several years under room temperature; (iv) high protein content in seeds (22-35%); and (v) simplicity of

[1]Agronomist, Asian Development Bank, P. O. Box 789, Manila, Philippines; and Soil Scientist, Farming Systems Program, International Institute for Tropical Agriculture (IITA), PMB 5320, Ibadan, Nigeria.

[2]The cowpea utilized as green pods is actually Vigna unguiculata (L.) Walp. subspecies sesquipedalis, also known as sitao, yard long bean, asparagus bean and kacang panjang, or vegetable cowpea.

preparation and multiplicity of edible forms (young leaves and shoots, green pods, green seeds and dry seeds).

CLIMATIC ADAPTATION

Cowpea is a hot weather crop well adapted to temperatures between 20°C and 35°C. For good germination, a minimum soil temperature of 20°C is required. It does not tolerate frost or temperatures below 10°C even for periods as short as 24 hours. Temperatures above 35°C will reduce yields through flower and pod abortion. Maximum dry matter production occurs at 27°C day temperature and 22°C night temperature.

Because of its drought resistance, cowpea can be grown in semi-arid regions with annual rainfall as low as 600 mm. However, for high yields, adequate moisture supply is necessary during vegetative and flowering stages. Most subsistence farmers grow cowpea toward the end of the rainy season or with residual moisture on soils with a high water-holding capacity. Excessive rain during crop growth promotes fungal and bacterial diseases that can be detrimental to seed yield. Since harvesting, drying and storage present major problems under high rainfall and relative humidity, cowpea for dry grain is seldom grown in the humid tropics. In this region, vegetable cowpea, harvested as green pods, is more suitable. Vegetable cowpeas are generally climbing types, although bush types have recently been developed in the Philippines.

SOIL ADAPTATION

Cowpea is well adapted to a wide range of soils, but cannot tolerate waterlogging. High seed yields are generally obtained on loamy soils. High fertility soils tend to cause more vegetative growth and lower seed yields. On sandy soils, moisture and nutrient stresses can hamper growth which can also sometimes be inhibited by heavy nematode infestations. Cowpea is fairly tolerant to acidity but intolerant to salinity.

PLACE IN FARMING SYSTEMS

In Africa, cowpea is generally intercropped with sorghum and millets and to some extent also with maize, cassava, yams, cotton and vegetables. The crop fits into subsistence conditions since (i) cereals and root crops are staple food crops and are planted at the onset of rains, (ii) without adequate crop protection, cowpea is a risky crop to be grown as a sole crop; and (iii) cowpea tolerates shading during the early part of its growth. Relay intercropped cowpea is frequently planted on ridges about 2 months after the cereal. As soon as the cereal is removed, the cowpea flowers towards the end of the rainy season and produces pods during the dry season. At this time, insect and disease pressures are relatively low, enabling the cowpea to produce low but stable yields.

In Southeast Asia, the climbing vegetable cowpea is generally supported with high stakes and grown on rice fields along the bunds. In the humid zone of Africa, vegetable cowpea is intercropped with yams, maize, and okra, the latter two crops being used as a means of support. Other means of support are hedges, fences and dead trees.

In the USA and South Africa, cowpea is planted as a sole crop. With good management and adequate crop protection, sole crop cowpea is more productive than intercropped cowpea, and it also facilitates mechanization.

NUTRIENT REQUIREMENTS

Potential Yields in the Tropics. In international uniform cultivar trials conducted by IITA, seed yields up to 4 mt/ha were obtained in Ethiopian highlands characterized by adequate soil moisture, high solar radiation, and moderately low night temperatures. Medium productivity (yield potentials of about 2.5 to 3.0 mt/ha) is observed in the lowland sub-humid regions of Ghana, Kenya, Nigeria, Benin, Togo, Zambia, Brazil and Thailand. Low productivity (seed yields of 0.5 to 1.5 mt/ha) occurs in the humid regions of Liberia, Nigeria, Sierra Leone, Zaire, Surinam, Indonesia and Sri Lanka.

Potential yields of intercropped cowpea vary from about 20 to 60 percent of the sole crop, depending on the extent of competition for water, nutrients, light, and space between cowpea and the major crop.

The potential yield of vegetable cowpea has not been properly assessed. With sole cropping, yields of 30 mt/ha of green pods can be obtained; when intercropped with maize, yields can drop to 30 percent of the sole cowpea yield.

Industry Farm Yields. Actual yields obtained in different countries range from 100 to 1,000 kg/ha. In Nigeria, which has over 77 percent of the cowpea growing area, the average yield is 189 kg/ha. The dismal yields are attributed to many factors: (i) insect damage, the major factor contributing to poor yield and poor seed quality--virtually all parts of the cowpea plant can be attacked by insect pests at some stages of growth; (ii) low plant population when grown in mixtures with cereals -- most cowpeas are intercropped with either sorghum or millets, are generally planted late and at low plant densities and may suffer from competition for light, moisture and nutrients; (iii) low-yielding varieties--most farmers grow traditional low yielding varieties with creeping plant type, these varieties are daylight-sensitive and susceptible to diseases and nematodes; (iv) low soil fertility -- cowpeas are widely grown on soils with low fertility (as a result of continuous cropping, inadequate fertilization, soil erosion and poor soil management) in the subhumid regions of tropical Africa, and farmers apply little or no fertilizers to cowpea; (v) inadequate weed control -- weeds can reduce cowpea yields up to 50% and are generally controlled by hand on small farms.

For the main traditional cowpea growing areas of tropical Africa, fertilizer application will not likely improve farmers' yields significantly unless combined with the use of insect control, improved varieties, and proper crop husbandry. Plant nutrients will play a more important role in cowpea production with sole cropping under a high level of management.

PRESENT LEVELS OF MINERAL NUTRITION

Very little information is available on mineral nutrition of cowpea in different regions. However, since cowpea is grown mainly by subsistence farmers as an intercrop, fertilizers are usually not used for the crop. Fertilizers are generally applied to the main cereal crops, and the relay intercropped cowpea utilizes the residual fertilizers. With the present low productivity of cowpea in most countries, the amount of nutrients available in the soils appears to be sufficient to sustain cowpea production, albeit at low levels. In the USA and South Africa, rates of up to 30-50 kg N/ha, 30-60 kg P_2O_5/ha, and 0-50 kg K_2O/ha are recommended for sole crop cowpeas.

The most common limiting nutrient for high yields in cowpea is phosphorus. Phosphorus deficiency is widespread in the savanna region of tropical Africa where cowpea is widely grown. Application of phosphorus improves nodulation and seed and protein yields. Inoculation of cowpea seed is seldom necessary in the major growing areas, as there is usually an abundance of indigenous Rhizobium strains capable of good nodulation. Addition of a small amount of starter N at planting in soils of low N status improves root growth, seedling growth, and seed yield. Despite the generally low sulphur status of West African savanna soils and S response in the greenhouse, no significant field responses to S application have been observed, presumably due to the low yields obtained. Potassium responses are only observed on continuously cropped sandy soils. Indications so far are that Ca, Mg and Mo have not been found to be limiting in cowpea production, except on acid soils in humid regions where cowpea is not well adapted and not widely cultivated.

SOIL TEST VALUES

Soil test values in relation to yields have been examined for P and S, the nutrients most likely to be limiting in the cowpea growing areas in African savannas. In field trials with improved varieties grown on Alfisols, the critical Bray (No. 1) extractable P-level was found to be 8 ppm P. About 5 ppm S was required in the soil solution for 95% of maximum grain production for cultivars TVu 76 - 2E and TVu 201 - 1D, and about 2.5 ppm S in soil solution for cultivar Sitao Pole, indicating varietal differences in the response of cowpea to sulphur.

PLANT TISSUE VALUES

Nutrient uptake by cowpea at different growth stages has received attention. The nutrient uptake of the determinate cowpea variety Prima grown under a high level of fertilizers is shown in Table 15.1. No study has been carried out to determine the nutrient composition of cowpea plant tissues at different soil fertility levels or at different yield levels for determinate and indeterminate cowpea varieties. Such a study is necessary in order to determine the nutritional status of the growing plant and its nutritional requirements for obtaining high yields.

Table 15.2 shows the elemental composition of the grain of 11 cultivars of cowpea. The amount of nutrients removed by grain depends very much on yield levels. At the present low yield levels in farmers' fields (200 kg/ha), the amounts of nutrients removed are so small that nutrient deficiency symptoms would not normally appear.

In southern Nigeria, the critical P-level of the uppermost fully expanded leaves at full-bloom stage is about 0.4% P. The critical S contents of uppermost fully expanded leaves at early flowering range from 0.3% S for variety Sitao Pole and 0.5% S and 0.6% S for cultivars TVu 201-1D and TVu 76-2E, respectively.

Table 15.3 shows the composition of leaves, green pods and grain at different growth stages.

VISUAL SYMPTOMS AS INDICATORS OF DEFICIENCY

Deficiency symptoms often observed in the field are N and P. Deficiency symptoms of K, S, and Mg have been observed only in localized areas and experimental plots.

Nitrogen deficiency is characterized by general yellowing and stunting of the plant. Lower leaves of N deficient plants will abscise as the symptoms become more severe. Such symptoms can be mistaken with those caused by root-knot nematode and can only be differentiated by pulling the plants and examining the roots. Roots infected by root-knot nematode are short, swollen, and full of

TABLE 15.1
Nutrient uptake over time by cowpea variety Prima[1]

Days After Planting (DAP)	N	P	K	Ca	Mg
			kg/ha		
10	1.4	0.1	1.0	0.2	0.2
17	5.5	0.5	4.7	1.0	0.9
24	16.2	1.8	14.5	3.6	3.8
31	41.7	4.0	31.5	8.2	7.4
38	65.6	6.3	48.0	14.1	13.1
(Flowering)					
48	108.1	10.1	74.9	22.4	21.9
52	112.2	12.5	98.8	21.2	28.2
(Podding)					
59	104.5	11.5	75.6	15.9	22.8
66	96.0	9.9	66.0	11.7	19.7
(Harvest)					

[1]Unpublished data of B.T. Kang and D. Nangju. The determinate cowpea was grown under a high level of management and received fertilizers at the rate of 90 kg N, 90 kg P_2O_5, and 90 kg K_2O/ha.

TABLE 15.2
Mean elemental composition of the grain of eleven cultivars of cowpea produced at IITA, Ibadan, Nigeria, and calculated quantities in harvested grain for two levels of production (After Fox, et.al. 1977)

			Quantity in Hypothetical Harvest	
Element	Concentration Mean %	Range %	Typical yield[1] kg/ha	Agronomically Possible yield[2] kg/ha
N	3.97	3.64 – 4.36	8.9	59.7
P	0.49	0.44 – 0.54	1.1	7.4
K	1.63	1.50 – 1.80	3.7	24.4
Ca	0.10	0.10 – 0.11	0.2	1.5
Mg	0.23	0.21 – 0.23	0.5	3.4
S	0.25	0.21 – 0.28	0.6	3.8
Zn	0.0044	0.9938 – 0.0051	0.01	0.07

[1]Based on seed yield of 224 kg/ha.
[2]Based on seed yield of 1,500 kg/ha.

galls, while the roots from N-deficient plants appear normal.

Phosphorus deficiency is characterized by general stunting, small and narrow leaves, early flowering, and a marked reduction in root nodulation.

In potassium-deficient plants, tips and edges of the older leaflets turn yellow; the yellowing spreads gradually toward the center and base of the leaflets. The condition is followed by necrosis and browning of tissue around the leaf margins and eventually dropping of the leaves.

Sulphur-deficient plants show similar symptoms as nitrogen deficiency, except that the chlorosis is more pronounced in younger leaves because S is less mobile in the plant. Magnesium deficient plants show a marked interveinal chlorosis; this condition is generally associated with acid soils. Calcium deficient plants show a distinct necrosis of the younger leaves; such symptoms are sometimes observed on acid soils. Cowpeas are sensitive to high levels of soil manganese; manganese toxicity is often observed on acid manganiferous Oxisols and soils that have been acidified with fertilizer. Severe manganese toxicity may induce Fe-deficiency and chlorosis, or plants may develop brown spots on the older leaves, eventually followed by leaf abscission and resulting in lower yields.

TABLE 15.3
Chemical composition of cowpea leaves, green pods and grain at different growth stages[1] (IITA)

Plant part	Protein Content	S/N[2]	Ca	P	Fe ppm
		%			
Cowpea leaves					
Recently opened	34.6	1.6	1.5	0.8	300
1 week old	32.6	2.1	1.6	0.5	260
2 weeks old	25.6	4.9	3.8	0.9	640
Green pods	23.2	3.9	0.3	0.6	80
Seeds	24.4	4.0	1.0	0.5	0

[1]Determined at about 5 percent moisture.
[2]S/N = sulfur/nitrogen ratio, as an indication of the level of sulfur containing amino acids.

IMPROVED VARIETIES AND THEIR NUTRIENT REQUIREMENTS

Improved varieties are continuously being developed at international and national research centers in the tropics, particularly at IITA. Most improved varieties have been bred mainly for resistance to diseases and

pests. Only after incorporation of disease and pest resistance can varieties be developed that are responsive
to fertilizers. The development of pest resistant varieties has been slow and difficult, and it is doubtful that
a variety resistant to all major pests can be developed
in the foreseeable future. Without that, high yields
cannot be obtained unless insecticides are applied during
flowering and pod development. With proper pest and weed
control, seed yields of about 1,500 kg/ha are agronomic-
ally feasible under field conditions using present improved varieties. Under these conditions, fertilizer
rates of 30 kg N, 30 kg P_2O_5, and 30 kg K_2O/ha at planting are recommended to ensure an adequate supply of nutrients. In soils high in organic matter and K, an application of 30 kg P_2O_5/ha alone is considered adequate.

FERTILIZER AND COWPEAS

Sources of fertilizers. Since most of the soils
where cowpea is grown are generally very sandy with low
effective cation exchange capacity (ECEC) and low moisture holding capacity, animal manure should be an excellent source of nutrients for correcting deficiencies.
However, in many countries, it is difficult to obtain
adequate amounts of animal manure to satisfy crop requirements, and hence it is seldom used by farmers. Chemical fertilizers are now gradually being used for sole
crop cowpeas. Among the chemical fertilizers, phosphate
fertilizers such as superphosphate, rock phosphate and
triple superphosphate are used most widely for cowpeas.

Method and rate of fertilizer application. Phosphate fertilizer can be broadcast, or drilled in bands a
few centimeters from the seedlings at, or just after,
planting. Both methods are equally effective. Foliar
application of phosphate fertilizer at flowering can be
effective in increasing cowpea yield, but is laborious
and expensive. Rates of phosphate applied vary with the
severity of phosphorus deficiency and the variety used,
but generally ranges from 25 to 60 kg P_2O_5/ha. Some of
the traditional varieties do not respond to phosphorus
application.

In poor soils, N applications at planting at the rate
of about 30 kg N/ha have been effective in improving
seedling growth and yield. If sulphur deficiency is also
suspected, ammonium sulphate or other S-containing phosphate fertilizers can be used.

In the USA, a complete fertilizer containing N, P
and K at the rates of 50 kg N, 46 kg P_2O_5, and 48 kg
K_2O/ha has been recommended. On strongly acid soils,
aluminum and manganese toxicities may limit nodulation

and cowpea growth and yield. Addition of low lime rates between 0.5 to 1.0 ton/ha to reduce aluminum and manganese toxicities will benefit the crop. There may be cultivar differences in tolerance to soil acidity.

SUGGESTED REFERENCES

FOX, R. L., B. T. KANG and D. NANGJU. 1977. Sulfur requirements of cowpea and implication for production in the tropics. Agron. J. 69:201-205.

NANGJU, D. 1978. Integrated approach to increased cowpea production in tropical Africa. AAASA Third General Conference and 10th Anniversary, April 9-15, 1978. IITA, Ibadan, Nigeria.

RACHIE, K. O. and L. M. ROBERTS. 1974. Grain legumes of the lowland tropics. Advances in Agron. 26: 1-132.

SELLSCHOP, J. P. F. 1962. Cowpeas, Vigna unguiculata (L.) Walp. Crop Abs. 4: 259-266.

SUMMERFIELD, R. J., P. A. HUXLEY and W. STEELE. 1974. Cowpea (Vigna unguiculata (L.) Walp. Fld. Crop Abs. 7: 301-312.

16
POTATO

O. A. Lorenz and D. N. Maynard

The potato (<u>Solanum tuberosum</u>) is the most important vegetable crop, and world production in 1979 was over 250 million metric tons. Potatoes are grown in all continents and serve as food for humans, feed for livestock and are important industrial sources of starch and alcohol. Popularity of potatoes as a basic food supply is well illustrated by the great Irish famine in the late 1840s when late blight disease wiped out the crop, forcing many Irish people to emigrate to America, as well as causing death by starvation of some one million persons.

Most potato production is in Europe and the Soviet Union, which combined account for over 80% of the world production (Table 16.1). Highest yields are reported from North America and Australia and the lowest yields from South America and North Africa. In the USA, potatoes are grown in every state. The leading states based on total production are Idaho, Washington, Maine and Oregon, which combined account for 56% of USA production. Based on the area planted, the five most important states in order are Idaho, North Dakota, Maine, Washington and Minnesota. Yields have increased steadily since about 1940 and now exceed 29 mt/ha. High yields are reported from Washington, Oregon and California.

The per capita consumption of potatoes in the USA is approximately 55 kg. About 40% is consumed fresh and 60% processed mostly as frozen french fries, but also as chips and strings or various dehydrated products.

[1]Professor, Vegetable Crops Department, University of California, Davis, California 95616, USA, and Professor and Chairman, Vegetable Crops Department, IFAS, University of Florida, Gainesville, Florida 32611, USA.

TABLE 16.1
Yield and production of potatoes in specified areas
(average 1976-78)

Area	Yield (mt/ha)	Production (1,000 mt)
North America	26.9	19,362
Europe	19.0	115,162
Soviet Union	12.0	84,884
South America	9.4	9,300
North Africa	11.8	1,773
Asia	14.8	11,877
Australia	21.1	719

Source: USDA Agricultural Statistics, 1979.

ECOLOGICAL ADAPTATION

Climate

The potato is principally a cool-weather crop, but it cannot withstand frost. Freezing temperatures kill both foliage and tubers. Growth is very slow at temperatures below 15°C, and optimum growing temperatures are from about 16° to 19°C, with best tuber development near 18°C. Temperatures above 22°C are definitely too warm for high yields, and near 30°C there is essentially complete inhibition of tuber production. With increasing temperatures the above-ground portion of the plant increases in relation to the tubers due to reduced translocation of carbohydrates to the tubers. Above about 19°C excessive respiration limits plant growth.

Since cool temperatures are so important, most of the world production is in the north temperate regions of North America, Europe and the Soviet Union. Temperatures in Africa are too warm for good production. The initiation of tubers is favored by short days, but there is complex interaction between daylength and temperature. Tuberization is affected by many environmental factors, including mineral nutrition, and depends largely on translocation and storage of carbohydrates in excess of that needed by the top growth for growth and metabolism.

Soils

Potatoes can be grown on various soil types, but it is essential that the soil is sufficiently retentive of moisture, well drained, well aerated and of good structure. If these conditions do not exist, tubers are likely to be misshapen, knobby and of poor general appearance. Compacted soils should not be used.

Potatoes can tolerate a wide range of soil reaction (acidity and alkalinity), from pH 5-8 or higher. Liming of acid soils to pH 6 or above may increase the severity of scab. Excellent yields are often produced on soils which are alkaline, as in the western USA.

Appropriate Place in Farming Systems

Since potatoes require a well-aerated soil, it is essential that they fit into a cropping system that favors good soil structure. It is desirable to include a rotation crop that supplies organic matter. Long rotations in which potatoes are planted at least 3 years apart reduce losses caused by such soil-borne organisms as Fusarium wilt, Verticillium wilt and scab.

In the USA potatoes grown for fresh market can be costly to produce but have potential for high return. Those grown for processing as in Idaho, Maine and Washington represent a much more stable situation with slightly less inputs and usually a lower but more secure market price.

MINERAL NUTRIENT REQUIREMENTS

Average Yields in Representative Countries

The yields of potatoes vary greatly from country to country due both to favorable climates and to production technology (Table 16.1). Highest yields are reported in North America and Australia, and the lowest yields occur in North Africa and South America. The average world yield for the 3-year period 1976-78 was 15.4 mt/ha, and the highest yields were reported in Switzerland at 38.1 mt/ha. Other countries with high yields were The Netherlands at 34.0 mt/ha, Belgium 23.7, United Kingdom 28.1, and USA 26.9. Lowest yields were the South American countries such as Bolivia, Brazil, Peru, and Uruguay, which had yields less than 10 mt/ha. The highest yields in the USA are reported from the state of Washington at slightly over 56 mt/ha. Oregon and California have potato yields approaching 40 mt/ha.

Nutrient Uptake

Potatoes have high rates of nutrient uptake (Table 16.2). Both foliar growth and tuber yield vary greatly; however, a 50 mt/ha crop will remove about 200 kg N, 40 P and 300 K. An average of five reports shows that each metric ton of tubers will remove 2.72 kg N, 0.52 P, and 4.28 K. These values approximate those for "Russet Burbank" reported by Kunkel, et al. 1973.

TABLE 16.2
Elemental accumulation in potatoes[a]

Tuber Yield (mt/ha)	Plant Part	Element Uptake (kg/ha)		
		N	P	K
39.7[b]	vines	51	3	65
	tubers	111	14	166
	total	162	17	231
56.0[c]	vines	108	8	246
	tubers	194	36	201
	total	302	44	447
18.2[d]	vines	34	3	81
	tubers	37	8	90
	total	71	11	171
63.5[e]	tubers	147	22	220
67.2[f]	tubers	201	51	280

[a]Source, growing area and cultivar:
[b]Lorenz (1947) - California (average of three crops) White Rose.
[c]Phosphate and Potash Institute (1979) - estimates from selected data.
[d]Carpenter (1963) - Maine (average of two crops) - Kennebec.
[e]Gunasena (1969) - United Kingdom - Craig Royal.
[f]Kunkel et al. (1973) - Washington (average of four crops) - Russet Burbank.

TABLE 16.3
Range of elemental concentrations in potato tubers. White Rose cultivar. Samples from many areas of California (Lorenz, 1963)

Elemental Level	Percent of Dry Weight		
	N	P	K
Low	0.90	0.10	1.35
Usual	1.60-1.90	0.20-0.35	1.60-0.90
High	2.25	0.45	2.60

The nutrient composition of potato tubers varies widely. Dry matter may be as low as 11% or as high as 30%, depending largely on cultivar and stage of maturity. Dry matter concentrations of 18 to 22% are common. Table 16.3 illustrates the great differences in elemental levels within the "White Rose" cultivar.

Soil Test Values Sought

Many researchers have found that soil analyses are of value in predicting fertilizer needs in potatoes. In California potatoes responded greatly to fertilizer application when the $NaHCO_3$ extractable PO_4-P (soil extractant used in soil testing) in the soil was less than 40 ppm, or when the NH_4 acetate extractable K was less than 100 ppm (Tyler et al., 1961). No response to fertilizer application was obtained when the values were 1-1/2 to 2 times higher.

Some areas provide specific fertilizer recommendations based on soil analyses, as shown in Table 16.4. Nutrient levels in the soil vary greatly depending on the soil extractant used.

Plant Tissue Values Sought

Much information is available for plant tissue analysis of potatoes. Petiole tissue of a recently matured leaf is used and related to plant age and yield (Table 16.5). Under conditions of very rapid growth or very long growing seasons, slightly higher values have been suggested for the critical nutrient ranges.

Leaf blade tissue serves almost equally well as petiole tissue. Also, analyses made for total N, P and K give diagnostic values very comparable to fractions extracted with 2% acetic acid and where analyses are made for NO_3-N (nitrate nitrogen), PO_4-P (phosphoric acid), and soluble K (potassium).

Visual Symptoms

In many cases it is possible to distinguish some of the nutrient deficiencies of potato by visual examination of the foliage as noted.

Nitrogen (N). General pale green or yellowing of leaves. Lower leaves are affected first and may curl and show "firing" of the margins. Reddish or purple colors may develop. Stems are thin, erect and hard. Leaves are small and plant growth is slow.

Phosphorus (P). Plants are stunted, grow slowly and maturity is delayed. Stems are thin and shortened. Foliage is crinkly and dark green, often becoming purplish. Severe deficiency can cause leaves to roll upward.

Potassium (K). Foliage is dark green, leaves are small and internodes short. Bronzing, and marginal leaf scorch develops on the older leaves. Chlorotic areas may develop throughout leaf.

TABLE 16.4
Soil analyses and fertilizer recommendations for potatoes

Area and Soil Extractant	Phosphorus (P) Soil Test (kg/ha)	Phosphate (P_2O_5) Recommended in Fertilizer (kg/ha)	Potassium(K) Soil Test (kg/ha)	Potash (K_2O) Recommended in Fertilizer (kg/ha)
Wisconsin – sand and sandy loam soil. (Bray P1 extractant, 0.025N HCl, 0.03N NH_4F).	0 – 170 170+	135 70	0 – 335 335+	270 200
Maryland and mid-Atlantic States. (0.05N, HCl, 0.025N H_2SO_4).	(kgP/ha) 0 – 30 31 – 45 46 – 100 100+	(kg/ha) 225 170 110 55	(kgK/ha) 0 – 78 79 – 150 151 – 181 187+	(kg/ha) 335 225 110 55
Ontario, Canada. (Na HCO_3 extractable P and NH_4 acetate exchangeable K).	(ppm) 0 – 9 10 – 20 21 – 40 41 – 80	(kg/ha) 180 100 30 30	(ppm) 0 – 60 61 – 180 181+	(kg/ha) 180 80 0

kg/ha = kilograms per hectare.
ppm = parts per million.

TABLE 16.5
Plant analysis guide for sampling time, plant part, and nutrient levels of potatoes

Time of Sampling	Plant Part	Nutrient	Nutrient Level	
			deficient	sufficient
Early season	Petiole of 4th leaf from growing tip.	NO_3-N, ppm PO_4-P, ppm K, %	8000 1200 9	12000 2000 11
Midseason	Petiole of 4th leaf from growing tip.	NO_3-N, ppm PO_4-P, ppm K, %	6000 800 7	9000 1600 0
Late season	Petiole of 4th leaf from growing tip.	NO_3-N, ppm PO_4-P, ppm K, %	3000 500 4	5000 1000 6

NO_3-N = nitrate nitrogen.
PO_4-P = phosphoric acid.
K   = potassium.
ppm = parts per million.

Magnesium (Mg). Symptoms of Mg deficiency often resemble those of K deficiency. Lower leaves are lighter green than normal and show yellowing between the veins. Dead tissue develops between the veins, which remain green. Older leaves may fall with prolonged deficiency. Supplemental Mg is often required for potato production. Potato vines showing early symptoms of Mg deficiency can be sprayed with a solution of epsom salts ($MgSO_4$), using 25 kg epsom salts/1000 liters of water/ha. Soil applications are made with the regular commercial fertilizer in which about 2% MgO is incorporated.

Calcium (Ca). The youngest leaves and growing point show the first symptoms. Severe deficiencies cause the terminal leaflets to roll upward with yellowish chlorosis of the leaflets. Stem elongation is restricted by death of the growing point.

Trace elements

Zinc (Zn). Foliage growth is restricted, resulting in rosetting or little leaf appearance. Leaves are thick and brittle. Lower leaves are chlorotic and with severe deficiency the entire leaflet may become yellow with dead tissue developing around the margins and tips. Zinc deficiency may occur in rare instances. In the eastern USA, where the soil test for Zn is below 0.8 ppm or on new land where leveling has exposed a high lime subsoil, Zn fertilizer at 10 kg Zn/ha is applied.

Boron (B). Youngest leaves are lighter green than normal. Leaves become thickened and roll upward. Stem tip may die or make distorted growth.

TIME AND METHOD OF FERTILIZER APPLICATION

High rates of N and K, when banded under the plant row, can result in fertilizer damage to potatoes. With low rates of fertilization it is best to apply the fertilizer in bands 5 cm to the sides and 5 cm below the seed piece. When the combined amounts of N and K exceed 350 kg/ha the excess should be broadcast and plowed under before planting. If needed, supplemental N can be side-dressed or applied during the growing season through the irrigation system.

PRESENT LEVELS OF MINERAL NUTRITION BY REGION

Fertilizers containing all of the major nutrients likely to be deficient in normal agricultural soils are universally recommended for potato production, but the quantities vary greatly between areas as shown in Tables 16.6 and 16.7.

TABLE 16.6
Fertilizer recommendations for potato in selected areas
of the world (de Geus, 1973)

| Area | Fertilizer Recommended | | |
	Nitrogen (N)	Phosphate (P_2O_5)	Potash (K_2O)
	kilograms per hectare		
Brazil	150	310	220
Ceylon (Sri Lanka)	112	124	56
India, Mysore	124	100	124
Indonesia	100	100	50
Mexico (Toluca)	60	120	60
Peru	160	160	100

TABLE 16.7
Fertilizer recommendations for potatoes in the USA

| State | Fertilizer Recommended (kg/ha) | | |
	Nitrogen (N)	Phosphate (P_2O_5)	Potash (K_2O)
New York (upstate)	135-170	135-340	55-340
New York (Long Island)	140-200	270-400	55-200
Florida	230	225	225
Massachusetts (high fertility soils)	100	150	150
Massachusetts (low fertility soils)	190	285	285
Maryland (low fertility soils)	140-170	225	335
Wisconsin (low fertility soils)	varies	135	240
Michigan (low fertility soils)	70	280	335
Maine (low fertility soils)	180	180	180
Washington (low fertility soils)	330	280	470

FERTILIZERS AND POTATO QUALITY

Without adequate fertilization with all major
nutrients, potato produces small tubers that do not meet
the accepted size or grade. With adequate fertilization,
larger tubers are produced and a high yield of U.S. grade
#1 results. With still higher fertilization, very large
tubers are produced and often many of these show growth
cracks, knobs, or other defects, which make them unac-

ceptable for the #1 grade. Even so, most experiments show that the largest total yield of U.S. #1 potatoes is normally produced by the fertilizer treatment which results in the highest total yield, even though the percentage of U.S. #1s may be less. In a few tests there have been definitely lower total yields of U.S. #1 grade from high N applications. In some tests from other states it has been reported that N application to a soil already adequately supplied resulted in decreases in the yield of grade #1s, increases in grade #2s, and no effect on the total yield.

Maturity of potatoes is influenced greatly by fertilizer treatment. Deficiencies of N and K are associated with early maturity. Potatoes grown with low levels of N may mature as much as one month earlier than those with adequate N. Conversely, high N applications are associated with delayed maturity.

Dry weight, starch, and specific gravity may be affected by fertilization. Starch accounts for about three-fourths of the total dry matter. In general, as soil fertility decreases, specific gravity and dry weight of tubers grown in these soils decrease. Tubers produced at low N levels usually are highest in specific gravity and dry matter. In this regard the relationship of N and maturity must be considered. An excess of N tends to stimulate the production of succulent foliage and delays maturity. During growth, the dry-matter content of tubers increases, and if the N applications are associated with delayed maturity, they are also associated with the production of low dry-matter tubers. Consequently, when sampled early, tubers grown on N-deficient soil are much higher in dry matter than those grown with sufficient N. An example of the effect of N on specific gravity and starch content of potatoes is shown in Table 16.8.

TABLE 16.8
Effect of nitrogen levels on specific gravity and starch content of White Rose potatoes.

N (kg/ha)	Specific Gravity	Starch (% of fresh weight)
0	1.096	17.0
56	1.089	15.6
112	1.083	14.4
168	1.075	12.8

Nitrogen fertilization often affects the color of potato chips; high N applications are often associated with dark-colored chips. This in turn is associated with higher levels of reducing sugars which can produce dark-colored chips. Also, high N produces higher levels of certain amino acids, which are also associated with dark chips.

Potassium fertilization also affects internal quality. In Pennsylvania specific gravity of the tubers was lowered by heavy K applications even though there was no effect on yield. Many times tubers deficient in K produce dark-colored chips, while those grown with sufficient K produce good, light-colored chips.

Even though there are many instances where fertilization has influenced specific gravity of potatoes, it is evident that cultivar and growing area or environment have a greater overall effect than fertilization. Some cultivars are inherently low in specific gravity and others high. Potatoes produced under cool conditions are higher in dry matter than those grown during hot weather.

BENEFITS FROM ADEQUATE NUTRITION

The great differences in potato yields in the various continents and areas (Table 16.1) are due to climate, seed stock, diseases, insects and fertilization. The very high yields in Belgium, The Netherlands, and the USA are made possible only by high rates of fertilization. In most areas economic production would not be possible if high rates of fertilization were not applied.

SUPPLEMENTAL READING

DE GEUS, J.G. 1973. Fertilizer Guide for the Tropics and Subtropics. Centre d'Etude de l'Azote. Zurich.

GARDNER, B.R. and J.P. JONES. 1975. Petiole analysis and the nitrogen fertilization of Russet Burbank potatoes. American Potato Journal 52:195-200.

KUNKEL, R., N. HOLSTAD and T.S. RUSSELL. 1973. Mineral element content of potato plants and tubers vs. yields. American Potato Journal. 50:275-283.

LORENZ, O.A. 1947. Studies on Potato Nutrition, III. Chemical composition and uptake of nutrients by Kern County potatoes. American Potato Journal. 24:281-293.

SMITH, Ora. 1968. Potatoes: Production, Storing, Processing. AVI Pub. Co. Inc., Westport, Connecticut.

TYLER, K.B., O.A. LORENZ and F.J. FULLMER. 1961. Plant and soil analyses as guides in potato nutrition. California Agricultural Experiment Station Bulletin 781.

17
CASSAVA

R. H. Howeler

Cassava (<u>Manihot</u> <u>esculenta</u>) is a tropical root crop known as "yuca", "tapioca", "manioc", or "mandioca". After rice it is the crop produced in largest quantities in the tropics, and in tropical America and Africa it takes first place in total production. It is estimated that it serves as a staple food for over 200 million people. Cassava is a traditional calorie source of the rural population in Latin America but has limited importance in urban areas, except where consumed as a flour, as in Brazil. Besides as a dry flour, the crop is sold as fresh roots for direct human consumption. In Brazil nearly 10% of total calories consumed is derived from cassava, while in Paraguay this is nearly 15%, mainly consumed as fresh cassava. After chipping and drying, cassava is also a major calorie source in animal feed rations, while the high-protein leaves are sometimes consumed by people or fed to livestock. Finally, the crop is used as a raw material for the production of starch or fuel alcohol.

ECOLOGICAL ADAPTATION

Cassava has a reputation as a rustic crop that grows reasonably well under adverse conditions of climate and soils. Although its production is essentially limited to the tropics, the crop has a considerable temperature range, growing from sea level to about 2000 meters altitude and tolerating during short periods temperatures down to 0°C. For good growth it has been suggested that soil temperatures should be above 18°C and rainfall should be above 1000 mm annually. However, cassava can be grown with much lower annual precipitation as long as sufficient soil moisture is available during the first 2 months of growth.

---

[1]Soil Scientist, Cassava Program, Centro Internacional de Agricultura Tropical (CIAT), Apartado Aereo 67-13, Cali, Colombia.

Once established, the crop tolerates 4 to 5 months of drought during which time it drops most of its leaves, sprouting vigorously again when enough soil moisture is available. Cassava appears to be more drought tolerant than most other crops and is therefore sometimes used as a famine crop.

It is generally recommended to grow cassava on well-drained sandy loam or clay loam soils since cassava will not tolerate excess water. However, when grown on ridges, extremely high yields (>80 t/ha/year) have been obtained on heavy alluvial soils of the Cauca Valley in Colombia. Harvesting, however, is greatly facilitated on light-textured soils.

Cassava grows best in soils of pH 6 to 7 but the crop is well adapted to a much wider pH range, e.g. producing excellent yields (40 t/ha) in a high organic matter Inceptisol with a pH as low as 3.5 and with 4-5 me Al/100 gm. Although well adapted to acid soils, the crop is relatively susceptible to high pH and associated problems of salinity, alkalinity and poor drainage, reducing its yields dramatically above pH 8.0 and 0.5 mmhos/cm electrical conductivity of the saturation extract.

Since cassava grows reasonably well on infertile soils, farmers often consider it a scavenger crop. Hence it is commonly grown as the last crop in a rotation before returning the plot to bush fallow in the slash and burn agricultural systems of tropical Africa, America and Asia. In general, cassava is grown as a subsistence crop without the use of fertilizers. However, now that cassava has been recognized to be one of the most efficient carbohydrate producers, interest has recently developed in its large-scale commercial exploitation as an animal feed or as a raw material for the production of starch or fuel alcohol. Countries with the greatest production include Brazil, Indonesia, Thailand, Zaire, Nigeria, and India.

MINERAL NUTRIENT REQUIREMENTS

Although cassava is reputed to grow well on poor soils, for maximum yields higher fertility levels are required. With adequate fertilization high yields of nearly 60 t/ha have been obtained in very acid and infertile soils. However, the crop is generally grown without fertilizers on infertile and often highly eroded slopes where no other crops can produce adequately. Under these conditions average yields may be as low as 3-4 t/ha. Although few reliable data are available, worldwide average yields were calculated to be 8.7 t/ha of fresh roots, with average yields of 6.5 t/ha in Africa, 11.0 t/ha in Asia and 11.5 t/ha in S. America. Thus the potential yield of the crop is nearly ten times higher than the average yield, and lack of fertilizer may be one important factor causing this production gap.

Fertilizer consumption statistics in cassava are essentially non-existent, but the scanty information available indicates little is used. However, fertilizer responses in cassava are as high or higher than those in other crops. Thus, numerous fertilizer trials conducted by FAO showed an average yield increase of 49% in cassava, compared with 43% in rice, 68% in maize and 50% in groundnuts in 11 countries in West Africa. In Latin America, yield increases due to fertilization were 110% in cassava compared to 101% in rice and 53% in maize, while in Asia this was 79% for cassava, 61% for rice, 76% for maize and 52% for groundnuts. Table 17.1 shows the average rates of N, P and K used in the FAO trials to obtain maximum yields of cassava compared with those of other crops. In West Africa cassava responded mainly to K and had the highest K requirement among eight crops tested; in Latin America cassava responded mainly to P, having a P requirement below that of beans and soybeans but above that of maize, rice and wheat; in Asia, cassava responded mainly to N, having an N requirement similar to that of maize and rice and much above that of groundnuts and soybeans. These data obtained by FAO in farm trials and demonstrations are in general agreement with data reported in the literature and summarized by Howeler (1981). In Latin America, cassava responded mainly to P, while in Africa and Asia the response was mainly to K and N, respectively.

DIAGNOSIS OF NUTRITIONAL DISORDERS

Nutritional disorders in cassava can be diagnosed by observation of visual symptoms or by soil or plant analyses. Unfortunately, visual symptoms of deficiencies in cassava, especially those of the major elements N, P and K, are not always very clear as the plant responds to an insufficiency of these elements mainly by reduced growth, rather than retranslocation of the limiting nutrient with a consequent production of deficiency symptoms. Cassava is well adapted to low levels of available nutrients because of its high efficiency of nutrient utilization, a slow internal distribution of nutrients, a low nutrient gradient between old and young tissue resulting in few deficiency symptoms, slow growth rates, a large root/shoot ratio, and less reduction in yield. Only in cases of very severe deficiencies of N, P or K are typical deficiency symptoms developed and then not necessarily in all cultivars. Because of this absence of clear symptoms, diagnosis of nutrient deficiencies depends largely on soil and plant tissue analyses.

TABLE 17.1
Average rates (kg/ha) of N, P_2O_5 and K_2O giving highest
yields in post-1969 fertilizer trials (source: FAO, 1980)

		No. of Trials	Average Rate for Highest Yield			Average Yield Increase (t/ha)
			N	P_2O_5	K_2O	
West Africa						
Cassava	(1)*	127	60	39	108	10.59
Cotton	(1)	132	111	57	33	0.58
Cowpeas	(1)	61	25	55	49	0.40
Groundnuts	(1)	134	25	50	54	0.92
Maize	(3)	996	92	70	32	1.64
Rice	(1)	41	74	31	00	1.31
Sorghum	(1)	99	45	45	45	1.48
Yam	(3)	128	56	41	52	6.17
Latin America						
Cassava	(1)	141	40	69	34	11.15
Beans	(1)	391	37	72	34	0.50
Cotton	(1)	510	46	55	37	0.36
Maize	(2)	563	63	61	31	1.57
Rice	(1)	361	37	59	31	1.16
Wheat	(1)	79	108	40	30	1.32
Soybean	(1)	111	–	85	19	0.88
Far East						
Cassava	(1)	69	119	50	38	6.37
Rice	(5)	796	129	56	26	1.77
Maize	(2)	346	111	63	31	1.52
Groundnuts	(2)	144	16	71	37	0.52
Soybean	(2)	121	21	63	31	0.41

*Number in brackets indicates number of countries in-
volved; for cassava these are Ghana, Brazil and
Indonesia.

Soil Testing

Data on critical levels of soil parameters (defined
as those corresponding to 95% of maximum yield) are
scarce, but those reported are summarized in Table
17.2. These values are not absolute but are merely a
guide in the interpretation of soil analyses data, since
critical values can vary with the cultivar used, soil
moisture conditions at time of sampling, soil texture,
and interrelations with other nutrients. In general,
cassava tolerates a wide range of soil pH, a high level
of exchangeable Al, and low levels of P and Ca; on the
other hand it is susceptible to soil salinity and to a
high level of Na saturation.

Plant Tissue Testing

Critical concentrations for deficiencies and toxicities in cassava plant tissue have been determined for most nutrients, and are summarized in Table 17.3. The plant tissue most indicative of nutrient stress is the youngest fully expanded leafblade (YFEL-blade); these leaves should be sampled at the time of maximum growth rate, i.e., approximately 3 to 4 months after planting. Sampling should not be done during periods of slow growth due to drought or low-temperature stress, nor should different plant parts be mixed into the same sample. Considerable differences in nutrient concentrations have been shown between leafblades, petioles, stems and roots, and between the upper and lower part of the plant. Also, nutrient concentrations vary during the growth cycle, and sampling of plant tissue should thus be strictly standardized to include only YFEL-blades collected at 3 or 4 months after planting. The critical values shown in Table 17.3 indicate that cassava leaves are very high in most nutrients and that the crop has a high yield internal nutrient requirement for near-maximum yield.

Deficiency Symptoms

Nutrient deficiency and toxicity symptoms in cassava have been determined in sand and nutrient solution cultures, and symptoms for most elements have been observed in the field. Color photographs and detailed descriptions were published by Asher et al. (1980) and can be briefly summarized as follows.

Deficiencies

Nitrogen (N). Reduced plant growth; in some cultivars uniform chlorosis of leaves, starting with lower leaves but soon spreading throughout the plant.

Phosphorus (P). Reduced plant growth, thin stems, short petioles; and small pendant leaves; under severe conditions some lower leaves turn yellow, flaccid, necrotic, and fall off.

Potassium (K). Reduced plant growth, small and chlorotic upper leaves but thick stems; under very severe conditions purple spotting, yellowing and border necrosis of lower leaves; fine cracks and premature lignification of upper stem; excessive ramification resulting in short prostrate plants.

Calcium (Ca). Reduced root and shoot growth; deformation and border chlorosis of upper leaves, leaftips and margins bending downwards.

TABLE 17.2
Critical levels of soil parameters for cassava

Parameter	Level	Method of Analysis
pH	4.7 and 7.8	1:1 soil-water ratio
Al	2.5 me/100 gm	1 N KCl
Al-sat	80%	Al/Al + Ca + Mg + K
P	7 ppm	Bray I-extract
	10 ppm	Bray II-extract
	8 ppm	Olsen-EDTA extract
	9 ppm	North Carolina extract
K	0.15 me/100 gm	NH_4-acetate
	60 ppm	North Carolina extract
Ca	0.25 me/100 gm	NH_4-acetate
Conductivity	0.5-0.7 mmhos/cm	saturation extract
Na-sat	2.5%	NH_4-acetate
Zn	1.0 ppm	North Carolina extract
Mn	5-9 ppm	North Carolina extract
SO_4-S	8 ppm	North Carolina extract

*Bray I = 0.025 N HCL + 0.03 N NH_4F
Bray II + 0.1 N HCL + 0.03 N NH_4F
Olsen-EDTA = 0.5 N $NaHCO_3$ + 0.01 Na-EDTA
North Carolina = 0.05 N HCL + 0.025 N H_2O_4
NH_4-acetate = 1 N NH_4-acetate at pH 7

Magnesium (Mg). Marked interveinal chlorosis of lower leaves; some reduction in plant height.

Sulfur (S). Uniform chlorosis of upper leaves, which soon spreads throughout the plant; very similar to N-deficiency.

Boron (B). Reduced plant height, short internodes, short petioles and small deformed upper leaves; purple-grey spotting of fully expanded leaves; gummy exudate on stem and petioles; suppressed lateral root development.

Copper (Cu). Deformation and uniform chlorosis of upper leaves, leaftips and margins bending upwards or downwards; petioles of fully expanded leaves long and droopy; reduced root growth.

Iron (Fe). Uniform chlorosis of upper leaves and petioles, which become white under severe conditions; reduced plant growth; young leaves are small but not deformed.

Manganese (Mn). Interveinal chlorosis of upper or middle leaves; uniform chlorosis under severe conditions; reduced plant growth; young leaves small but not deformed.

Zinc (Zn). Interveinal yellow or white spotting of young leaves, which become very narrow and chlorotic at the growing point; necrotic spotting also on lower leaves; reduced plant growth.

Toxicities

Aluminum (Al). Reduced root and shoot growth; yellowing of older leaves under severe conditions.

Boron (B). Necrotic spotting of older leaves, especially along leaf margins.

Manganese (Mn). Yellowing of older leaves with purple-brown spotting along the veins; leaves become flaccid and drop off.

IMPROVED GERMPLASM

Although cassava has been grown for thousands of years, it is only very recently that serious attempts have been made to improve its yield potential and stability through breeding. As yield levels increase so do the nutrient requirements of the crop. According to nutrient extraction data (Howeler, 1981), every ton of fresh cassava roots harvested removes on average 2.5 kg N, 0.5 kg P, 4 kg K, 0.6 kg Ca and 0.3 kg Mg. Thus, a 30 t/ha root yield removes as much as 120 kg K and 75 kg N. For sustained high yields of cassava it is therefore necessary to fertilize with at least these levels of N and K to prevent soil depletion. The P requirement depends on the level of available P and the P fixing capacity of the soil, rather than the quantity of P extracted by the plant.

Plant breeding can also be a tool to improve crop tolerance to adverse soil conditions. Through field screening of large numbers of varieties, germplasm can be identified that has special tolerance to soil salinity, acidity, or low levels of P; these can be used as parents in breeding to improve yield as well as adaptation to adverse soil conditions.

ROLE OF NUTRIENTS IN CROP MANAGEMENT

Cassava is exceptionally tolerant to acid soils, and liming is generally not necessary. If used, lime is applied at low rates mainly to supply Ca and/or Mg. In many acid soils 500 kg/ha of dolomitic lime is sufficient to supply these nutrients. For small-scale cassava production the use of compost or cow and chicken manure are recommended as a well-balanced source of major and minor nutrients. For large-scale production these sources are generally not available or not economical, and chemical fertilizers are used. Nitrogen can be applied as urea, ammonium sulfate, or ammonium nitrate; no significant differences have been found among these sources. Phosphorus is generally applied as single or triple superphosphate, basic slag or ground phosphate rocks. The latter are excellent and cheap sources of P for acid soils. Cassava has been shown to be highly dependent on mycorrhizal associations for P uptake from low-P soils (Howeler, 1981). It is speculated that P fertilizer requirements may be reduced through mycorrhizal inoculation, especially in those soils having a low or inefficient population of native strains of these fungi. Potassium can be applied as potassium chloride or sulfate, or as potassium-magnesium sulfate; the latter two sources are more expensive but are used on soils low in S or Mg.

Lime, basic slag and rock phosphate should all be broadcast and incorporated before planting. The more soluble fertilizers such as urea, triple superphosphate and potassium chloride are best applied in bands under or close to the 'stake' at planting or 1 or 2 months after planting. Care should be taken that the stakes are not in direct contact with the fertilizers, otherwise fertilizer burn may cause poor sprouting. In well-drained soils it is recommended to apply part of the N and K at planting and 2 to 3 months later as the plants approach their maximum rate of growth.

Cassava is susceptible to Zn deficiency, especially during early growth. Planting stakes can be dipped in a solution of 1-2% $ZnSO_4.7H_2O$ for 15 minutes at planting time. If this is not sufficient to prevent Zn deficiency, a foliar spray with 1% $ZnSO_4.7H_2O$ may eliminate the symptoms, while a soil application of 5-10 kg Zn/ha as $ZnSO_4.7H_2O$ or ZnO may eliminate Zn deficiency for several years. Application of other micronutrients is generally only required in extreme cases of soil infertility, or in calcareous or organic soils.

Localized placement of fertilizers will stimulate the growth of cassava but not of weeds, resulting in more rapid cover and fewer weeds. Rapid closing of the canopy also protects the soil from erosion. In general, vigorously growing plants are less susceptible to insects or diseases, and cassava has a remarkable ability to recuperate from foliage loss due to insect or disease attack. Little is known about the effect of specific elements in enhancing tolerance to insects and diseases. Moderate levels of K may reduce the incidence of cassava bacterial blight. Similarly, K application reduced the incidence of anthracnose in trials in Colombia. Liming also appeared to decrease bacterial blight in highly acid soils.

TABLE 17.3
Critical nutrient concentration for deficiency and toxicity in cassava plant tissue

Element	Plant Tissue	Concentration*
N def	YFEL** blade	5.1%
	shoots	4.2%
	YFEL blade	5.7%
	YFEL blade	>4.65%
P def	leaves	0.25%
	YFEL blades	>0.4%
	shoots	0.49-0.68%
K def	YFEL blades	1.1%
	YFEL petiole	0.8%
	stem	0.6%
	shoot and root	0.8%
	YFEL blade	1.2%
	YFEL petiole	2.5%
Ca def	shoots	0.4%
Mg def	shoots	0.26%
	YFEL blades	0.29%
S def	YFEL blades	0.32%
Zn def	YFEL blades	43-60 ppm
	YFEL blades	37-51 ppm
B def	shoots	17 ppm
B tox	shoots	140 ppm
Mn def	shoots	100-120 ppm
Mn tox	shoots	250-1450 ppm
Al tox	shoots	70->97 ppm
	roots	2000-14000 ppm

* Range corresponds to values obtained with different cultivars.
**YFEL = Youngest Fully Expanded Leaf.

CROP MATURATION

Cassava does not have a clearly defined stage of maturity and is generally harvested between 8 and 18 months after planting. Fertilization may increase the total yield at any particular time, but there is no indication that it would shorten the growth cycle. Cassava adapts to low-fertility conditions by increasing its root/shoot ratio and harvest index. This implies that fertilization, especially N fertilization, will stimulate shoot growth more than root growth, resulting in a decrease in harvest index. Excessive fertilizer may increase the leaf area index beyond its optimum of 3-4, thus resulting in a yield decrease. Hence, the crop is susceptible to over fertilization.

TABLE 17.4
The average highest percent yield increase due to fertilization and Value/Cost Ratio (VCR) for cassava as compared to other major crops in various countries. (Source: FAO, 1980)

Country	Crop	No. Trials	Average Best	
			% Response	VCR
Brazil	Cassava	66	111.6	5.63
1970-76	Maize	510	83.1	2.74
	Cotton	490	84.2	3.73
	Beans	391	91.1	4.89
	Rice	385	76.6	4.81
	Soybean	124	102.4	2.38
	Sugarcane	105	66.6	3.64
Colombia	Cassava	16	124.5	5.89
1962-70	Maize	102	95.2	6.38
	Beans	47	67.2	5.96
	Forage	41	153.6	1.68
	Potatoes	33	266.4	11.40
	Wheat	15	72.8	4.43
Ghana	Cassava	134	71.0	19.90
1961-75	Maize	775	121.2	9.59
	Groundnuts	134	52.1	18.70
	Cotton	92	82.1	18.31
	Cowpeas	61	65.1	15.90
Indonesia	Cassava	56	176.4	4.19
1969-76	Rice	378	62.6	3.12
	Sorghum	312	217.0	2.46
	Groundnut	135	60.0	4.88
	Soybean	117	59.9	3.12
Nigeria	Cassava	28	53.5	11.26
1961-77	Maize	478	64.1	5.12
	Yams	348	43.5	22.60
	Rice	277	41.8	13.78

MEASUREMENT OF BENEFITS FROM ADEQUATE NUTRITION

Economic analyses of 300 cassava fertilizer trials conducted by FAO (1980) resulted in average value-cost ratios (VCR) ranging from 4.2 in Indonesia to 19.9 in Ghana. Table 17.4 shows the percent response and VCR for cassava compared with other crops. Although cassava has the reputation not to "need" fertilizers, it is clear from these data that cassava is highly responsive to fertilization and has a higher VCR than many traditionally fertilized crops.

Similar calculations for fertilizer trials in the Colombian Llanos also showed highly positive net returns from fertilization. However, in on-farm trials where crop management practices are seldom optimum, fertilizer responses were much lower, and in general were not economical. Thus, care should be taken not to recommend fertilizers to farmers based solely on experiment station results.

UTILIZATION OF INCREASED PRODUCTION

Cassava has several alternative uses in rather independent markets. Thus highest prices are generally obtained in the fresh root market for human consumption, but this market is limited and retail costs are high due to the perishability of the crop. Any increase in production will soon saturate this market. However, with simple drying or processing techniques fresh roots can be transformed into cassava flour for human consumption or pellets or chips for animal consumption. In this form cassava can be stored and transported over long distances both to local and to international markets, where it competes mainly with grains such as maize or sorghum. Finally, it can be used as the raw material for production of starch and fuel alcohol, in which case it competes with maize and sugarcane, respectively. There is a tremendous potential for increased production in most of these markets, but cost of production of cassava dictates its relative competitiveness. Considering the high value-cost ratios obtained in fertilizer trials with cassava, it is likely that with adequate but not excessive fertilization, yields can increase markedly, thus decreasing production costs per ton of roots produced and making it more competitive with other crops in a number of potential markets.

SUPPLEMENTAL READING

ASHER, C.J., EDWARDS, D.G., and HOWELER, R.H., 1980. Nutritional disorders of cassava (Manihot esculenta Crantz). University of Queensland, St. Lucia, Q., Australia. 48 pp.

FAO, 1980. Review of data on responses of tropical crops to fertilizers. 1961-1977. Food and Agriculture Organization of the United Nations. Rome, 101 pp.

HOWELER, R. H., 1981. Mineral Nutrition and Fertilization of Cassava (Manihot esculenta Crantz). CIAT, Series 09EC-4, Cali, Colombia. 52 pp.

NORMANHA, E.S., 1961. Adubacao da mandioca (Fertilization of cassava). FIR 3(8): 18-19.

18
SWEET POTATO

Walter A. Hill

Sweet potato (<u>Ipomoea batatas</u>) ranks seventh in world crop statistics after wheat, rice, maize, potato, and cassava and ranks twelfth in dry matter production and fifth in harvested weight in the world when compared to the 20 major foods (milk, beef, fish, swine, and 16 crops). Total production per unit area and food value of sweet potato exceed that of rice. World production of sweet potato roots is dominant in Asia with China accounting for 80% of the world's production (Table 18.1). Nations that produce more than one-half million mt per year include China, Vietnam, Indonesia, India, Korea, Philippines, Japan, Brazil, Burundi, Rwanda, Bangladesh, Uganda and USA.

In Papua New Guinea and the Philippines, sweet potato is a staple food crop. In countries such as Brazil, Colombia, Paraguay, Nigeria, Burundi, and Rwanda, sweet potato and other root crops contribute as much or more caloric input to the human diet as do cereal crops. Sweet potato roots are high in carbohydrate, and beta carotene or provitamin A (orange flesh varieties) and are a good source of Vitamin C, potassium, calcium, iron, and dietary fiber. Depending on cultivar, growing conditions, and method of preparation a 100 g portion of sweet potato contains 22.5-34.3 g carbohydrate, 5000-9200 IU provitamin A (orange flesh varieties), 8-23 mg vitamin C, 120-300 mg K, 30-40 mg Ca, 0.7-0.9 mg Fe, 0.3-0.9 g dietary fiber, 60-66 g water, 0.1-3.3 g fat, 0.4-1.0 g ash, 29-59 mg P, 10-48 mg Na, 0.03-0.10 mg thiamine, and 0.03-0.07 mg riboflavin. Protein content in sweet potato roots ranges from 0.5 to 23% depending on cultivar and cultural practices. The crop has been an important protein source in China and Papua New Guinea. In a number of countries sweet potato leaves and tender shoots are eaten as a green vegetable prepared similarly as spinach, mustard, turnip and collard greens. The leaves are also

---

[1]Associate Professor of Plant and Soil Science, Tuskegee Institute, Tuskegee, Alabama 36088, USA.

TABLE 18.1
Sweet potato production in 1980*

Countries	Production (million metric tons)	Yield (mt/ha)
World	107.2	8.0
Africa	5.0	6.5
Burundi	0.9	9.9
Kenya	0.3	10.0
Rwanda	0.9	7.7
Uganda	0.7	4.8
North and Central America	1.3	6.0
Cuba	0.3	3.8
Dominican Republic	0.1	7.1
Haiti	0.3	3.1
USA	0.5	12.0
South America	1.7	8.7
Argentina	0.3	9.0
Brazil	1.0	9.1
Paraguay	0.1	8.2
Peru	0.2	10.0
Asia	98.6	8.2
China	87.1	8.2
India	1.6	7.1
Indonesia	2.0	6.8
Japan	1.4	21.5
Vietnam	2.4	6.0
Europe	0.1	10.4
Oceania	0.6	5.3
Papua New Guinea	0.4	4.5
Solomon Islands	0.1	9.9
Tonga	0.1	12.7

*FAO. 1980. FAO Production Yearbook.

rich in protein (12.7%) and minerals (10% ash), and contain total nutrients equivalent to beef or pork.

Sweet potato roots are processed into canned, flaked (dehydrated), frozen, and pureed products. Roots are also processed into flour, starch, hot cakes, gruel, noodles, pies, cakes, french fries, chips, candy, ice cream, milkshakes, jellies, and syrup. Carotene, pectin and tempeh can also be produced from sweet potato.

Sweet potato roots alone or mixed with other feeds are favorable feeds compared to corn and are used for dairy, beef, goat, sheep, pigs, and poultry. Substitution for maize may be especially important in locations with low maize yields. Sweet potato chips are more nutritious than maize in hog rations and soybeans in beef rations.

The high starch and sugar content also make sweet potato a potential feedstock for ethanol. Potential commercial production of 200 proof ethanol from sweet potato is 1776 liters/ha. The residue after distillation potentially can be used as an animal feed.

ENVIRONMENTAL REQUIREMENTS

Sweet potato is important in the the tropics, subtropics and parts of the temperate zone. In the subtropics and tropics more than one crop can be produced per year. Its drought tolerance permits production in semi-arid conditions, and appropriate cultivar selection and cultural practices permit its growth on a wide range of soils. Cultivars should be developed for wet season plantings in the tropics.

Sweet potato is grown under intensive farming, subsistence farming and dual purpose production for both foliage and storage roots. It is well adapted for multiple cropping systems and is grown in mono-, relay, rotation, shifting-cultivation and inter-cropping systems with staple and cash crops including rice, soybeans, maize, beans, peanuts, barley, oil palm, cola, taro, coconut, vegetables, tobacco, and sugarcane.

WORLD PRODUCTION

Average root yields in major producing countries are shown in Table 18.1. These yields are considerably below the 85 t/ha experimental yields obtained in North Carolina.

NUTRITIONAL REQUIREMENTS

Leaf analysis. Common nutrient ranges found in mature leaves of sweet potato at mid-season are: 3.2-4.2% N, 0.2-0.3% P, 2.9-4.3% K, 0.73-0.95% Ca, 0.4-0.8% Mg, and 40-100 ppm Mn. The deficiency-sufficiency ranges of selected nutrients in the petiole of the sixth leaf from the growing tip of sweet potato at mid growth are given in Table 18.2. These values are used as plant analysis guides in some regions in the USA. Direct comparisons of results in different regions are difficult because of growing conditions and sampling and analytical techniques.

Foliar symptoms. Nutrient deficiency symptoms associated with leaves and roots are given below. Deficiency symptoms begin to occur when nutrient concentrations in shoots (including stems and leaves) and leaf blades are 0.08% S, 0.10-0.12% P, 0.05-0.16% Mg, 0.5-0.75% K, 0.2% Ca and 1.5-2.5% N (Spence and Ahmad, 1967).

TABLE 18.2
Deficiency-sufficiency ranges of selected soluble nutrients in sweet potato leaves (Geraldson et·al., 1973)

Nutrient	Deficient	Intermediate	Sufficient
NO_3-N, ppm	1,500	2,500	3,500
PO_4-P, ppm	1,000	1,500	2,000
K, %	3	4	5

Nitrogen (N). Growth of leaves and internodes is severely restricted. Leaves are smaller than normal and pale green to pale yellow. Chlorosis first appears on older leaves and later on younger leaves. Purple pigmentation can occur on stems, petioles, leaf margins and veins. In the late stages abscission of the older leaves occurs. Storage roots are similar in size to control plants. Bright purple to red pigmentation occurs on roots. Fibrous roots are very much reduced in development.

Phosphorus (P). In late stages, growth is restricted and leaves smaller than normal. Leaf blades are slightly darker. Abscission of older leaves occurs and purple pigmentation is darker on petioles and main veins on the under surface of leaves. Moderate storage and fibrous root formation occurs. Storage roots become very dark purple and fibrous roots become severely necrotic.

Potassium (K). In early stages, internodes and petioles are shorter and leaves smaller than normal. Leaf blades are darker green than normal. Brown spots may occur under the surface of the leaves. In later stages, interveinal chlorosis occurs with dark green bands at the base of the blades and along secondary veins. The chlorotic bands eventually become dark brown and necrotic. No storage roots form. There is moderate to small formation of coarse fibrous roots.

Calcum (Ca). Small chlorotic patches appear scattered over the leaf surface, these patches becoming necrotic. Leaves are leathery to the touch and there is little pigmentation on the petioles. In the later stages, leaf size is smaller than normal. Death of the main growing points and abscission of old leaves occur. Storage roots do not form. There is restricted fibrous root formation and marked necrosis.

Iron (Fe). Vein islets on young leaves become pale green in color giving the leaf a mottled appearance. Leaf size is normal. Purple pigmentation is greater on stems and on veins on the under surface of leaves. Severe chlorosis can occur in young leaves with the entire blade of each being yellow to white in color, the vein network being pale green. Root development is similar to controls but root pigmentation may be redder or paler than controls.

Sulfur (S). Growth is not restricted. Leaves are pale green to pale yellow, the younger leaves yellowing first. In young leaves very narrow green bands persist for some time on main and secondary veins. In very late stages smaller leaves are produced and a narrow purple band occurs on the leaf margins. Storage roots are radish-shaped or rounded and extensive root formation occurs.

Magnesium (Mg). Internodes are shortened and leaves are smaller than normal. There is interveinal chlorosis over the whole surface of old leaves and eventually in young leaves. A pink flush eventually appears in old leaves. Margins of young leaves curl up. Necrotic patches develop on severely chlorotic leaves. Root initials are formed at the nodes at the intermediate stage of foliar symptom development. No tubers are formed. There is moderate formation of coarse roots which show marked necrosis.

Soil Tests

Soil tests can be used to determine soil P and K deficiency levels. Deficiency levels and corresponding recommended fertilizer rates for sweet potato are given in Table 18.3. For large-scale commercial production, rates from 34 to 224 kg P_2O_5 and 72 to 252 kg K_2O/ha are recommended. Phosphorus and K recommendations in tropical regions generally range from 50 to 100 kg P_2O_5/ha and 84 to 169 kg K_2O/ha. Potassium rates recommended in temperate regions range from 72 to 252 kg K_2O/ha. When soil tests are "very high" in P and/or K, P and K fertilizers are not recommended.

Compared to other tropical crops, sweet potato is highly tolerant of low levels (< 0.01 ppm) of soil solution P. The N and K ratio in plant tissue has been shown to influence significantly root and/or foliage yield and quality. Evaluation of studies from various countries indicates that the highest percent yield increases over unfertilized controls occurs with N/K_2O rates of 40:90, 34:132, and 80:200.

FERTILIZATION PRACTICES

In most parts of the tropics sweet potatoes are seldom fertilized. Fertilizer is not always affordable or available to small farmers. Limited work has been done

to evaluate mineral nutrition of sweet potato in traditional rotation, relay, or inter-cropping systems. For
this reason the sweet potato breeding program at the International Institute for Tropical Agriculture (IITA) in
Nigeria does not apply fertilizers in breeding line and
cultivar field evaluations.

TABLE 18.3
Soil deficiency levels and soil test fertilizer recommendations for sweet potato in Alabama (Cope et al.,
1980)*

Phosphorus	Potassium				
	Very High	High	Medium	Low	Very Low
	Pounds	P_2O_5 or K_2O per acre)			
Very high	0-0	0-80	0-120	0-160	0-200
High	40-0	40-80	60-120	80-160	40-200
Medium	80-0	80-80	80-160	80-160	100-200
Low	120-0	160-80	120-120	160-160	200-200
Very low	120-0	160-80	120-120	160-160	200-200

* Pounds per acre x 1.12 = kg/ha.

On Puerto Rico Ultisols, highest yields were obtained when pH was above 5, and the Al/exchangeable bases ratio was <0.2. An Ultisol in the central hilly region of
Puerto Rico with a pH of 4.7 apparently supplied enough
for sweet potato production. In the mainland USA liming
is commonly recommended for sweet potato if the soil pH
is <5.8. Results from a 13-year experiment in Louisiana
indicated that when soils at pH 5.2 were treated with
sulfur (358 kg/ha) the pH varied between 4.4 and 5.0.
The sulfur treatment restricted growth of "Centennial"
sweet potato in the early stages of growth and yields
were somewhat decreased. A definite chlorosis pattern
appeared to be less involved in causing the foliage abnormalities. When the same soil was limed at 1100 kg/ha,
the soil pH increased from 5.2 to 7.0, marketable root
yield was reduced, and the incidence and severity of soil
rot infection was increased. These results suggest that
a moderately acid soil should be maintained for high
yields and quality. If magnesium is low and a soil test
indicates that lime should be applied, then dolomitic
lime should be used.

FERTILIZING THE CROP

A 15 t/ha sweet potato crop removes 70 kg N, 20 kg
P_2O_5, and 110 kg K_2O per ha. Though some work indicates
fertilizer timing and placement do not influence root
yields, others have shown that yield and quality are influenced by these practices. Two split applications of N

or complete (N-P-K) fertilizers are commonly recommended for sweet potato and are applied at planting (33 to 50%) and 4 weeks after planting (50 to 67%). Three split applications of N and two split applications of K are recommended on North Carolina sandy soils. The N is applied preplant (34 kg/ha), at the last cultivation (45 kg/ha), and 4-5 weeks after the last cultivation (22 kg/ha). The K is applied preplant (34 kg/ha) and at the last cultivation (168 kg/ha). Delayed application or late season supplemental application of N usually reduces root yield, but applying the recommended amount in five split applications via drip irrigation has produced greater marketable yields and the same total yields as two split applications of side-dressed N.

Commercial fertilizer N recommendations for a specific soil type, variety or climate are usually based on marketable and/or total yield response curves. On a worldwide basis, recommended fertilizer N rates for sweet potato range from 0 to 146 kg N/ha. In temperate regions where high energy dependent, mechanized cultural practices are used, the recommended rates range from 22 to 146 kg N/ha. In tropical regions the rates applied range from 0 to 60 kg N/ha. In Japan N applications for sweet potato from all sources range from 40 to 100 kg N/ha, averaging 65 kg/ha. In the USA the recommended rate of N varies from state to state, but ranges from a low of 22 to 54 kg/ha in Louisiana to 101 to 146 kg/ha in North Carolina. Rates recommended in Taiwan range from 20 to 90 kg N/ha.

Vine cuttings or sprouts are used as seedstock for production. If sprouts are used application of fertilizer to the beds is recommended. Suggested application rates vary from 0.1 kg/m^2 of 6-12-6 to 1 kg/m^2 of 10-10-10, depending on bedding practices used. Fertilizer increases size, vigor, and number of slips. Additional N fertilizer is recommended after each pulling. Starter solutions are recommended for soils with low fertility levels. For example, Georgia recommends mixing 1/3 kg of 10-52-17 in 190 liters of water and applying 500 ml per plant at transplanting.

CROP MANAGEMENT

Sweet potato plants are spaced at 20 to 45 cm in rows that are 60 to 122 cm from center to center. The recommended spacing in each area depends on climate, variety, soil type and cultural practices. In general, as plant population increases the number and mean weight of tubers decrease and total yields do not significantly change. If spacing is too wide there may be an increase in the number of large roots produced. Closer spacings are often recommended for early yielding varieties (135-150 days) and wider spacings for late yielding varieties. Generally, wider spacings are required on infertile sandy soils than on fertile, finer-textured soils.

EFFECT OF FERTILIZERS ON NUTRITIONAL VALUE

Increasing fertilizer N rate up to 100 kg N/ha increases firmness of canned roots but decreases percent fiber and flesh color (carotenoid pigments) of fresh and canned roots. Increasing the fertilizer N rate from 100 to 200 kg N/ha increases weight loss during storage and decreases the reducing sugars, sucrose, total sugars and ascorbic acid, but root intercellular space and cortex thickness are not influenced by N rate. Percent protein in foliage and roots increases with increasing fertilizer N level. Late applied, supplemental N increases Ca and Na content, but does not influence P, K, Fe, Mg, N, carotene and ascorbic acid content of roots. Root cracking, firmness, and weight loss after 112 days of storage are not influenced by form or rate of late applied N, but intercellular space is decreased. Good storage quality of sweet potato roots has been obtained by applying fertilizer in an N:K ratio of 1:1:4.

Yield and quality of sweet potato foliage are important where foliage is used as greens for human consumption. Shoot development, lateral branching of shoots and foliage production of sweet potato are directly related to fertilizer N rate, with vine yields increasing with each increment of N up to 100-120 kg N/ha. Oxalate and carbohydrate contents are decreased and percent protein is increased with increasing N rate from 0 to 120 kg N/ha.

VARIETAL DIFFERENCES

Considerable variation among sweet potato genotypes in response to soil fertility level has been demonstrated in both tropical and temperate regions, and must be considered in the adoption of fertilizer practices. Bush types, long vine types, and intermediate length types have been described. One hundred sweet potato cultivars have been classified into N-responsive, N-indifferent, and N-depressive types according to the effects of N fertilizer applications on final root yields. High-yielding cultivars have been developed that produce up to 50 t/ha on experiment station fields without addition of fertilizer. A number of studies in tropical countries have shown no response of sweet potato to fertilizer application. It has been suggested that plants with small foliage capable of high yields with close spacing and intensive fertilizer application be used in intensive production, and that subsistence farmers use plants with long vines capable of adequate yields in the absence of fertilizer application.

FUTURE PROSPECTS

Sweet potato is considered as a staple, vegetable, or dessert food depending on level of consumption, method of preparation and perspective. Sweet potato is considered as a staple "survival" or "poor man's" food where high sweet potato consumption has been associated with war or famine. When other crops are available consumers often shift away from sweet potato as a staple food. Such factors make it difficult to predict the requirements for increased consumption of sweet potato. However, potential solutions to increased sweet potato consumption most likely include the following: development of specialized types of sweet potato for human food, animal feed and industrial users; development of low cost drying processes; development of new, pre-prepared, packaged products that require a minimum of cooking; improvement in yield; reducing seasonal supply fluctuations; lowering production costs through mechanization; developing varieties with low sugar content and flatulence; increasing protein, vitamin A, and other important nutrient contents; and use of media campaigns to publicize the nutritional value and versatility of sweet potato.

SELECTED REFERENCES

COPE, J.T., C.E. EVANS, and H.C. WILLIAMS. 1980. Soil test fertilizer recommendations for Alabama. Auburn University Experimentation Circ. 251.

CULBERTSON, R.E. and J.H. BOULWARE. 1951. Sweet potatoes in Japan. Report 145, Department of the Army, Washington, D.C.

GERALDSON, C.M., G.R. KLACAN, and O.A. LORENZ. 1973. Plant analysis as an aid in fertilizing vegetable crops. In: Soil Testing and Plant Analysis. L.M. Walsh and J.D. Bouton, eds. Soil Science Society of America, Madison, WI, USA.

HILL, W.A. 1982. Nitrogen fertility and uses of sweet potato - past, present and future. Proceedings of the L.T. Kurtz Colloquium on Science in a Changing Agriculture, Univ. Illinois, Dept. Agronomy, pp. 89-112.

HILL, W.A. 1985. Effect of nitrogen on quality of three major root/tuber crops. Chapter 43. In: Nitrogen in Crop Production. R.D. Hauck, (ed.) American Society of Agronomy, pp. 627-659.

JONES, L.G., et al. 1977. Effects of soil amendment and fertilizer applications on sweet potato growth, production, and quality. Louisiana State University, Agriculture Extension Station Bull. No. 704.

MILLER, C.H. and H.M. Covington. 1982. Mineral nutrition studies with sweet potatoes during a three year period. North Carolina Agriculture Research Service Tech. Bull. No. 273.

ONWUEME, I.C. 1978. The Tropical Tuber Crops. John Wiley, New York.

SPENCE, J.A. and N. AHMAD. 1967. Nutrient deficiencies and related tissue composition of the sweet potatoes. Agronomy Journal 59:59-62.

VILLAREAL, R.L. and T.D. GRIGGS (eds.) 1982. Sweet Potato - Proceedings of the First International Symposium. Asian Vegetable Research and Development Center, Tainan, Taiwan.

WILSON, L.G., C.W. AVERRE, J.V. BAIRD, E.A. ESTES, K.A. SORENSON, E.O. BEASLEY, W.A. SKROCH. 1980. Growing and marketing quality sweet potatoes. North Carolina Agriculture Extension Service AG-09.

19
ONIONS

D. N. Maynard and O. A. Lorenz

Onions (Allium cepa) are grown universally and constitute an important part of the diet for people in virtually every country. Pecause they can be stored for long periods, or dehydrated and stored for even longer periods, they serve as important food sources in seasons of the year when fresh production is not possible.

Onions are of worldwide importance since countries in Asia, Europe, North America, and South America are listed among the major producers. China is the largest onion-producing country with annual production exceeding 2.5 million mt (M mt). Production in the USA, India, and USSR is in excess of 1.5 million mt (M mt) for each country; Japan and Spain produce over 1 M mt annually. Petween 0.5 and 1 M mt are produced each in Turkey, Fgypt, Brazil, and Italy.

Onions may serve as a staple or as a condiment in the diet, depending upon culinary use and ethnic origin. Per capita consumption of both fresh and dehydrated onions in the USA was 6.3 kg in 1978. This might serve as an average consumption figure with higher or lower utilization in other countries.

The unique flavor and odor of onions are due to allyl sulfide. Otherwise fresh, dry onions are 89% water; a fair source of energy; and are relatively low in protein, mineral, and vitamins. Recently developed fresh onion hybrids may have dry matter contents as high as 18%.

ECOLOGICAL ADAPTATION

Temperature and daylength are the most critical environmental factors in onion production. Control of

[1]Professor and Chairman, Vegetable Crops Pepartment, IFAS, University of Florida, Gainesville, Florida 32611, USA; and Professor, Vegetable Crops Department, University of California, Pavis, California 95616, USA.

moisture at maturation and curing is important for successful drying of the crop.

Seed germination and growth are optimum at 20-25°C; relative growth rates are restricted at 30°C and above. These temperature regimes are consistent with those found in onion producing areas of the USA, growth in relatively cool weather and maturation and harvest in relatively warm weather.

Bulb formation is generally promoted by long days, but varieties that bulb at varying daylengths have been developed. The minimum daylength when bulbing (the growth stage when bulbs are initiated) occurs is called the "critical daylength." Short-day varieties bulb at 12-14 hours, intermediate-day varieties at about 14 hours, and long-day varieties at 14-16 hours. The "critical daylength" must be exceeded in each case for bulbing to occur. Accordingly, it is essential that the correct variety be selected for a particular growing area.

Premature seed-stalk initiation, called bolting, limits bulb production in certain cases and is induced by several factors, including temperature. In winter crops in mild climates, an extended period below 8°C will induce bolting. When sets are used as propagules, storage at 5-10°C will induce bolting of the succeeding crop.

Onions can be grown successfully on soils ranging from light, sandy loams to organic soils. Heavy clays, stony soils, and saline soils are unfavorable and should be avoided.

NUTRITIONAL REQUIREMENTS

Numerous interacting conditions influence onion yield, e.g. variety, plant population, daylength, temperature, water, and availability of essential elements. Maximum potential yield can be achieved only with a favorable blend of these factors. Assuming that most, if not all, of these conditions are met in onion production in developed countries, average yield would easily exceed 35 mt/ha. On the other hand, yields in developing countries average less than 10 mt/ha. Certainly part of this gap could be closed with efficient fertilizer use.

Yields of 50 mt/ha are considered to be very good in England. However, in small plots at the National Vegetable Research Station in Wellsbourne, yields of 130 mt/ha have been achieved. High plant densities, with irregular irrigation, and a high level of fertilization are necessary to achieve such yields.

Average Yields

Average yields are considerably lower than potential yield (Table 19.1). For example, onion yields in North America and Oceania are three times greater than in the USSR and Asia. Limited fertilizer use certainly accounts for a major portion of this difference.

TABLE 19.1
Onion yields and production (1977-79 Average)

Area	Yield (mt/ha)	Production (1,000 mt)
Africa	10.9	1,318
North America	29.5	1,844
South America	13.1	1,556
Asia	10.3	8,395
Europe	17.9	4,444
Oceania	29.0	166
USSR	9.7	1,576

Source: FAO Production Yearbook 33:155-156 (1979).

Present Fertilization Recommendations

Extensive research in California showed a fairly close approximation between growth rate and elemental accumulation in onions. More than 65% of the fresh weight and 74% of the dry weight were produced from bulbing to harvest. During this same period, 68% of the total N, 75% of the P, and 47% of the K was accumulated by the onion plant. Total elemental removal by onion plants was not always related to fertilizer application rates (Table 19.2).

In Idaho, elemental removal by a 56 mt/ha onion crop was 68 kg N, 13 kg P, 81 kg K, 10 kg Ca, 13 kg Mg, and 8 kg Na (sodium). Removal of Zn, Mn, Fe, Cu, and B was less than 1 kg/ha.

Recommendations for N fertilization of onions in various countries are summarized in Table 19.3. Considering the different growing conditions, the recommendations are remarkably similar.

General fertilizer recommendations for several locations in North America are quite variable among and within locations (Table 19.4). This variability suggests that, in these areas, onion fertilization has been more extensively researched, and that extremes in fertilizer use efficiency have been determined.

Soil Tests

Soil test values in California for P, K, Zn, and Mn obtained with the indicated soil extraction procedures are shown in Table 19.5. When test results are in the deficient range, applications of 123-168 kg P/ha, 112-124 kg K_2O/ha, 11 kg Zn/ha, or 11 kg Mn/ha are recommended.

TABLE 19.2
Yield and nutrient elemental application and removal by "Southport White Globe" onions

Trial	Yield (mt/ha)	N Applied (kg/ha)	Removed	P Applied (kg/ha)	Removed	K Applied (kg/ha)	Removed	Na	Cu	Mg Removed (kg/ha)
1	55.4	225	148	29	26	0	110	4	40	8
2	53.6	287	168	22	29	0	90	5	50	9
3	60.2	287	149	49	29	94	111	3	50	9
4	35.1	143	85	29	17	0	63	2	26	5
5	45.5	208	—	—	—	—	—	—	—	—
Average	50.0	230	136	37	25	28	95	4	39	7

Source: Zink, Hilgardia 37:203-218, 1966.

TABLE 19.3

Nitrogen fertilizer recommendations for onions

Country	Basic Application (kg/ha)	Supplemental Application (kg/ha)
Congo	90	-
Germany	40	20-40
India	60-80	-
Indonesia	100	-
Italy	50	50
Netherlands	100	-
Peru	80-100	-

Source: J.G. DeGeus. 1967. Fertilizer Guide for Tropical and Subtropical Farming. Centre d'Etude de l'Azote, Zurich.

TABLE 19.4

Fertilizer recommendations for onions in North America

Area	Soil	N (nitrogen)	P_2O_5 (phosphate)	K_2O (potash)
			(kg/ha)	
USA				
California	Mineral	224-336	123-168	0-224
Florida	Mineral	213-314	179	213-314
	Organic	67	224	470
	Marl	94-161	81	114-181
Indiana	Organic	67	269	269
Kentucky	Mineral	56-112	56-224	112-224
Maryland	Mineral	84-112	28-224	28-224
Michigan	Organic	180	0-336	0-448
New York	Old Organic	112-140	45-112	45-168
	New Organic	112	112	168
	Mineral	90-112	56-168	56-168
Oklahoma	Mineral	80-112	59-90	34-78
Canada				
Ontario	Organic	112	50	150
	Mineral	112	112	112

194

TABLE 19.5
Deficient and sufficient soil test results for onions in
California, USA

Element	Soil Extraction Agents	Range (ppm) Deficient	Sufficient
P	Water soluble, for organic soils	0-0.6	>1.0
P	Sodium bicarbonate, for mineral soils	0-8	>12[a]
K	Ammonium acetate	0-80	>100
Zn	DTPA or dithizone	0-0.5[b]	>0.8[c]
Mn	DTPA	0-0.8	>1.0

[a]20 for fall-winter-seeded crops. ppm = parts per millio
[b]0-0.4 for soils high in organic matter.
[c]0.6 for soils high in organic matter.

Source: Voss, Division of Agricultural Sciences, University of California, Bulletin 4097, 1979.

In Michigan (USA), fertilizer recommendations for onions (Table 19.6) are based on P extractable with Bray's P_1 solution and on K extractable with 1.0 N neutral ammonium acetate.

In Maryland and most states in the southeastern USA, the double-acid extraction (0.05 N HCl, 0.025 N H_2SO_4) procedure is used. Based on this extraction, Maryland fertilizer recommendations for onions are shown in Table 19.7.

From the foregoing, it is apparent that soil tests provide useful information on which to base fertilizer applications. However, caution must be exercised in use of a soil test report without knowledge of the specific soil extraction procedures.

Plant Tissue Tests

Results of tissue tests made during the growing season provide information on which to base supplemental fertilizer applications, especially applications of N, K, and micronutrient-containing fertilizers. In addition, the tissue test results from one year together with soil test results provide a basis for fertilization in the next year. Deficient and sufficient concentrations of N, P, K, Zn, and Mn in onion leaves at midseason have been determined in California (Table 19.8).

Tissue test results have not been established widely; however, those obtained in California should be applicable to other areas for diagnostic purposes.

TABLE 19.6

Soil test results and onion fertilization in Michigan

Yield Potential Soil (mt/ha)	Soil P (kg/ha)	Recommended P_2O_5 (kg/ha)	Soil K (kg/ha)	Recommended K_2O (kg/ha)
Mineral 44.8	< 21	336	< 76	280
	22-24	280	67-11	224
	44-77	224	112-167	168
	78-111	168	168-223	112
	112-167	112	224-335	56
	168-223	56	>335	0
	224	0		
Organic 67.2	< 21	336	<140	448
	22-44	280	150-223	392
	44-77	224	224-307	336
	78-111	168	308-391	280
	112-167	112	392-475	224
	168-223	56	476-559	168
	>224	0	560-643	112
			644-727	56
			>728	0

Source: Michigan Extension Bulletin E-550, 1976.

TABLE 19.7

Soil test results and onion fertilization in Maryland

Soil P (kg/ha)	Recommendation P_2O_5 (kg/ha)	Soil K (kg/ha)	Recommendation K_2O (kg/ha)
34-68	224	40-94	224
69-116	112	95-180	112
118-230	56	181-358	56
>230	28	>360	28

Source: Maryland Extension Bulletin 236, 1980.

TABLE 19.8

Deficient and sufficient elemental concentrations in onion leaves[a]

Element	Level	
	Deficient	Sufficient
Total N (%)	0-2	>2.5
Total P (%)	0-0.1	>0.2
Total K (%)	0-2	>2.5
Total Zn (ppm)	0.15	>20
Total Mn (ppm)	0-10	>20

[a]Samples of third (tallest) leaf, taken at midseason.

Source: Division of Agricultural Science, University of California, Bulletin 4097, 1979.

Visual Deficiency Symptoms

Visual diagnosis of elemental deficiencies in onions is not as dependable as in some other crops because of leaf structure and growth habit. Nonetheless visual diagnosis together with results of tissue and soil tests should prove useful.

Nitrogen (N). Erect upright growth of light green or yellowish leaves.

Phosphorus (P). Wilting and tipburn of oldest leaves followed by necrosis. Slow growth, delayed maturity, a large proportion of thick necks.

Potassium (K). Dark-green tops, followed by yellowing of oldest leaves, tipburn and necrosis.

Copper (Cu). Tipburn; thin, light-colored outer scales of mature bulbs. Early maturation.

Manganese (Mn). Leaves tipburned, are light-colored, and curled. Growth is restricted and bulbing delayed. High percentage of thick necks.

Zinc (Zn). Restricted growth. Yellow striping and bending of leaves.

Improved Varieties

Improved onion varieties and hybrids are made available to farmers by government agencies and the seed industry. Most new varieties are higher yielding than those that they are meant to replace. In order to obtain the maximum benefit of a newly introduced variety, cultural systems--including fertilization management--may require modification. In most but not all cases this involves increased fertilizer use.

FERTILIZER REQUIREMENTS

Animal manures applied and incorporated well before planting at 20 to 30 mt/ha may substitute for all or part of the required fertilizer when commercial fertilizer is unavailable. A combination of animal manures and commercial fertilizers may yield better results than animal manures alone.

Onions do not tolerate high soil acidity, consequently, acid soils should be limed before planting to achieve a pH of 6.0-6.5. Highly alkaline soils may be made more acid by S applications.

Fertilizer Placement

For maximum efficiency of fertilizer use, all of the P, K, and micronutrients (as shown by soil test results) and part of the N should be applied before planting. Because the onion has a sparse root system, best results have been obtained by placing the preplant fertilizer 5 to 10 cm below the seed row. When K is required, a portion may be broadcast or sidedressed. The remainder of the N can be applied in 1 to 3 applications as a sidedressing, topdressing, or in irrigation water.

Fertilizer requirements increase with plant populations. The increase is not proportional, however, since efficiency of fertilizer use increases with higher plant populations. Consequently, an increase of 30% in plant population may only require a 20% increase in fertilizer.

Relation of Time of Planting and Plant Population to Fertilization

Onions planted and grown when temperatures are near the lower limit of growth may require additional N fertilization since the mineralization of organic soil N is restricted at low soil temperatures. Also since root growth is restricted, band placement of fertilizer is preferable to broadcast applications when soil temperatures are low.

Crop Maturation and Harvest

The market quality of onions is affected greatly by levels of all essential elements. At deficient levels, bulb size is reduced and often the bulbs are too small for market acceptance. Conversely, in some cases, extremely large bulbs do not readily meet consumer acceptance. High N fertilization results in more "doubles" or "splits" which lower the percentage of #1 size bulbs.

High N fertilizer applied at the time of bulbing reduces bulb formation whereas low N fertilization enhances it. Supplemental N fertilizer applications should be made prior to the time of critical daylength to encourage early vegetative growth. At the bulbing stage of development, a high P and K to low N ratio is desirable.

With onions, dry matter content is the most common quantitative measurement of internal quality, and especially important with onions grown for dehydration where dry matter yield is of extreme economic consideration, both because of the yield of the actual dry product, and also because of the cost of dehydration. There has been considerable speculation as to the effect of fertilization on dry matter. This was investigated in a number of fertilizer experiments in California where very large yield and maturity responses to both N and P were obtained. Only very small differences in dry matter were in any way related to fertilizer treatment and even these differences were not consistent.

On the other hand, the variety of onion is the most important factor contributing to dry matter content. Some varieties may be as low as 6% dry matter, and others may be as high as 18%. The area of production has some effect on dry matter but not nearly as much as variety. In California, onions are produced under extremely varied conditions of daylength and temperature. There is some indication that onions maturing under very hot conditions are lower in dry matter than when matured under cooler conditions; for example, where onions were grown experimentally at Shafter (warm), at Davis, and Tulelake (cool), California. Usually those grown in Tulelake were the highest in dry matter and those at Shafter the lowest.

Benefits from Adequate Fertilization

Increased onion yields can be expected when fertilizers are applied to soils deficient in one or more of the essential elements. Highest yields are attained only when high genetic (inherited) potential of a variety is combined with suitable environmental factors. Adequate fertilization is only one of these factors.

Results of experiments conducted in South America show increased yields and net return and value: cost ratios of 3.1 to 37.5 from fertilizer applications (Table 19.9).

TABLE 19.9

Economics of fertilizer use in onions

Country	Fertilizer N-P$_2$O$_5$-K$_2$O (kg/ha)	Yield (kg/ha)	Yield Increase (kg/ha)	(%)	Net Return ($/ha)	Value Cost Ratio
Ecuador	Control	5 562	–	–	–	–
	45-0-0	6,492	930	17	69	4.8
	45-45-0	7,414	1,852	333	143	5.6
	45-45-45	7,448	1,886	34	136	4.3
Ecuador	Control	3,348	–	–	–	–
	45 0-0	4,691	1,343	40	122	7.6
	45-45-0	5,418	2,070	62	186	6.9
	45 45-45	4,967	1,619	48	129	4.1
	90-90-90	5,826	2,478	74	177	3.1
Colombia	Control	6,667	–	–	–	–
	45-0-0	11,000	4,333	665	633	37.5
	45-45-0	9,334	2,667	40	371	13.6
	45-45-45	10,000	3,333	50	463	13.5
	90-90-90	10,343	3,667	55	476	7.5

Source: J.G. DeGeus. 1967. Fertilizer Guide for Tropical and Subtropical Farming. Centre d'Etude de l'Azote, Zurich.

SUPPLEMENTAL READING

BREWSTER, J.L. 1977. The physiology of the onion. Hort. Abstracts Part I. 47:17-23 , Part II. 47: 103-112.

DeGEUS, J.G. 1967. Fertilizer Guide for Tropical and Subtropical Farming. Centre d'Etude de l'Azote, Zurich.

JONES, H.A. and L.K. MANN. 1963. Onions and Their Allies. Interscience, New York.

VOSS, R.E.   1979   Onion production in California. Division of Agricultural Sciences, University of California, Bulletin 4097.

ZINK, F.W.   1966.   Studies on the growth rate and nutrient absorption of onion.   Hilgardia 37:203-218.

20
SUGARCANE

J. A. Silva

Sugarcane (Saccharum officinarum) has the potential to produce 100 to 175 million calories per hectare per year. It is also the world's major source of sucrose sugar. In high income countries sucrose sugar may provide 15% or more of the total calories consumed. Sugar is a concentrated food source and is almost pure energy. It is used as a sweetener in many products, including candy, soft beverages, bread, cake, pastry and ice cream, and also as a preservative in canned and frozen foods.

Molasses, a by-product of sugar manufacture, is fermented to produce alcoholic beverages as well as ethanol for fuel. With the present high cost of fossil fuel, production of alcohol fuel has increased to the extent that a large proportion of the sugarcane grown in Brazil is used to make ethanol. Molasses is also used as an animal feed and in the production of yeast, acetic acid, citric acid, butyl alcohol, and carbon dioxide.

Energy can be supplied by sugarcane in the form of dry matter as well as alcohol. The fiber (bagasse) remaining after the juice is extracted is burned to generate steam for the production of electricity. In some countries sugarcane is also grown for biomass production and varieties are selected for rapid accumulation of dry matter rather than sucrose. Bagasse is utilized in the manufacture of furfural, paper, particle board, and acoustical wall board in many countries. Pith, separated from bagasse, is fed to animals and is another source of food from sugarcane.

The sugar produced from sugarcane is the major source of income for many countries and considerable effort is being spent to encourage its production. In some countries non-centrifugal sugar in the form of gur, jaggery, panela, papelon, etc. is the primary product for trade and barter by the small farmer who grows and pro-

---

[1]Professor of Soil Science, University of Hawaii, Honolulu, Hawaii, USA.

cesses his own sugarcane. Sugarcane by-products such as molasses and bagasse also represent important sources of income. The production of sugarcane and the manufacture of sugar are major business enterprises which provide many jobs and support various inter-related industries. They create a need for fertilizers and other agricultural chemicals, the manufacture and repair of farm equipment, engineering expertise, development of transportation facilities, and a host of other services.

Thus, it is apparent that sugarcane plays an important role not only in meeting man's food needs, but also in achieving his energy and economic goals.

ECOLOGICAL ADAPTATION

Sugarcane, a perennial grass which requires a tropical or subtropical climate, grows only from about 35° north to 35° south of the equator. It has a wide temperature range from over 38°C (100°F) to below freezing. However, cane growth is greatly reduced by temperatures below 21°C (70°F) and growth stops at temperatures below 10°C (50°F). The optimum temperature for cane growth is between 27° and 33°C (80 to 90°F).

Sugarcane may be found growing at sea level as well as at 1,000 m or more under conditions which range from hot deserts to the humid tropics with monsoon rains and even to rather temperate areas where it grows during the warm summer. In temperate climates the above-ground stalks are killed by freezing temperatures in winter and the stubble that survives in the ground gives rise to new stalks from underground buds when temperatures rise in spring. Sugarcane will also tolerate a range of soil conditions. It grows best in rich, fertile soils, but will also survive in infertile sandy or clay soils. Sugarcane requires abundant water for good growth but has the ability to survive periods of drought. Although growth is greatly reduced or stopped completely, it will resume when water becomes available.

MANAGEMENT PATTERNS

In managing the crop, it is necessary first to grow sufficient leaves and stalks to provide the systems to produce sugars and store sucrose. Therefore, there should be a vegetative stage during which maximum growth is the objective and when reducing sugars will be used mainly for the production of leaves and stalks that are the foundation of the crop. This should be followed by a ripening stage during which growth is reduced to promote storage of sugars as sucrose. The methods for achieving these two stages vary with geographical location and growing conditions.

Areas With Climatic Stress

Where growth of sugarcane is restricted by climate due to low temperature or drought during part of the year, the growing season and thus the vegetative stage are short. Fertilization must be completed early to promote maximum growth and to allow nutrient depletion by the time the climatic stress occurs. Where temperature gradually decreases to 10°C (50°F) or lower, there will be a gradual reduction in growth and photosynthesis. Sucrose storage will increase because little of the reducing sugars produced will be used for growth. In areas with a definite dry season, growth is gradually reduced with increasing moisture stress which results in increased sucrose storage.

Areas Without Climatic Stress

Where conditions are favorable for the growth of sugarcane the year round, special care is needed in maintaining the vegetative stage. The fertilizer and water provided must be adequate for crop growth as well as properly timed to allow the cane to ripen before harvest. The grower must induce the ripening stage by allowing nutrients, especially N, to become limiting a few months prior to harvest or by reducing irrigation to impose a moisture stress which will reduce growth and increase sucrose storage.

STATUS OF WORLD PRODUCTION

Sugarcane is grown for sugar production under a wide range of environmental conditions as well as with varying levels of management, so that yields vary with sunlight, temperature, length of growing season, water supply, fertilization, sugarcane variety, and general management. Regions with the highest yield potential are those which allow continuous growth and have abundant sunlight (400-500 gram calories per cm^2 per day), warm day temperatures (27-33°C or 80-90°F) with cool night temperatures (10-18°C or 50-65°F), and fertile soils provided with adequate irrigation. Generally, sugarcane growth under rainfed conditions is reduced by cloud cover which restricts sunlight and may also keep day temperatures too cool and night temperatures too warm for maximum sugar production.

Sugar production for 1979 is shown in Table 20.1 for countries with centrifugal plus non-centrifugal sugar production of about 500,000 mt or more. Centrifugal sugar is the major product of most countries, except India and Pakistan, where non-centrifugal sugar (gur, shakkar) makes up 55% and 76%, respectively, of total production. The world total centrifugal cane sugar production in 1979

204

TABLE 20.1 Sugar production in selected countries, 1979a

Region and Country	Cane Yield mt/ha	Cent. Sugar 1,000 mt	Non-cent. Sugar 1,000 mt	Total Sugar 1,000 mt	Sugar Yield mt/ha	mt Cane Per mt Sugar	Crop Duration mo.
Africa							
Egypt	84.2	668	--	668	6.42	13.11	12-16
Mauritius	79.2	730	--	730	9.12	8.68	12-18
South Africa	82.8	2,079	--	2,079	9.36	8.85	11-15
N. C. America							
Cuba	58.9	8,048	--	8,048	6.13	9.61	12-16
Dominican Rep.	57.8	1,200	--	1,200	6.74	8.58	12-14
Mexico	64.3	3,078	80	3,158	5.87	10.95	12-14
U.S. Mainland	59.9	1,387	--	1,387	5.40	11.09	8-10
Hawaii	215.4	948	--	948	23.58	9.13	18-30
S. America							
Argentina	46.2	1,411	--	1,411	4.61	10.02	12-14
Brazil	54.8	7,020	200F	7,220	2.84	19.30	12-18
Colombia	87.5	1,105	984	2,089	7.41	11.81	
Peru	130.3	715	15	730	13.52	9.63	15-22

Asia							
India	50.2	6,378	7,800F	14,178	4.54	11.06	9-16
Indonesia	100.0	1,307	255	1,562	9.76	10.24	12-16
Pakistan	36.3	660	2,152	2,812	3.73	9.73	9-12
Philippines	45.8	2,269	39	2,308	5.44	8.42	11-14
Thailand	42.2	1,845	400F	2,245	4.68	9.02	12-18
Oceania							
Australia	79.2	2,963	--	2,963	11.10	7.14	12-18
Fiji	66.2	490	--	490	8.17	8.10	12-18

aOnly countries with total (centrifugal plus non-centrifugal) sugar production of 490,000 metric tons (mt) are included. Data are from FAO. 1980 Production Yearbook, Vol. 34; USDA Agricultural Statistics, 1981, and Hawaiian Sugar Planters' Association, Hawaiian Sugar Manual, 1980.

F = FAO estimate.

was 51 million metric tons (M mt). Brazil, Cuba, and India were the largest producers with 6.9 M mt, 6.4 M mt, and 5.2 M mt, respectively. However, if both centrifugal and non-centrifugal sugar are considered, India's production is 10.9 M mt, which is the highest. The wide range in cane and sugar yields per hectare is due to the yield potential of the country, management practices followed, and duration of the crop.

The combination of good management and adapted, high-yielding varieties exploits the environmental yield potential. For example, in Hawaii, average raw sugar yields of well-managed, irrigated plantations are often 14 mt/ha while individual fields have yields 17 mt/ha or more on a 12-month basis. The range of cane quality denoted by the tons of cane per ton of sugar reflects not only the environment, but also management of the crop.

Fertilizer Used for Sugarcane

Fertilizer use varies among the sugarcane growing area (Table 20.2), with N rates ranging from 0 on Florida muck soils to 600 kg/ha on soils of the high rainfall areas of Hawaii. Potash rates show a similar range, from 0 to 800 kg/ha, while P has a smaller range, 0 to 400 kg/ha. Hawaii uses the highest amounts of all three nutrients due to the longer cropping period and the high level of production sought. With the exception of muck soils with high organic matter, sugarcane soils generally do not supply adequate N for a crop, so it is always applied. Phosphate and potash fertilizers are not applied to soils which have adequate levels of P and K according to soil analysis or have not responded to applications of these nutrients. The amounts of P and K fertilizers applied vary with the availability of these nutrients in the soil due to P sorption (fixation) and to K leaching. Fertilizer rates are also related to the cost of fertilizer and the value of the expected yield increase.

METHODS FOR ASSESSING NUTRIENT REQUIREMENTS

The nutrient needs of sugarcane can be assessed by soil analysis, plant tissue analysis and/or deficiency symptoms. A combination of these three methods gives the most complete inventory of the nutrient status of a crop. Although deficiency symptoms indicate that a particular nutrient or nutrients are deficient, they are usually expressed after the deficiency has caused a reduction in growth. Soil and plant tissue analysis, on the other hand, can identify potential or incipient deficiencies which can be corrected before growth reduction occurs. Soil analysis performed prior to planting identifies the need for pH adjustment and/or for P fertilization. This is essential since both lime and P must be applied prior to or at planting. Soil analysis also

TABLE 20.2
Fertilizer used for sugarcane in selected countries[a]

Region and Country		N kg/ha	P_2O_5 kg/ha	K_2O kg/ha	Crop Duration mo.
Africa					
Malagasy Rep.	Plant	100	0	125	11-14
	Ratoon	150	0	125	11-12
Mauritius		125	110-320	145-220	12-18
South Africa		90-150	variable	100	12-16
N. C. America					
El Salvador		140	60	45	12
Florida	Muck	0	15-55	370	8-10
	Sand	85	15-55	370	8-10
Hawaii	Irrig.	330-450	0-250	450-550	18-24
	Rainfed	350-600	260-400	500-800	21-30
Louisiana	Plant	90-140	45	90	8
	Ratoon	150-170	45	90	8
S. America					
Argentina		60-95	0	0	12-14
Brazil		40-60	70-100	50-100	12-18
Peru		90-105	0	0	15-22
Venezuela		65	95	75	11-12
Asia					
India	North	110-170	0	0	9-12
	South	340	0	0	12-16
Indonesia		120-130	70-80	100-150	12-16
Philippines		165	100	185	11-14

a Amounts of nutrients are the totals for the duration of the crop.
Data from Smith, D. 1978. Cane Sugar World. Palmer Publication,
New York, and from personal communications.

indicates the status of K, Mg, and other nutrients, which aids in formulation of a fertilization program for the crop. When sugarcane is growing, analysis of tissue samples reveals the crop's nutrient status and indicates the nutrients needed to maintain optimum conditions for growth.

Soil Analyses

Many methods have been developed for the extraction and determination of soil nutrients. The method selected should measure the form of the nutrient taken up by sugarcane and should be calibrated for local soils.

A representative soil sample should be collected from the field to a depth of 30 cm because sugarcane is deep-rooted. Generally about 25 subsamples per hectare should be collected and composited. Separate samples should be collected from dissimilar areas such as hills and valley bottoms or from two different soils in the same field.

Nitrogen. Most sugarcane soils have inadequate N to produce an acceptable crop. Available tests for soil N have not shown a consistent ability to predict the quantity of soil N available throughout the crop. Therefore, N fertilizer recommendations are generally made on the basis of field experiments or past experience.

Phosphorus. Many extraction methods are used for soil P. Critical levels indicating the level of soil P above which application of P is not likely to give a yield response, and below which application of P is expected to produce a yield response, are given in Table 20.3. The bulk density of tropical soils can range from 0.5 to 1.9 grams per cubic centimeter and this may cause misinterpretation of soil nutrient levels expressed as ppm. Researchers in Hawaii have overcome this problem by expressing soil nutrients in terms of pounds per acre foot. In this way, the nutrient level is adjusted for bulk density and a single critical level can be applied to all soils. Critical levels for several nutrients are expressed in terms of pounds per acre foot in Table 20.3.

Cations. Critical levels for K, Ca, and Mg extracted with N ammonium acetate, pH 7.0, are 500, 800, and 50 pounds per acre foot, respectively (Table 20.3). Soil Ca is generally adequate if soil pH is greater than 6.0, and is often not determined in this situation.

Sulfur. The critical level for ammonium acetate-extracted S is 50 pounds per acre foot (Table 20.3).

Silicon. Although Si is not generally considered an essential element, marked increases in cane and sugar

TABLE 20.3 Critical levels for soil nutrients

Nutrient	Critical Level[a]	Extractant
P	30 ppm P_2O_5	0.02N sulfuric acid[b]
P	85 lb P/acre ft.	0.02N sulfuric acid[c]
P	15 ppm P_2O_5	0.5 M sodium bicarbonate, pH 8.5[b]
P	85 lb P/acre ft.	0.5 M sodium bicarbonate, pH 8.5[c]
P	50 ppm P_2O_5	1% citric acid[b]
P	15 ppm P_2O_5	0.03M ammonium fluoride + 0.025M hydrochloric acid, pH 3.5[b]
K	100 ppm K	N ammonium acetate, pH 7.0[c]
K	500 lb K/acre ft.	N ammonium acetate, pH 7.0[c]
Ca	800 lb Ca/acre ft.	N ammonium acetate, pH 7.0[c]
Mg	50 lb Mg/acre ft.	N ammonium acetate, pH 7.0[c]
S	50 lb S/acre ft.	N ammonium acetate, pH 7.0[c]
Si	100 lb Si/acre ft.	0.5N ammonium acetate, pH 4.8[c]
Si	10 ppm Si in solution	Water-saturation extract[d]

[a]To convert lb/acre ft to kg/ha-30 cm, multiply lb/acre ft by 1.12.

[b]Du Toit, J.L., B.D. Beater, and R.R. Maud. 1962. Proc. 11th Congress Int. Soc. of Sugarcane Technologists, p. 101-111.

[c] Experiment Station, Hawaiian Sugar Planters' Assoc. Personal communication.

[d]Fox, R.L. Personal communication.

yield have been obtained from applications of calcium silicate to highly weathered soils in high rainfall areas of Hawaii and Mauritius. Therefore, in Hawaii, certain soils are routinely assessed for their Si content using 0.5N ammonium acetate, pH 4.8. The critical level accepted in Hawaii is 100 pounds Si per acre foot. A saturation extract with water has also been used to assess the Si status of soil, and the critical level is about 10 ppm (Table 20.3).

Tissue Analyses

Several methods of tissue analyses are used throughout the sugar world because procedures have been developed for different cropping systems. They all involve the sampling and analysis of recently matured leaf, sheath, or stalk tissues. Interpretation and use of the analytical information vary with the sugar growing area.

CROP LOG SYSTEM. This is the most comprehensive system of tissue analysis used for sugarcane, and it was developed by Dr. H. F. Clements in Hawaii, where sugarcane is grown for 24 months or more. Data recorded on the log include growth measurements, sheath moisture and nutrient concentrations of leaves and sheaths, sunlight, temperature, and rainfall measurements, fertilizer and irrigation applications, as well as relevant crop information such as starting and harvest dates, yields, quality, etc. Thus the progress of the crop can be followed and a permanent record is obtained for future reference.

Sampling and Analysis. Samples are collected every 35 days starting when the crop is about 3 months old and consist of leaves and sheaths +3, +4, +5, and +6, counting the spindle as +1, from 5 normal stalks in a sampling area.

Interpretation and Use. An example of a crop log for a plant crop of variety H59-3775 grown for 25 months under unirrigated (rainfed) conditions is shown in Figure 20.1. In the top graph are plotted the average minimum and maximum temperatures (°F) as well as the average sunlight for the sampling period (plotted as vertical bars) in terms of gram calories per square cm per day (g cal/cm^2/day). This information provides a record of the general climatic conditions prevailing during the crop. In the second graph is plotted the Growth Index, which is the fresh sheath weight (in grams) per stalk. The Growth Index depicts the pattern of crop growth. The Nitrogen Index is plotted in the third graph and is the total N content of leaf blades +3, +4, +5, and +6, expressed as percent dry matter. The x's near the line are values of the Normal Nitrogen Index (NN) which is the leaf N value expected for a high yielding crop of

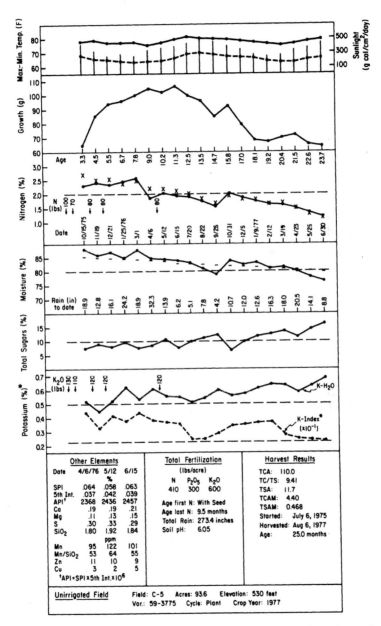

Figure 20.1. An example of the Crop Log developed by Dr. Harry Clements in Hawaii.

sugarcane at the age and moisture content for the particular sample. A general equation used for calculating NN is the following:

$$NN = 0.07287X_1 - 0.03481X_2 - 3.5851$$

where NN is the Normal Nitrogen Index, X_1 is sheath moisture, and X_2 is age of the crop in months at the time of sampling. Equations for specific varieties in particular areas have been developed and used by some growers. If the sample N value drops below the NN value, application of N fertilizer is required, especially during the first year of the crop. Note the low N levels for the 3.3, 4.5, and 9.0 month samples and the N applications made. Additional information shown in this third graph is the age in months at the time of sampling (at the top), the time and amount of N applications (indicated by arrows), and the date samples are collected (on the bottom).

The Moisture Index plotted in the fourth graph is the moisture content of the sheath sample. The horizontal bars near the solid line are the values of the Normal Moisture Index (NM) which is the level of sheath moisture at each age associated with maximum yields and therefore, ideal conditions. It is calculated for a specific variety for a particular location with the age, maximum temperature and/or sunlight associated with a given sample. Although equations vary with variety and location, an example equation for variety H59-3775 grown on an unirrigated plantation is given below.

$$NM = -0.26949X_1 - 0.08068X_2 + 92.4796$$

Where NM is the Normal Moisture Index, X_1 is age in months at the time of sampling, and X_2 is average maximum temperature ($°F$) for the sample period. Comparison of NM with the sample Moisture Index indicates the moisture status of the crop relative to a crop which produced high yields. Rainfall in inches for the sampling period is indicated at the bottom of the graph and irrigation applications would be indicated by arrows at the top of the graph for an irrigated field. Note the effects of low moisture during the September 26th sample period when the cane was 14.7 months old. Growth and N dropped sharply while total sugars increased above 10% and $K-H_2O$ also increased. When moisture increased in the following sample period, these indices returned to normal levels.

The total sugars or Primary Index representing the reducing sugar and sucrose content of the sheaths expressed as percent dry matter is an indicator of the plant's fitness into its environment and is shown in the fifth graph. If plants are growing rapidly, it is below 10%, but if the plant is under stress due to drought, nutrient deficiency, etc., the level rises over 10%.

The potassium status of the plant is shown in the sixth graph as two lines. One (solid line) is K-H$_2$O, which is the K content of the sheaths expressed as percent of the sheath tissue moisture content, while the other (broken line) is the K-index, which is sheath K expressed as a percent of the sugar-free dry weight of the sheaths. The critical level for K-H$_2$O is 0.5%, while that for the K-index is 2.25%. The scale for this graph should be read with a decimal after the first digit for the K-index. K-H$_2$O is less affected by varietal differences, so the critical level applies to all varieties, whereas the critical level for the K-index can differ with variety. Applications of K fertilizer are indicated by arrows in the top part of the graph. Potassium applications are generally made when K-H$_2$O falls below the critical level or when N is applied. It is desirable to have a fertilizer K$_2$O:N ratio of 1.1 to 1.6 when N is applied to provide sufficient K for the increased growth from the N application.

The P status of the plant measured in three samples taken during the period of rapid growth is indicated by three indices which are shown in the lower left corner of the log. The first, the SPI (standardized P index), is the P content of the sheaths expressed on the dry weight basis and adjusted to a standard moisture and total sugar content. The second is the 5th internode P, which is the P content of the 5th internode below the last green leaf expressed on the dry weight basis. The third, the API (amplified P index), is obtained by multiplying the SPI and 5th internode P values together as whole numbers, i.e., SPI = 0.064; 5th internode P = 0.037; therefore, 64 x 37 = 2,368. The critical level for the API is 2,400. The API of one crop has been used to indicate the need for P fertilization of the next crop. If the next crop is to be a plant crop, the API readings are divided by 2 because the plant crop has a less extensive root system initially and would have lower API readings during the first 12 to 15 months than a ratoon crop. If API/2 exceeds 2,400, no phosphate fertilizer is required. If it is between 1,800 and 2,400, then 225 kg P$_2$O$_5$/ha should be applied, and if it is less than 1,800, 450 kg P$_2$O$_5$/ha are required. If the next crop is to be a ratoon crop, the API readings are used without modification, and an API above 2,400 indicates that no phosphate is needed on the next crop. If API is below 2,400, 255 kg P$_2$O$_5$/ha is needed by the ratoon crop, except in soils with high capacity to "fix P" where values below 1,000 indicate that 450 kg P$_2$O$_5$/ha should be applied.

Other nutrients may be determined in these sheath samples taken during the period of rapid growth and the data are recorded in the table with the P indices at the bottom of the log. Critical levels established for several nutrients are presented in Table 20.4.

TABLE 20.4
Critical levels[a] for the Crop Log[b]

Nutrient	Tissue	Sample Period (mo.)	Critical Level
H_2O	Leaf sheath	3–24+	Compare with normal Moisture Index
N	Leaf blade	3–24+	Compare with normal Nitrogen Index
P (API)	(Leaf sheath & 5th internode)	(9–15) (3 samples)	2,400
K_2O	Leaf sheath	3–24+	0.500%
K–index	Leaf sheath	3–24+	2.25%
Ca	Leaf sheath	3–24+	0.20%
Mg	Leaf sheath	(9–15) (3 samples)	0.075%
S	Leaf sheath	(9–15) (3 samples)	0.200%
Mn	Leaf sheath	(9–15) (3 samples)	25 ppm
Mn/SiO_2	Leaf sheath	(9–15) (3 samples)	5

Zn	Leaf sheath	(9-15) (3 samples)	10 ppm
Cu	Leaf sheath	(9-15) (3 samples)	
Mo	Leaf sheath	(9-15) (3 samples)	0.05 ppm
Fe	Leaf sheath	(9-15) (3 samples)	40 ppm or more
Cl	Leaf sheath	(9-15) (3 samples)	219-289 ppm

aCritical level is defined as the nutrient level below which growth reduction occurs and response to application of the nutrient would be obtained.

bFrom Clements, H.F. 1980. Sugarcane Croplogging and Crop Control Principles and Practices. The University Press of Hawaii, Honolulu.

Pertinent fertilization and crop data are also recorded on the bottom of the log, giving a complete record of the field.

LEAF BLADE SYSTEM. This system has been developed on crops grown for 12 to 15 months and generally has one to three samples taken about midway in the crop's growth.

Sampling and Analysis. Several variations of this method are used throughout the sugar world. Most of them utilize the third leaf blade, which may be defined a little differently in the various systems, but is generally the leaf with the top-most visible dewlap (the junction of the blade and sheath). This leaf is the +1 leaf using the method of numbering leaves according to Kuijper and is comparable to leaf +3 in the Crop Log System. Older leaves are numbered +2, +3, etc. while younger leaves are numbered 0, -1, -2, etc. in Kuijper system. The leaf sampled is called the third leaf or the top visible dewlap (TVD) leaf. Samples are collected when the cane is 4 to 6 months old and 60 or more leaves are collected from locations uniformly spread throughout the field. The middle 15 to 20 cm portion of the leaf is taken, the midrib removed, and the laminae are dried at 80°C, ground, and analyzed for N, P, K and micronutrients. Standard methods of analyses should be used.

Interpretation and Use. The nutrient levels found in the tissues are compared to established critical levels and the fertilization program modified accordingly. Most methods have established critical levels for N, P, and K, while critical levels for micronutrients have been established for the TVD method. The critical levels given in Table 20.5 are generally applicable to both the third leaf and TVD leaf samples. The critical levels for a particular sugar growing area may be refined and the amounts of fertilizer required to correct a deficiency determined by installation of microplots in the field which received variable amounts of N, P, or K and which are sampled along with the field. Nutrient levels of these plots, which are related to crop growth within the plots can be used to interpret nutrient levels in samples from the field.

VISUAL SYMPTOMS OF DEFICIENCY

One means of identifying nutrient problems of a growing crop of sugarcane is to interpret the growth and color patterns of the plant. Characteristic symptoms produced by nutrient deficiencies can suggest the cause of a growth problem. However, when a nutrient deficiency has progressed to the stage where it causes symptoms, growth reduction has already occurred and corrective measures generally will not completely overcome the growth loss. Nonetheless, deficiency symptoms can be helpful in monitoring a growing crop of sugarcane.

Nitrogen (N). The symptoms of N deficiency are the gradual yellowing of all leaves, with the youngest leaves remaining green the longest since N is mobile and moves to those leaves. There is a general retardation of growth and the stalk diameter is small, while the older leaves die prematurely. Roots grow long, but generally are of small diameter.

Phosphorus (P). A striking characteristic of P-deficient stool is the lack of tillers. There are usually only a few primary stalks with small internodes per stool, which results in a very open stand. The length and diameter of the stalks are reduced, and these short, slender stalks taper rapidly to the growing point. Leaves are generally narrow and greenish-blue, in marked contrast to the wide, dark green leaves of normal plants. Older leaves dry at the tips and margins and gradually die. Root systems are very poorly developed with limited growth of secondary rootlets. This results in reduced root/soil contact and reduced nutrient and water uptake.

Potassium (K). One of the early symptoms of K deficiency is dieback of the leaf tips followed by death of the leaf edges which progresses downward. Older leaves become orange-yellow with numerous chlorotic spots which gradually become necrotic and coalesce so that a general browning with death of the leaf results. This is referred to as "firing." The upper surfaces of midribs often have reddish discolorations which are confined to the epidermal cells. Young leaves remain dark green longest because mobile K moves to these leaves and maintains more normal color. Growth is depressed and slender stalks are produced.

Calcium (Ca). Symptoms of Ca deficiency first appear on the leaves as minute chlorotic spots with dead centers which later become dark reddish-brown. As the leaves age, the spotting becomes more intense. Plants are weak, with thin stalks having soft rind; growth is retarded. With severe deficiency the spindle dies and growth stops completely.

Magnesium (Mg). Young leaves are light green while older leaves are yellowish-green with small chlorotic spots that become dark brown. These spots later coalesce and give the leaves a rusty or freckled appearance. The spotting is most pronounced on the older leaves. In severe cases of Mg deficiency, stalks may show internal browning.

TABLE 20.5

Critical levels for third leaf and top-visible dewlap (TVD) leaf samples[a]

Nutrient	Critical Level
N	2.0%
P	0.20%
K	1.10%
Ca	0.16%
Mg	0.08%
Mn	20 ppm
Zn	<10 ppm
Cu	4 ppm
Mo	0.05 ppm
B	1 ppm

[a]N, P, K values from Samuels, G., S. Alers Alers, and P. Landrau, Jr. 1957. J. Agr. Univ. of Puerto Rico 4(1):1-10.

[b]All other values from Evans, H. 1959. Proc. 10th Cong. Int. Soc. Sugar Cane Tech., 1959. p. 473-408.

Sulfur (S). Sulfur-deficient plants generally have an off-color or yellowish-green appearance which is much like N deficiency. However, unlike N deficiency, the youngest leaves are more chlorotic than the older leaves because S is not mobile in the plant. Stalks are short and thin, leaf area is reduced, and there is often a suggestion of redness over the chlorosis.

Silicon (Si). Cane growing in highly weathered soils of high rainfall areas where soil Si is low frequently shows characteristic leaf freckling which is due to small yellowish-brown spots on the leaves. The freckling starts as small, clear areas on the young leaves. The areas gradually become chlorotic with a brown or reddish cast and increase in number as the leaf ages. The tips of older leaves become necrotic and leaves die prematurely as the necrosis spreads toward the sheath. Stalks are short and slender with few tillers. Whether these symptoms are due to a deficiency of Si or Ca, or due to a toxicity of Mn, Fe, or some other nutrient has not been definitely established. However, these symptoms are usually alleviated by application of calcium silicate.

NUTRITION AND CROP MANAGEMENT

Supplying a crop of sugarcane with adequate nutrients at the right time is essential for the production of a good sugar crop. The strategy in fertilization follows the growth pattern of sugarcane discussed earlier. That is, ample nutrients are supplied to the crop to maximize growth during the vegetative stage, and then nutrients are allowed to become limiting in the ripening stage to maximize sucrose storage. This means that all of the P and sufficient N and K are applied early to allow the plant to make rapid initial growth, then additional amounts of N and K are applied to sustain this rapid growth without being applied so late or in such quantity that rapid growth continues into the ripening period. This pattern varies with the length of crop and environmental conditions during the ripening and harvest periods.

Fertilizer practices are separated into two groups based on the length of the cropping period, i.e., crops grown for 12 to 16 months and those grown for 18 to 24 months or longer.

N, P, and K for Crops 12 to 16 Months Long. Crops are grown for 12 to 16 months where a normal seasonal change markedly reduces growth or stops it completely, e.g. cold and freezing temperature, or a dry season. The amounts of fertilizer applied vary widely, but generally exhibit a similar pattern of application. An example schedule is shown in Table 20.6. The N may be applied at planting or 1 month later, depending on conditions for growth at planting. If it is cold, N may not be applied until it warms up sufficiently for cane to grow. In the ratoon crop, N is applied when growth starts due to favorable temperatures or sufficient rainfall. Generally, a second application made when the cane is 2 to 3 months old will carry the crop to maturity. This may be based on leaf N levels if tissue analysis is practiced. Nitrogen is usually not applied after 4 months to reduce vegetative growth near harvest and to induce ripening. All of the P fertilizer is applied at planting with the seed, or at the beginning of a ratoon crop, below the surface near the stool. The amounts applied may be based on soil analysis and/or experience. The amounts of K applied are likewise based on soil analysis and/or experience and are usually applied in two applications (Table 20.6, K_2O-1). K is sometimes applied only at planting with the seed or spread on the surface near the stool for a ratoon crop (Table 20.6, K_2O-2). The ratio of K_2O:N applied ranges from 1.1 to 1.6.

N, P, and K for Crops 18 to 24 Months or Longer. Cane grows all year in areas where there are no natural stresses to stop growth, and the crop must be forced to ripen either by withholding irrigation or by allowing nutrients to become limiting. In many areas chemical ripeners are also used. The amounts and time of fertilizer application vary between irrigated and rainfed culture and between plantations. Examples of fertilizer schedules are given for irrigated and rainfed conditions in Table 20.7.

TABLE 20.6
Example of fertilizer schedule for sugarcane grown for 12 to 16 months

Age (mo.)	Amount of Fertilizer (kg/ha)			
	N	P_2O_5	K_2O	
			(1)	(2)
0	40 – 90	0 – 100[a]	50 – 90[a]	0 – 180[a]
1				
2	60 – 70		50 – 90	
3				
Total	100 – 160	0 – 100[a]	100 – 180	0 – 180[a]

[a]Amount based on soil analysis and/or experience.

TABLE 20.7
Example of fertilizer schedule for sugarcane grown for 18 to 24 months or more

Age (mo.)	Amount of Fertilizer (kg/ha)					
	Irrigated			Rainfed		
	N	P_2O_5	K_2O	N	P_2O_5	K_2O
0	50- 75	0-200[a]	50- 75[a]	50- 0	200-400[a]	60- 0[a]
1				40- 60		60- 80
2	75		75-100	60- 80		100-120
3						
4	75		75-100			
5				70- 90		90-120
6						
7	75-100		75-100	70- 90		90-120
8						
9	50-100		75-100			
10				70- 90		90-110
11				50- 75[b]		75-100[b]
12						
Total	325-425	0-200[a]	350-475	360-410 / 410-485[b]	200-400[a]	490-550 / 565-650[b]

[a] Amount based on soil analysis and/or experience.

[b] Optional application, depends on tissue N and K levels, projected season and age at harvest, variety, etc.

Under irrigated conditions, all of the P is applied with the seed at planting and some of the N and K may also be applied at the same time. Later applications of N and K may be applied by hand, machine, air, or in irrigation water. In ratoon crops, P is applied below the surface near the stools, while N and K may be applied underground or on the soil surface near the stool by any of the above methods. The initial amounts of N applied are based on experience while the initial amounts of P and K are based on soil analysis and/or experience. The number of applications may range from 2 to 5, depending on the philosophy of the plantation and on the levels of N and K in plant tissues. Where the crop log is used, N is generally applied when the Nitrogen Index falls below the Normal Nitrogen Index, while K is applied when $K-H_2O$ drops below 0.50%. The objective is to keep the tissue N and K values at or above the critical levels to maintain optimum conditions for growth during the first year. The last application is usually made at 9 months; however, applications may be made as late as 12 months if the need arises. Generally 1.1 to 1.6 kg of K_2O are applied per kg of N to provide adequate K for the increased growth resulting from N application. Where trickle or drip irrigation is practiced, N, P, and K may be applied in the irrigation water and applications of small amounts of fertilizer may be made every few weeks.

Under rainfed conditions, 5 to 7 applications of N and K may be made due to the higher rates of leaching and generally greater porosity of soils in these areas. The amounts of fertilizer applied and time of application are based on the same considerations discussed for irrigated conditions; however, due to the lack of control of water, ripening depends largely on nutrients becoming limiting before harvest. The amount and time of the last, optional, N and K application are critical and depend on tissue N and K levels, projected season and age at harvest, as well as variety. This application should not be made after 12 months unless the field is scheduled to be harvested when it is older than 24 months. For instance, in areas of slow growth due to cool temperatures, where cane is harvested at 36 months or older, N and K may be applied in the second year of growth, if needed.

Other Nutrients and Lime for All Crops. When soil analysis indicates that Mg, Ca, S, or Si are low, the appropriate materials should be applied before planting or ratooning and worked into the surface soil. Mg may be applied at the rate of 100 kg Mg/ha in the form of Epsom salts. Dolomitic limestone may be used if pH is low. If the soil Ca level is low but pH is adequate, a Ca source such as calcium sulfate should be added to supply about 800 lb Ca per acre foot. With low pH as well as Ca, sufficient liming material should be applied to raise soil pH to 6.0-6.5. Although sugarcane will grow in a wide range of pH, this range is believed to be optimum. The amount of material required will vary with the buffering capacity of the soil and the neutralizing value of the liming material. Both of these characteristics should be determined for efficient liming.

When soil S levels are inadequate, a sulfur source should be applied at the rate of 100 to 200 kg S/ha. If it is desired to use elemental S to lower pH, the amount required should be determined from an acidulation curve developed for the specific soil. Sulfur rates on the order of 1,000 to 2,000 kg/ha, depending on the soil, should be applied and tilled into the soil prior to planting to allow microbial production of sulfate to lower pH.

Rates of calcium silicate applied to increase the soil Si level vary with the form and content of silicate, particle size, and silicate solubility of the material used. Applications on the order of 2,000 to 4,000 kg/ha of locally-produced calcium metasilicate (49% SiO_2) have been found adequate for soils with low Si in Hawaii.

NUTRITION OF SUGARCANE VARIETIES

Many of the major sugar-growing areas have active breeding and selection programs which produce high-yielding varieties that are adapted to specific environmental conditions. Varieties have been developed for a 12- to 14-month crop period and also for an 18- to 24-month or longer crop period. The growth patterns of these varieties differ and fertilization is varied accordingly. Requirements for specific nutrients can differ among sugarcane varieties with some requiring higher amounts of certain macro- and micronutrients than others. Some varieties are also more sensitive to adverse ripening and harvest conditions so must be fertilized more carefully to avoid cane and juice quality reduction. With the large number of varieties grown and the continuous production of new varieties, it is not practical to discuss nutrient needs of particular varieties. However, the specific nutrient requirements of promising new varieties should be determined so they can be properly fertilized when released to growers.

SUPPLEMENTAL READING

CLEMENTS, H.F. 1980 Sugarcane Crop Logging and Crop Control Principles and Practices - The University Press of Hawaii, Honolulu.

21
SUGARBEETS

Albert Ulrich and F. Jackson Hills

The sugarbeet plant (<u>Beta vulgaris</u>) is highly prized as a field crop in the temperate zone as a source of white sugar (sucrose) which gives food a pleasant taste and a major increase in energy content. The crop is also valued for its tops and beet pulp for feeding beef cattle and dairy cows. On the farm, the crop fits well into many crop rotations, and helps provide a relatively stable income to the grower.

The world production of all sucrose in 1979 was 90.7 million mt; 60% was from cane and 40% from beet. Of this total, 24% was produced in Europe, 21% in North America, 20% in Asia, 14% in South America, 9% in the USSR, 8% in Africa and 4% in Oceania. In California, the record average yield in 1976 was 64.1 mt of roots per ha. This yield produced 9629 kg of sugar; 3513 kg of dry tops; 3520 kg of dry beet pulp; and 121 kg of monosodium glutamate. Under favorable conditions, sugar yields on a yearly basis are approximately the same for cane in Hawaii and for beet in California.

The sugarbeet grows well in fertile, well-drained soils with a pH range of about 6.0 to 8.3. It tolerates soils and waters with a moderate degree of salinity and boron content and does well on calcareous soils. On acid soils, lime is required for good growth. In the temperate zone sugarbeets require a minimum period of 5 to 6 months for growth in a climate which is relatively free of frost and day temperatures below 38°C (100°F). Excessive heat may be avoided in some areas by fall planting and spring harvesting. Sugarbeets in a warm climate need approximately 91 cm (36 inches) of water to meet the water requirements of an average crop. The judicious use of nitrogen from 0 to 280 kg/ha (0 to 250 lb/acre) is required for best crop production on soils containing

[1]Respectively, Plant Physiologist (Emeritus), University of California at Berkeley, Berkeley, California 94720; Extension Agronomist, University of California at Davis, Davis, California 95616, USA.

an ample supply of other nutrients. The crop must also be protected against pests and pathogens, as well as be under good management, and the sugar produced must be fairly priced.

Sugar yields from sugarbeets free of diseases and pests vary enormously from year to year with the controlling factors in some instances being climatic and in other instances, nutritional (plant), or a combination of both. On the nutritional side, a deficiency of one or more of the plant nutrients will depress beet yields and nearly always lower processing quality, except for a deficiency of N which will nearly always result in roots with a higher sucrose concentration and superior processing quality. On the other hand an excessive supply of N will quite frequently result in a high yield of roots with a low sucrose concentration and poor processing quality. Unfavorable results such as these can often be prevented by pre-sampling the soil for nitrate-nitrogen and by sampling the plants at 4-week intervals in order to determine the N status of the crop from thinning to harvesting. In essence, it is possible to detect, correct and prevent mineral nutrient deficiencies before they occur by using an appropriate soil and plant analysis program. In California, there are approximately 50 laboratories ready to provide technical services for sugarbeets and other crops for efficient crop production. Plant samples properly dried and ground can be analyzed effectively by relatively simple analytical procedures and equipment for nitrate-N, soluble phosphate-P, total potassium and chloride, or by automated procedures involving highly sophisticated instrumentation for those nutrients and for Mg, Ca, Zn, Mn, Fe, Cu, and Na and possibly for Mo, SO_4-S and B.

DETECTION OF NUTRIENT DEFICIENCIES

Visual Diagnosis

The most effective way of detecting a mineral nutrient deficiency of growing sugarbeets is by a careful inspection of the plants at regular intervals from the time of emergence to harvest. In this way, the early warning signs displayed by an occasional plant can be detected long before large losses in yield have taken place. Successful identification of a deficiency symptom depends on the fact that each mineral deficiency has a characteristic leaf pattern which an experienced person can readily recognize with the aid of color photos for comparison, and by following a key. This identification must be made shortly after the plant symptoms have appeared, otherwise they may be modified by other causes such as wind, pests, disease, drought or smog.

KEY TO SUGARBEET NUTRIENT DEFICIENCY SYMPTOMS

Uniform Yellowing

Nitrogen (N). Overall yellowing of leaves. Premature senescence of older leaves, followed by formation of small green center leaves. Cotyledons of seedlings yellow, center leaves, green. Petioles, cut slantwise, give negative test for nitrate with diphenylamine reagent. [Diphenylamine reagent is prepared as follows: Add 0.2 gm of diphenylamine to 100 ml of nitrate-free, concentrated sulfuric acid. Store reagent in a fully labelled pyrex glass-stoppered bottle and keep in the dark. Dispense small amounts as needed to glass-stoppered eyedropper bottles for field use. CAUTION: Sulfuric acid is highly caustic. Keep away from children and animals, and do not spill on clothing. If reagent contacts skin, wash immediately with large amounts of water. Follow this with a dilute solution of baking soda. Always pour unused acid or reagent into water, NOT the reverse.]

Sulfur (S). Similar to N deficiency except young leaves become a light green or yellow. Brown blotches of an irregular pattern without regard to veination or location become necrotic on older blades. Petioles give a positive test for nitrate with diphenylamine reagent.

Molybdenum (Mo). Resembles S deficiency closely, except pitting develops along the veins. Stem petioles give a positive test for nitrate with diphenylamine reagent.

Stunted Greening

Phosphorus (P). Difficult to diagnose. Plants appear normal but are smaller in size, may have a deep green color ranging from a dull grayish-green to almost bluish-green. Confirm with plant tissue analysis.

Leaf Scorch

Potassium (K). Marginal tanning, scorching of mature leaves, center leaves remain normal, for "high" sodium plants. Interveinal scorching, crinkled leaf surface of mature leaves. Center leaves remain green, often hooded, for "low" sodium plants. Scorched blades of low and high sodium plants contain 0.3-0.6% K. Do not use petioles for K analysis since those from low Na plants will be exceptionally high in K, even from leaves with pronounced K deficiency symptoms.

Magnesium (Mg). Easily confused with K deficiency. Blades of mature leaves become chlorotic, interspersed with interveinal necrosis. Base of blade remains green, often forming a green triangle at the base.

Growing Point Damage

Calcium (Ca). First signs are a crinkling, downward cupping (hooding) of young leaf blades with chlorotic margins that fail to expand fully. Progressively, the young blades become the blackened stubs (tip-burn) of petioles, which recede to become the typical growing-point damage of the beet plant and the cause of dark rings of dead cambium in the root.

Boron (B). The first sign of deficiency is a white-netted chapping of the upper blade surface or a wilting of tops. Young leaves wilt the most, instead of least, as in dry soil. Other symptoms are: pronounced crinkling of blades, transverse cracking of petioles, exudation of syrupy blobs from older blades, and ultimately death of growing point followed by root rot.

Yellowing with Green Veining

Zinc (Zn). Light greening or yellowing first appears in the larger, younger leaves near the center of the plant, followed by pitting, collapse and drying of interveinal tissue, leaving the veins green and prominently displayed. Leaves become erect.

Manganese (Mn). A chlorosis (loss of green color) begins in blades of young leaves that can be seen through transmitted light as a netted veining not readily distinguishable from that of Cl, Fe and Cu deficiency. The chlorosis gradually increases to a uniform yellow. As the deficiency continues to increase, a gray metallic, sometimes purplish lustre develops on the upper surface, followed by a gray-to-black freckling along the veins, that in time, coalesces to shrunken black necrotic areas along the larger veins. Upward curving of blade margins occurs frequently.

Iron (Fe). The blades of young leaves change from green to a light green, and finally to a uniform bleached yellow. The veins remain green at first, but finally they too, become bleached. Eventually, the bleached blades become necrotic, but if iron is reabsorbed before necrosis occurs, the fine veins become green and prominently netted, a characteristic Fe deficiency symptom.

Chlorine (Cl). Symptoms appear first as a chlorosis on the blades of younger leaves near the center of the plant. When viewed against bright light, the leaf blades show a netted, mosaic pattern, branching out from the main veins. As the symptoms develop, the interveinal areas appear as flat, yellow-green depressions which become dry between the raised green veins, a unique deficiency symptom.

Copper (Cu). Early symptoms are similar to those of Fe, Cl and Mn, but thereafter the symptoms progress to a fine, green netted veining, followed by a "bleaching" of the leaf blade tissues. These symptoms differ from the spotted necrosis of Fe deficiency, black spotting of Mn deficiency and raised veining of chlorine deficiency.

CHEMICAL DIAGNOSIS

Unless the observer is very familiar with the deficiency as diagnosed visually, it is always desirable to confirm the diagnosis by a chemical analysis of the affected leaf material. If the analytical results fall within the deficiency range given in Table 21.1, then steps should be taken to correct the deficiency if there is sufficient time remaining before harvest for the treatments to become effective.

The lower and upper analytical limits given for a deficiency in Table 21.1 are for leaf material with severe and mild symptoms, respectively. The critical values given in the table are tentative values for field sampling and apply only to the nutrient status of the plant at the exact moment of sampling. The loss in yield from a deficiency will be proportional to the time the plants are at or below the critical concentration. Plants distinctly above the critical concentration will not grow faster than those just above it, although they will grow longer before becoming deficient after the soil supply has become depleted.

Correction

Once a deficiency symptom has been identified, the next step is to confirm the diagnosis and evaluate the response to the deficient nutrient. This can be done by comparing adjacent untreated and treated plots, four rows wide and 15 to 20 meters long. Two or more rates of the nutrient should be applied and replicated from four to six times in a randomized complete block design. In the case of N, this may simply involve the addition of 90 kg N/ha in a convenient form followed by an irrigation or adequate rainfall. Within a few days, the leaves will become green and start to grow, accompanied by a marked decrease in sugar content of the beet root and a very large increase in petiole nitrate-N in the petioles of the young, recently matured leaves.

TABLE 21.1
A plant analysis guide for sugarbeets

Nutrient	Plant Part	Critical Concentration Field Samples[a,b]	Range Showing Deficiency Symptoms[c]	Range Without Deficiency Symptoms[d]
Boron	Blade	21 ppm	12–40 ppm	35–200 ppm
Calcium	Petiole	0.1%	0.04–0.10%	0.2–2.50%
	Blade	0.5%	0.1–0.4%	0.4–1.5%
Chlorine	Petiole	0.4%	0.01–0.04%	0.8–8.5%
Copper	Blade	—	<2 ppm	>2 ppm
Iron	Blade	55 ppm	20–55 ppm	60–140 ppm
Magnesium	Petiole	—	0.010–0.030%	0.10–0.70%
	Blade	—	0.025–0.050%	0.10–2.50%
Manganese	Blade	10 ppm	4–20 ppm	25–360 ppm
Molybdenum	Blade	—	0.01–0.15 ppm	0.20–20.0 ppm
Nitrogen (NO_3–N)	Petiole	1,000 ppm	0–200 ppm	350–35,000 ppm
	Storage root	1,000 ppm	0–500 ppm	800–4,000 ppm
Phosphorus (PO_4–P)	Petiole	750 ppm	150–400 ppm	750–4,000 ppm
	Blade	—	250–700 ppm	1,000–8,000 ppm
	Seedling:			
	Petiole	1,500 ppm	500–1,300 ppm	1,600–5,000 ppm
	Blade	3,000 ppm	500–1,700 ppm	3,500–14,000 ppm
	Cotyledon	1,500 ppm	200–700 ppm	1,600–13,000 ppm
Potassium (Na more than 1.5%)	Petiole	1.0%	0.2–0.6%	1.0–11.0%
	Blade	1.0%	0.3–0.6%	1.0–6.0%

Potassium (Na less than 1.5%)	Petiole[e]	—	0.5-2.0%	2.5-9.0%
	Blade	1.0%	0.4-0.5%	1.0-6.0%
Sodium	Petiole	—	—	0.02-9.0%
	Blade	—	—	0.02-3.7%
Sulfur (SO₄-S)	Blade	250 ppm	50-200 ppm	500-14,000 ppm
Zinc	Blade	9 ppm	2-13 ppm	10-80 ppm

[a] The critical concentration is that nutrient concentration at which plant growth begins to decrease in comparison with plants above the critical concentration.

[b] All critical concentrations except for P in seedlings and for N in roots are based on a young mature leaf sample.

[c] Leaf material for chemical analysis must be collected shortly after appearance of leaf symptoms, otherwise deficient plants may accumulate nutrients in the leaf without restoring chlorotic tissues to normal. Petioles of old yellow leaves from N-deficient plants often give a higher nitrate test, even though petioles from the light green mature and young green leaves on the same plant give a very low nitrate test. All values here are for the element on the dry basis. Use a color atlas (Ulrich and Hills, 1969) to help identify what deficiency, if any, has occurred.

[d] The upper value reported is the highest value observed to date for normal plants. Abnormally high values are often associated with other nutrient deficiencies, for example, blades low in Fe may contain up to 4% Ca, 900 ppm Mn. etc.

[e] Because of the influence of Na on K content of the petioles, blades must be used for K analysis when the petioles contain less than 1.5% Na.

Leaf samples (stem petioles from the young mature leaves) should be taken for nitrate-N analysis from all plots just before fertilization and again 2 weeks after irrigation (or rainfall). Thereafter, plant samples should be taken every 2 to 4 weeks until harvest in order to evaluate the effectiveness of the N fertilization. The best N program will be one that supplies ample N for early vegetative growth, followed by a period of N depletion of 4 to 8 weeks prior to harvest to allow for sugar accumulation (Figure 21.1).

A similar series of corrective measures may be taken for other nutrients found to be deficient visually and confirmed chemically. Unfortunately, other deficiencies are not as easily detected or corrected as that of N.

Phosphorus deficiency, as mentioned earlier, is not readily detectable, since the plants appear normal, except for a darker green foliage and smaller leaf size, which occurs frequently during cold weather. Under these conditions, the petioles will contain less than 750 ppm of acetic acid soluble phosphate-phosphorus (dry basis). A side dressing of 55 to 225 kg of P_2O_5 per ha should be tried experimentally as outlined for N, including the taking of plant samples at 2-week intervals. Trials with P combined with 25 kg N/ha, placed about 10 cm below the seed, should be tried at planting when the sodium bicarbonate soil test for P is less than 10 ppm and less than 1.5 ppm by the water extraction method on soils containing 10% or more of organic matter (Table 21.2). Cotyledons (first pair of leaves) with less than 1500 ppm of phosphate-phosphorus will indicate P deficiency.

TABLE 21.2
Sugarbeet responses expected for different concentrations of P in the first 30 cm of soil

ppm PO_4-P, dry soil extracted by:		
Bicarbonate	Water	Expected Response
0-10	0-1.5	Root yield
11-20	1.6-2.0	Seedling growth
20+	2.0+	None

A K deficiency is readily detectable visually from leaf scorch symptoms and chemically when leaf blades with clearly defined deficiency symptoms contain less than 0.6% K. A side dressing of 55-225 kg K_2O/ha from potassium chloride or potassium sulfate should correct the deficiency.

Magnesium deficiency is often confused visually with K deficiency, but chemically the leaf blades contain only 250 to 500 ppm of Mg (dry basis). A side dressing of 50-100 kg per ha of magnesium sulfate should correct the deficiency.

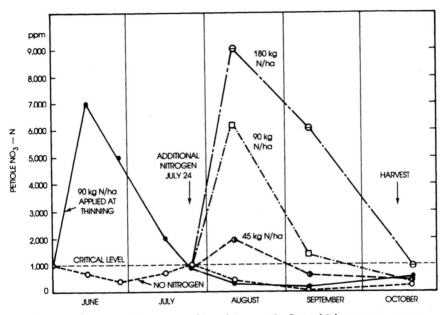

Figure 21.1. Crop monitoring and fertilizer program evaluation of sugarbeets side-dressed with ammonium nitrate, Davis, California. Relationship of petiole nitrate-N (dry basis) to nitrogen fertilization and sampling date.

Calcium deficiency is directly or indirectly associated with tip-burn, which occurs primarily on plants growing vigorously with an ample supply of nutrients, including N, during periods of cool weather in late spring or early summer. Tip-burn has no material effect on yields or on sugar concentration of roots, since under most field conditions the subsequent new growth is normal, although at times the growing point is damaged and rot of the storage root may occur. The addition of gypsum at the rate of 1,120 kg/ha or its equivalent in Ca from calcium chloride is recommended as a trial in early spring where symptoms appear frequently in late spring or early summer.

Boron deficiency when detected can be corrected by the addition of 2-3.5 kg B/ha but the trials should be made cautiously, since other crops are often harmed by amounts of B normally tolerated by sugarbeets.

Sulfur deficiency can be corrected by applying 35 to 70 kg/ha of S-containing fertilizers or by incorporating broadcasted gypsum ($CaSO_4 \cdot 2H_2O$) into the soil prior to planting, or by adding gypsum by hand as a side dressing or dissolving it in the irrigation water.

Manganese deficiency may occur in overlimed acid soils or on some alkaline soils. Strangely, however, no deficiency of Mn or of Fe has been observed on sugarbeets in highly calcareous California soils where Fe chlorosis on fruit trees is often severe. An application of 15 kg Mn/ha from manganese sulfate, plus plant sampling, is suggested on a trial basis.

Copper and chlorine deficiencies have not yet been observed in sugarbeets under field conditions.

Symptoms of Fe deficiency have been observed in Europe, but effect on yield has been minimal.

PREVENTION OF SIGNIFICANT NUTRIENT DEFICIENCIES

After a nutrient deficiency has been identified and corrected, the next step is to prevent its recurrence on the same field or its appearance on neighboring fields. This can be done by monitoring each field on a regular basis from thinning to harvest through a plant analysis program. The program should also include tests for nitrate-N and to a lesser extent, for P, K, Mg and Zn.

Field trials

Field trials with sugarbeets have shown a general need for N and occasionally for other plant materials, but it has not been possible to extend these results to neighboring fields or even to adjoining fields with accuracy, because of differences in soil type and fertilizers and cropping history of each field. To overcome this difficulty and to reduce costs and time in getting results from extensive field trials, special soil and plant tissue tests have been developed for sugarbeets.

Soil tests on fields scheduled for sugarbeets

Nitrogen. A soil test based on the amount of N as nitrate present in the soil to a depth of 0.9 m (3 feet) has been found to be useful in California for estimating the amount of fertilizer N required for maximum sugar yield. The test is done by taking a representative soil sample from each quarter of a field to a depth of 0.9 m just after seedling emergence, or no earlier than just before planting, and then determining the average kg of nitrate-N per ha, to 0.9 m depth. Nitrate-N so determined in field trials conducted statewide has been correlated with the root yield of nonfertilized plots to give the regression, metric tons (mt) per ha = 47.04 + 0.08 S_n, where S_n is kg NO_3-N/ha 0.914 m. Inserting the average NO_3-N from properly taken field samples in the equation, then subtracting the calculated yield from the historical yield when the field was well fertilized and multiplying the difference by 8.0 estimates the amount of fertilizer N needed per ha. In field trials it was found that it takes, on the average, 8 kg of fertilizer N to increase the root yield by 1 mt per ha for plots fertilized with N for maximum sugar yield. The factor 8.0 in the metric system is equivalent to 16 lbs of N per acre per ton increase in root yield in the English system.

To illustrate the use of the equation, if soil nitrate-N is 112 kg/ha 0.914 m, the expected yield of the field would be 47.04 + 0.08 (112) or 56 mt/ha. If the historical yield has been 67.2 mt/ha when well fertilized, the difference, 67.2 - 56 or 11.2 mt is multiplied by 8.0 kg N/mt to give about 90 kg of fertilizer N/ha.

In the English system the equation becomes: Tons/acre = 21 + 0.04 S_n, where S_n is pounds of NO_3-N/acre to depth of 3 feet. For a yield of 30 tons/acre and a S_n value of 100 lb NO_3N/acre, the fertilizer N required is 16 30-[21 + 0.04 (100)] , or 80 lb N/acre.

Quite likely the yield equation will require adjustments for various sugar producing areas, but nevertheless, fields testing low in nitrate-N will require more N fertilization than those testing high. But, regardless of the outcome of the soil test and the amount of fertilizer applied, plant samples should be taken to see how well the N program has met the needs of the crop, especially for early season rapid growth as well as the period of 4 to 8 weeks of N depletion prior to root harvest. This schedule improves beet quality and increases root sugar concentrations.

Phosphorus. Soil samples from fields to be planted that are properly taken, analyzed, and interpreted can serve as a guide to the need for P fertilization of sugar beets. Soil values for P by two different methods of extraction and the kind of response expected are given in Table 21.2. The sodium bicarbonate method of analysis is recommended for mineral soils, and the water extraction method for soils containing 10% or more of organic matter. Although the values given in Table 21.2 are based on considerable experience, they should be viewed as tentative and subject to revision with more experience. Guidance to the need for P fertilization can also be gained from plant analysis of the previous crop. If the values were adequate to high throughout the season, the following sugarbeet crop would be unlikely to respond to P fertilization.

Other nutrients. Soil tests for other nutrients have only been slightly successful. Responses to K have been observed when both the replaceable and Nenbauer K values have been very low or when symptoms have been observed.

PLANT TESTS

Plant tests have been used to prevent the appearances of nutrient deficiency symptoms by monitoring each sugarbeet field from thinning to harvest in a well-planned plant analysis program. Fundamentally, the monitoring system is based on the critical nutrient concept in which it is assumed that an element required for plant growth must be obtained within the plant and that growth proceeds unrestricted by the presence of that element (nutrient) until its concentration has decreased to a specific concentration, the critical nutrient concentration. At this point the growth rate becomes less than for comparable plants above the critical concentration. For practical purposes, the critical concentration is taken at the point on the calibration curve where the yield is 10% below the maximum (Figure 21.2). The critical values given in Table 21.1 are for petioles or blades taken from young mature sugarbeet leaves. These values are presently tentative and thus may be raised or lowered slightly, based on field experience.

Since a petiole or blade from a young, mature leaf is of the same physiological age at all sampling dates, it becomes possible to determine the nutritional status of the crop at the moment it is sampled during the growing season from field thinning to harvest. When plants are found to be below the critical concentration, they are considered deficient in that nutrient and are not growing as rapidly as those above the critical concentration. Whether to apply fertilizer and thereby prevent a deficiency will obviously depend on many factors, including the time remaining before harvest and the feasibility of applying the materials in time to become effective. In essence, the longer the deficiency lasts and the earlier in the growing season the deficiency occurs, the larger the response to be expected from appropriate fertilization, when all other growth factors remain favorable. Conversely, when the nutrient concentrations are

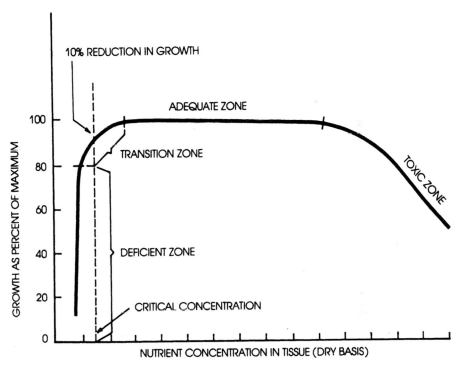

Figure 21.2. Calibration curve. Growth of sugarbeets in relation to the nutrient concentration of tissue (dry basis). The critical concentration is usually taken at the point where growth is 10% less than at maximum. Deficiency symptoms usually appear below the critical concentration and not at all above it. The sharper the transition zone, the more useful calibration is for purposes of plant analysis. Toxicity symptoms appear in the toxic zone as a nutrient becomes excessive.

above the critical concentration, the plants are well
supplied with nutrients at the moment of sampling. Adding more nutrients will not make the plants grow faster
as long as the untreated plants remain above the critical
concentration. It is only when the nutrient concentrations fall below the critical concentration that the rate
of growth decreases. Predicting when such a deficiency
will occur and how long it will last will naturally depend on the balance between nutrient supply from the soil
and the nutrient demand of the plant. It is this balance
between supply and demand that is determined by plant
analysis and it is this information that permits us to
fertilize a sugarbeet crop effectively, according to its
needs instead of by guess.

Nitrogen. Nitrogen is the key fertilizer element
for the quantity and quality production of sugarbeets for
processing at the sugar refinery. When N is in short
supply, root yields will be unusually low, but processing
quality will be high, although total sugar produced per
hectare will be low and most likely uneconomical to the
grower. Conversely, when the N supply is abundant on
fertile soils, root yields will be high but the sugar
concentration and processing quality will be relatively
low. This may make processing uneconomical at the factory and consequently uneconomical to the grower. Within
these limits, the judicious use of N is essential for efficient beet sugar production.

To achieve this objective, a soil analysis before
planting will estimate the initial N supply of the soil;
and depending on the yield goal, it will estimate the N
to be added as fertilizer. To evaluate how well this
fertilizer application will meet the needs of the crop
and to decide whether an additional application of N
needs to be made, the N status of the crop should be monitored by collecting and analyzing leaf petiole samples.
In most instances, this evaluation would be based on four
sets of plant samples; one taken about 1 month prior to
mid-season, 2 weeks later, another midway between the
second sampling and harvest, and finally just before
harvest. If the first two petiole samples show that the
crop will be deficient (below the 1000 ppm nitrate-N
critical level, Fig. 21.1) for more than 10 weeks prior
to harvest, an additional application of about 90 kg N/ha
is indicated. If the plant samples show that the
petioles remained above the critical level until 4 to 8
weeks before harvest, the conclusion would be that the
crop was well fertilized.

In a study at Davis, California (Table 21.3) the addition of P, or P and K, with N had no significant effect
on quality, root size or sugar yield. These results
agreed with the petiole phosphate-P and K values, since
these values were well above their critical concentrations at all sampling dates and therefore no response was
anticipated and none was observed.

TABLE 21.3
Effect of nitrogen, phosphorus and potassium fertilizers on sugarbeet yields, Davis, California

Kg of Nitrogen (N), Phosphorus (P$_2$O$_5$) or Potassium (K$_2$O) per ha		Beet Yield, mt per ha	Percent Sugar	Sugar Yield, kg per ha
First Application May 27[a]	Second Application July 24			
		Harvest October 20		
0	0	47.3	15.9	7,490
90N	0	52.9	15.5	8,160
90N	45N	58.1	15.1	8,770
90N	90N	63.9	14.7	9,420
90N	180N	68.2	14.0	9,510
224P$_2$O$_5$, 90N	180N	66.6	13.6	9,040
224P$_2$O$_5$, 224K$_2$O, 90N	180N	63.7	14.1	8,990
Sig. dif.	(19:1)	5.2	0.7	1,120

[a]Applied shortly after thinning.

SUMMARY

Routine soil and plant analysis. The use of soil and plant analysis as guides to fertilization of sugarbeets has increased tremendously on a routine basis in the last 10 years. Soil analysis has indicated how much N is in the soil at planting and the follow-up with plant analysis has indicated how well the fertilizer program actually met the needs of the growing sugarbeet crop up to the time of harvest. With this information, along with weather forecasts, harvesting has been advanced or delayed whenever feasible to achieve best sugar production.

Detection (trouble shooting). When leaf symptoms appear and nutritional deficiencies are suspected, the cause of the symptoms may be determined visually by comparing leaves to color photos or chemically by comparing analytical results to those of leaves with known deficiency symptoms.

Correction (field trials). Once a deficiency symptom has been identified visually or chemically, a well-planned field trial should be set up currently or on the next crop to verify the diagnosis.

Prevention (monitoring). Regardless of how a fertilizer program has been established, it should be evaluated through a systematic plant analysis program. This evaluation usually involves taking four sets of samples-- (i) shortly after thinning (ii) early mid-season, (iii) late mid-season, and (iv) just before harvest. The results along with pertinent crop information will aid in formulating a better fertilizer program for the current and following crop.

SUPPLEMENTAL READING

DRAYCOTT, A. P. 1972. Sugar-beet Nutrition. Applied Science Publishers Ltd. London. 250 pp.

HILLS, F. J., R. SALISBERY, and A. ULRICH. 1978. Sugar-beet fertilization. Bulletin 1891. University of California, Division of Agricultural Science, 16 pp.

JOHNSON, R. T., J. T. ALEXANDER, G. E. BUSH, and G. R. HAWKES. 1971. Advances in Sugarbeet Production: Principles and Practices. Iowa State University Press, Ames, Iowa, USA 470 pp.

ULRICH, A. and F. J. HILLS. 1969. Sugar beet nutrient deficiency symptoms, a color atlas and chemical guide. Priced Publication 4051. Berkeley: University of California, Division of Agricultural Sciences. 36 pp.

22
TOMATO

Allen V. Barker

Tomatoes (<u>Lycopersicon</u> <u>esculentum</u>) are grown around the world. In dollar value, tomatoes rank second, behind potatoes, among vegetables produced in the USA and income approaches a billion dollars per year. In the USA, more than half of the income is from fresh tomatoes; however, tonnage of processed tomatoes far exceeds that of the fresh fruit.

The tomato appears to have originated in tropical America, probably Mexico or Peru. The cherry tomato is viewed by some people as being the original type from which cultivated forms developed; however, larger fruited forms are found growing wild in South America. It apparently was taken to Europe during the early 1500s and was grown extensively in Italy long before it was grown in North America and England, where it was first considered as a curiosity or an ornamental plant. The cultivation of tomato for market dates from about 1800 in Europe and about 1830 in America. In its early history of cultivation, the fruits were classified by color, size, and shape, and in the 16th and 18th centuries four to seven types were recognized. Today, there are hundreds of named cultivated varieties and other lines. The processing industry started about 1850 in Pennsylvania.

Tomatoes are produced now in all areas of the world, however, little or no growth occurs with soil or air temperatures below 13 to 18°C. Where the outdoor temperatures do not satisfy this requirement, greenhouse production increases in importance as in northern Europe.

ECOLOGICAL ADAPTATION

The tomato is a warm-season plant requiring a relatively long season for profitable yields. It is reason-

---

[1]Head, Department of Plant and Soil Sciences, Massachusetts Agricultural Experiment Station, University of Massachusetts, Amherst, Massachusetts 01002, USA.

ably resistant to heat and drought and grows under a wide range of climatic conditions. It is cold-tender, however, and will not withstand a hard freeze. A frostfree period of at least 115 days is needed for profitable production because the crop requires 3 to 4 months from seeding to harvest of the first ripe fruit.

The crop does best when the weather is clear and dry and temperatures are uniform and moderate, 18 to 30°C. Below 0°C, the plants usually freeze, and above 35°C, fruits do not increase in size. High temperatures and high humidity favor the development of foliar diseases. Hot, drying winds cause the flowers to drop. High and low temperatures have an effect on market quality, especially color, of the fruit. High light intensity favors the formation of ascorbic acid (vitamin C) in the fruit. Light and temperatures may affect the formation of carotene.

Soil Adaptation

Tomatoes may be grown on many kinds of soils from light sands to heavy clays. Where earliness is important sandy soils are preferred, for they drain and warm quicker in the spring than the heavy soils. Provided the growing season is long, loamy and heavier soils are preferred, especially when large yields, as in production of processing tomatoes, are important. In all cases a well-drained soil is essential. Deep, loamy soils ample in organic matter and fairly well limed are ideal. The acidity of the soil should be in the range of pH 6.0 to 6.5, however, the tomato is tolerant of more acidic conditions. Yields are improved normally if soils in the range of pH 5.0 to 5.5 are limed. Very alkaline soils are undesirable for tomato production.

Level or gently rolling fields are preferred over rolling ones. Mechanical harvesters may dictate certain sizes or shapes of fields, as row lengths of less than 200 m may hinder harvester efficiency. Good wind protection is required for direct seeding. Whereas rows can be oriented to minimize damage from wind and sand blasting, windbreaks can be helpful if sand blasting is a serious problem. Strips of cereal grains may help reduce this problem.

Stony and cloddy soils and trashy or weedy soils should be avoided, especially with mechanical harvesting. Organic matter in sands should be 1.0% or higher, and in sandy loams and heavier soils organic matter should be more than 1.5% to improve soil physical properties and increase water holding capacity.

MINERAL NUTRIENT REQUIREMENTS

When nutrient needs are met, tomato fruit yields exceeding 50 mt/ha are probable under favorable soil and

climatic conditions with adequate pest control. Maximum
potential yields under ideal conditions may be much
higher. In California yields of 100 mt/ha of marketable
fruit are known. The current average yield in California
is about 70 mt/ha. Reduced losses of green, cracked, and
damaged fruit could increase the yield potential up to
140 to 150 mt/ha. Highest yields are normally those of
processing tomatoes. For the fresh market, yields are of
the order of 30 mt/ha under favorable soil fertility and
good environment conditions. Yields of tomatoes in
greenhouses approach 200 mt/ha with multiple cropping.

Average Yields in Representative Regions

The world average yield of tomatoes is 20 mt/ha.
Yields vary considerably among regions of the world and
among countries of a region, as shown in Table 22.1.

TABLE 22.1
Yield of field-grown tomatoes by region (from FAO
Production Yearbook 1979)

Region	Area harvested (1000 hectares)	Yield		
		Low	High*	Average
		(metric tons per hectare)		
World	2407	0.7	51.6	20.5
Asia	706	2.6	51.6	15.9
Europe	490	5.2	41.8	28.3
USSR	395	**	**	16.2
Africa	354	0.7	24.3	13.5
N. America	314	2.3	42.1	31.3
S. America	135	6.1	26.5	21.2
Oceania	10	**	**	21.3
Australia	8	**	**	16.2

* On harvested areas on 500 ha or larger.
** Not available.

The total world production of tomatoes in 1979 was
49,201 million metric tons (mmt). The highest average
yields are found in North America and Oceania. Europe
(13,817 mmt), Asia (11,252 mmt) and North America (9,847
mmt) are the leading regions in total tomato production.
The USA is the largest producing country with 7,663 mmt
in 1979 followed by the USSR with 6,400 mmt.

Mineral Nutrition for Acceptable Yields

Nitrogen (N). Nitrogen is likely to be deficient in
soils because of leaching and limitations on rate of mineralization of N from organic matter, which varies greatly in soils. A plant must have adequate N to produce a
frame to support the plant and foliage for photosynthesis

and shading of the fruit. Too much N early in the season
may make the plant too vegetative, and late in setting
and maturing fruit. Yellow foliage and small plants
result from insufficient N. Late applications in addition to delaying fruiting or to splitting sets of fruits
may give lower soluble solids, more disorders in fruit
(blotchy ripening, gray wall, yellow eye, sprouting
seeds), detinning in cans (high nitrates), and poorer
suitability for machine harvesting. Tomatoes may be
sidedressed with N for hand picking, but for machine
harvesting, the N supply should be depleted as most of
the fruit mature.

Phosphorus (P). Phosphorus stimulates vigorous root
growth and promotes sturdy stems and foliage. In regions
where the growing season is short, P fertilization is
more important than in regions where the season is long,
for the plant has less time and a smaller root system to
absorb P.

Potassium (K). Tomatoes utilize large amounts of K
which is required for a number of physiological and metabolic processes and greatly affects plant growth and development and fruit quality.
Fertilizer ratios of $N-P_2O_5-K_2O$ for tomatoes may
range from 1-2-1 or 1-2-2 for sandy soils to 1-3-1 or 1-
4-1 for loamy and heavier soils. Fifty to 100 kg N/ha
are typical applications with P and K in proportions as
described in Table 22.2.
Starter solutions of water-soluble fertilizers, such
as diammonium phosphate or a commercial fertilizer (10-
40-10) with a high P analysis, promote earlier maturity
and heavier yields.

TABLE 22.2
Plant nutrients applied by top tomato growers in Ohio

| NUTRIENT | 1955 | Amounts applied | |
		1965	1970
		kg/ha (Ratio)	
N	40 (1)	70 (1)	75 (1)
P_2O_5	120 (3)	175 (2.5)	190 (2.5)
K_2O	120 (3)	150 (2.1)	230 (3.0)

About 1 kg of fertilizer per 100 to 200 liters of
water makes a satisfactory starter solution for watering-
in plants.

Soil Tests, Tissue Tests, and Deficiency Symptoms

Soil tests should include pH, magnesium, phosphorus,
and potassium.

Liming. Land should be limed to pH 6.0 to 6.5 to correct soil reaction and to provide Ca and Mg. Dolomitic limestone is recommended for soils that test low in Mg. Soil reaction should not be allowed to fall below pH 5. Blossom end rot of fruits is linked with Ca deficiency. The frequency of this disorder increases in dry weather. Adequate irrigation is essential to prevent blossom end rot when rainfall is low. Magnesium deficiency is manifested by interveinal chlorosis (mottling) of lower leaves. A metric ton of tomato fruit will contain about 0.5 kg of Mg, and about 2.4 kg of Ca.

Nitrogen (N). Fertilizer rates for N are based on yield responses, tissue tests, or on visual symptoms. Tissue tests should be made early in the growth cycle. The petiole (stem) from a recently matured leaf, such as the fourth leaf back from the tip, is taken. Guides for deficient and sufficient levels of nitrate -N in tomato leaf petioles are given in Table 22.3. Nitrogen deficiency symptoms appear as yellowing of the lower leaves, although yellowing is permissible as the fruits mature and exhaust N from the soil and lower portions of the plant. About 3 kg of N are in each metric ton of tomato fruit.

TABLE 22.3
Guides for nutrient concentration in leaf petioles of tomatoes grown for processing (adapted from University of California Bulletin 1979)

Time of Sampling*	Nutrient Concentrations **					
	Deficient			Sufficient		
	N	P	K	N	P	K
	% dry wt					
Early bloom	0.8	0.2	3	1.2	0.3	6
Fruit at 2.5 cm diameter	0.6	0.2	2	1.0	0.3	4
First color in fruit	0.2	0.2	1	0.4	0.3	3

* Petioles of fourth leaf from tip.
** NO_3-N, PO_4-P, and total K.

Phosphorus (P). Soil tests for P have a fair correlation with plant yield response to fertilization and nutrient uptake. Yield increases are likely with tomatoes grown in soils testing less than 6 ppm available P (0.5 M, pH 8.5 sodium bicarbonate extraction) and are unlikely if the test shows values greater than 12 ppm. Tissue test values for adequate P nutrition of tomato are shown in Table 22.3. Phosphorus deficiency is shown by stunting of the plants and off-green colors of the leaves. Lower leaves show the symptoms first. Leaves

first become a gray or blue-green color after which a
reddening of the leaf blade occurs. The reddening is
prominent between the veins and appears more pronounced
on the lower sides of the leaves. With severe deficiency
the lower leaves may yellow and become necrotic. Often P
deficiency symptoms disappear as the roots of the plant
expand and are able to extract more P from the soil.
Stunting of plants may persist. Any appearance of P
deficiency symptoms usually results in reduced crop production. A metric ton of tomato fruit contains about 0.4
kg of P.

Potassium (K). Crop yield responses correlate fairly well with K soil tests. Yield responses are likely
with K fertilization for soil test values below 50 ppm
(1 M ammonium acetate extraction) and are unlikely for
values above 80 ppm. Tissue test values for adequate K
nutrition are given in Table 22.3. Potassium deficiency
symptoms appear on the foliage as marginal burn of the
leaflets beginning at the leaf or leaflet tips and progressing back and into the blade in some cases. Gray
wall and blotchy ripening of fruit are associated with
inadequate K. Lesions on stems of tomatoes fertilized
heavily with ammoniacal fertilizers can be corrected by
the addition of K fertilizers. About 4 kg of K are in
each metric ton of tomato fruit.

MINERAL NUTRITION OF IMPROVED VARIETIES AND HYBRIDS

New cultivars may have different nutritional requirements than older ones. Breeding and selection occur normally under high regimes of soil fertility, and
the nutritional requirements of new varieties and hybrids
must be determined to assure that the full production potential is realized and that fruiting disorders related
to nutrition are identified and corrected by appropriate
fertilization. The development of cultivars with increased efficiency for nutrient use is possible, for
genetic variability in uptake, transport, and utilization
of nutrients--and in tolerance to high concentrations of
elements in a medium--has been observed. In tomato, differences in nutrition or tolerance with respect to Ca, P,
B, Na and ammonium have been identified.

NUTRIENT REQUIREMENTS IN THE FIELD

When relatively small amounts of fertilizers are
used, application by the side of the row at planting increases nutrient use efficiency over broadcasting. With
heavy fertilization (1,000 kg/ha or more), split applications are desirable with some fertilizer being plowed
down before planting and with sidedressed application at
planting or soon afterward; for instance, after thinning

of direct-seeded crops. Generally, heavy N applications should be split, possibly with some being applied during fruit development for hand picking. Under most circumstances, no advantage is gained by more than one application of P or K fertilizer.

Farm manures may be used to provide a portion of the plant nutrients with some preference given to rotted manure because of its soil-building properties and mineralization rate. Manures should be considered as N and K fertilizer as they are low in P. A chemical fertilizer with a ratio of 1:3:1 should be applied in conjunction with any farm manures. Tomatoes may follow green manures or soil-building crops. About 30 to 60 kg N/ha applied before plowing aids in the decomposition of soil-building (green manure) crops. Chemical fertilizers and soil-building crops are the main resources used to maintain soil fertility for tomato production.

Higher tomato plant populations are used when plants are trellised or staked than when they are grown on the ground, and higher rates of fertilization are required accordingly. Tomatoes harvested by machine in a once-over or destructive harvest would not be fertilized late in plant development or as heavily as plants that are grown for hand picking and a prolonged period of productivity. If rain should leach N deep into the soil, or if too little N was applied at planting, about 25 kg N/ha may be applied within 4 weeks after transplanting, or at a similar stage of development with direct seeding for machine harvesting.

UTILIZATION OF INCREASED PRODUCTION

Generally the consumption of tomatoes is rising. The use of tomatoes in the home is high. Fast food restaurants, pizza restaurants, and manufacturers of prepared foods use large amounts of catsup, sauces, and paste. The consumption of tomato juice has declined over the years in the USA with a rise in the consumption of frozen orange juice.

The trend toward increased overall consumption of tomatoes seems to ensure a market for higher tomato production and to maintain price stability.

TOMATO PRODUCTION IN TROPICAL CLIMATES

Problems and Progress

Leading tomato producers are in the temperate regions. In the tropics, tomatoes are normally produced in mountain regions or in lowlands during cool seasons. Tomato seeds available to most farmers in the tropics are imported temperate varieties, and as a result tomato production in the tropics is characterized by seasonality and low yields. Unadapted varieties is one of the constraints limiting production of tomatoes in the tropics.

Soil temperatures above 39°C reduce root expansion of tomato. In soils warmed by the tropical sun, high temperatures eliminate root growth in the top 5 to 10 cm of soil where air, moisture, and organic matter contents are otherwise favorable for root activity. For optimum fruit-setting, tomatoes require night temperatures of 15 to 20°C. In tropical regions, temperatures below 20°C are rare in the lowlands even during the cooler months. Moisture conditions are high also during the periods of high temperature. Major yield-reducing diseases of tomato also occur in the tropics.

Heat Tolerance of Tomatoes

In California, Texas, the USSR, the Philippines, and at the Asian Vegetable Research and Development Center (AVRDC) in Taiwan, breeding and screening are underway to identify heat tolerant tomatoes. Heat tolerance is genetically controlled, but heritability is confounded by environmental factors, and is complex. Many genotypes which may be heat tolerant may be lost in screening techniques used to identify other desirable traits such as resistance to bacterial wilt. Until high-yielding, heat tolerant cultivars are available, management will be very important in tomato production in the tropics. Mulching with organic materials, such as rice straw or husks or dry grass, helps to reduce surface soil temperatures as well as to improve distribution of water in the soil profile and to reduce nutrient leaching losses. Plastic mulches tend to elevate soil temperatures and should be avoided in the tropics. Multiple cropping of tomato with paddy rice increases the possibility of profitable tomato production. Commonly the first rice crop is shifted to summer tomato, or the second rice crop is shifted to the winter crop. Pruning and training of indeterminate tomato plants, or adopting determinate cultivars can program the duration of tomato cropping from 45-60 days to 120-150 days, allowing much flexibility in sequential cropping. Tomatoes have been successfully interplanted on intensive vegetable farms with sugarcane, leeks, or other vegetables of long duration. Tomato is a very labor-intensive crop, requiring about 5,000 man-hours per ha. Fresh tomato production gives even labor distribution throughout the season. Adopting the tomato in multiple cropping patterns will be most suitable for small-scale farmers with a surplus of family labor. Processing tomatoes require intensive management and involves short harvest times and long intervals between harvest. Although the total labor may be reduced to 1,750 person-hours per ha, the dependency on hired labor is greatly increased.

Fertilizer Requirements of Tomatoes in the Tropics

In temperate regions, the tomato plant responds to heavy rates of fertilization. In Japan, fertilizer application rates by outstanding growers are 340 kg N/ha, 410 kg P_2O_5/ha, and 320 kg K_2O/ha with average yields of 105 mt/ha. Although these rates of fertilization are much higher than those used in the USA, similar or higher rates are used in Korea, The Netherlands, and Italy. Tomatoes require high nutrient concentrations in the soil solution, and high salt concentrations may eventually become a problem.

In the tropics, uncertain environmental conditions and market economics may not favor intensive fertilization. Quality of fruit may not be adequately paid for. Fertilizer rates for tomatoes in the tropics usually range from 40 to 120 kg N/ha, 30 to 90 kg P_2O_5/ha, and 30 to 90 kg K_2O/ha. In relative terms, fertilizer response is striking for tomatoes in the tropics. But because of low crop yields, efficiency of use may be low. Improved management of natural soil and water resources and better adapted varieties are needed before fertilizer rates can be matched to the genetic capability of tomatoes to respond to fertilization in the tropics.

23
CABBAGE AND CAULIFLOWER

Ernest L. Bergman

Cabbage (Brassica oleracea, Capitata Group) and cauliflower (Brassica oleracea, Botrytis Group), a part of the large group of cole crops, are grown worldwide, and are important fresh market and processing crops depending on varied head shapes and sizes, different leaf colors (such as red cabbage) and forms (Savoy cabbage).

ECOLOGICAL ADAPTATION

Areas of cabbage production range from near the Arctic Circle to the Equator, even though it is a cool-season crop. Development of new cultivars and research in production practices were instrumental in increasing this range of adaptability. Cultivar selection, combined with choice of growing season as dictated by local climatic conditions, made it possible to grow cabbage in any one or more of the four seasons. For instance, a cabbage cultivar being used as a summer crop at one location in the temperate zone might grow better as a fall, winter, or even spring crop in a subtropical or tropical region.

Cabbage and cauliflower production can be very versatile. Home gardeners and small growers enjoy growing these crops for home use and for local markets. Depending on cropping systems, these two crops are produced in large hectarage grown alone as monoculture; propagated by direct seeding, or transplanted; and manually or mechanically harvested. Cabbage can be planted as a second crop after grain harvest or in a labor-intensive system by intercropping with other vegetables. In the USA, approximately 38,000 ha of commercially grown cabbage are harvested annually for fresh consumption, and about 17,000 ha of cauliflower are produced for both fresh market and processing. Overall, as indicated from area

[1]Professor of Plant Nutrition, Department of Horticulture, The Pennsylvania State University, University Park, PA 16802, USA.

harvested, cauliflower production is slightly increasing annually while that of cabbage is decreasing.

Temperatures. Requirements for growing both cabbage and cauliflower vary with cultivars but cauliflower seedlings bolt (produce flower stalks) quicker when subjected to 12°C followed by hot weather, than cabbage at 2°C. Generally, an average temperature of 15-20°C per month is best for both crops. Seeds of both crops will germinate in soil temperature as low as 8°C, with the best range being 10-16°C. For seed germination a minimum soil temperature of 5°C is needed; however, both crops will withstand up to 36°C, but the best range is between 7 and 29°C. At very low soil temperature, seed will take longer to germinate than at 25°C for instance.

Soil. Any kind of soil, be it high in minerals (clay, silt, or sand) and/or organic matter (peat, muck) will produce acceptable cabbage and cauliflower yields. Both crops prefer a mildly acid to neutral soil (pH of 6.5 to 7.0), but they will grow on soils that are moderately acid or alkaline. When compared to other vegetables, cabbage has a salt tolerance similar to melons and potatoes, but a lower tolerance than tomato, and a higher tolerance than lettuce and onions. Overall, as long as climate, cultivar, nutrition, and cultural practices are in balance, there should be no problems with either cabbage or cauliflower production in most arable soils.

NUTRIENT REQUIREMENTS

In order to determine the mineral nutrient supplying power of the soil to support the crop, estimates of percent base saturation (BS) and cation exchange capacity (CEC), its chemical and storage capacity, are very useful. The hydrogen ion concentration (measured in pH values) and base saturation which includes Ca, Mg, and K contribute to the cation exchange capacity of the soil. Under soil conditions high in sodium (alkali and saline soils) this element is included as a part of base saturation, otherwise it is omitted and not taken into consideration except under greenhouse conditions. Base saturation is expressed in milliequivalents per 100 g soil, and percent saturation is the concentration of each element expressed as percent of cation exchange capacity. A balanced soil will generally fall within these ranges as to content of basic elements: calcium (Ca) 60-80%; magnesium (Mg) 10-15%; and potassium (K) 3-5%. These values are important in cabbage and cauliflower production and will be dealt with further when the individual elements are discussed.

Nitrogen (N) deficiency. Of all the essential elements N has probably the most effect on production since plants react drastically to deficient levels and have severely reduced growth and yield, while too much will have an antagonistic effect on the elements. Hence, for all fertilizer applications N must be the basis, especially when applied in combination with other elements or as a component of manure. To date, there is no soil test available for N which gives consistently satisfactory results, and therefore the amount applied is based on the specific crops. All cole crops have the highest need for N of any vegetable. Under most soil and climatic conditions, 160 kg N/ha fertilizer will help to produce acceptable cabbage and cauliflower yields, as well as ensure acceptable quality. If frequent irrigation is used, and/or the soil is sandy, N must be increased to 185 kg/ha.

There are many N-containing fertilizers on the market and generally a complete fertilizer (N-P-K) is cheapest. Where the soil is high in K and no K is needed, a mono- or diammonium phosphate application is sufficient. Ammonium nitrate, ammonium sulfate, anhydrous ammonia or urea will work if only N is needed; however, continuous ammonium sulfate application will lower soil pH (increase acidity), and possibly decrease Ca and increase N uptake. Animal manures, depending on the animal, or if fresh, stored, or dry, are highly variable in nutrient content.

When transplanting cabbage or cauliflower seedlings from the seed bed to the field, 0.75 kg/100 liters of a high analysis fertilizer starter solution, applied through the transplanter, gives the seedlings a good start. If hand-planted, other means of application must be used, but about 250 milliliters of solution per plant are sufficient in any case.

Cabbage and cauliflower plants deficient in N have a uniform yellowish color, with the leaves closest to the roots (oldest) being lighter in color than younger ones near the top of the plant. The root system of these plants is generally very large. Cabbage heads may not be as firm as with adequate amount of applied N, and cauliflower plants may button prematurely. Bolting (producing flower stalks) is generally not influenced by N since it is brought about by early cold temperatures followed by heat in the young plant stage. Compactness of inflorescence seems to improve with N.

Phosphorus (P) deficiency. There are many areas in the world where cabbage and cauliflower are grown where soils are deficient in P. Deficiency can be due to naturally low P level in soils, by continuous cropping without addition of P, or by low soil pH so that P is not in a readily available form for plant uptake. An intermediate soil pH range is the best insurance for having soil

P available to the plant. Phosphorus is most important for cabbage and cauliflower at the seedling stage, especially when the soil is cold, since low soil temperature hinders uptake. The best insurance against P deficiency is soil testing. In clay soils for instance, 240 kg P/ha as indicated by the Bray #1 laboratory test is a good value to strive for.

Under low fertility soil conditions, cabbage and cauliflower will show deficiency symptoms at the early stage of growth. Their oldest leaves will show purpling on the underside first, and the margins of these leaves (tips) will then turn purple followed by the whole leaf taking on this color. When temperature increases and enough P is in the soil, the purpling will disappear. When P is deficient, growth is retarded and plants have a poor root system. When transplanting on a low P soil, a starter solution high in P should be applied. Soil liming is essential, and if additional P is needed above that applied in a complete fertilizer (Table 23.1), diammonium phosphate should be used. If K is high or if P alone is needed superphosphate or triple superphosphate are excellent under practically all soil pH conditions. Rock phosphate works well in acid soils but is less available to plants unless soils are strongly acid.

Potassium (K) deficiency. All cole crops are heavy K feeders. This makes these crops valuable to use in rotation with other crops having low K in order to get the soil back into nutritional balance. Cabbage shows positive yield response to K application and the K concentration in tops will increase nearly proportionally. Since the top portions leave the farm, K taken out of soil by the harvested crop, as much as 300 kg/ha, must be replaced. Under well-balanced conditions, 4-5% of CEC in a field with 85% Ca brings best yields. An oversupply of K will depress Mg uptake.

Potassium deficiency appears first on oldest cabbage leaves and is expressed as spots on the leaf surface. Shiny green leaves turn dull green, leaf margins turn a yellowish green (under severe conditions to dark brown), followed by withering away. Leaf veins are not affected but mature heads are loose and smaller. However, head size is an inherited trait of the cultivar and is influenced much more by plant spacing than by mineral nutrition. On red cabbage, the red color will fade and when K is deficient loose head formation and poor red color can be observed. In K-deficient cauliflower, leaf tips turn brown, leaves turn inward and can have a crinkled surface.

If complete fertilizer (Table 23.1) did not produce desired results, muriate of potash (KCl) or potassium sulfate (K_2SO_4) can be used. Generally, animal manure is high in K and if applied should be plowed under since using manure (especially fresh poultry manure) as top dressing can injure the crop.

TABLE 23.1
Fertilizer recommendations based on soil phosphorus (Bray #1) and potassium (% of CEC by Pennsylvania Soil Testing) for cole crops[1]

| % K in Soil | Fertilizer to be Applied (kg/ha)[2] | | |
	Nitrogen (N)	Phosphorus (P)	Potassium (K)
A. If P below 70 kg/ha			
0 - 2.0%	160	160	240
2.1 - 3.0	160	160	185
3.1 - 3.5	160	160	120
3.6 - 4.0	160	160	95
4.1 - 4.5	160	160	47
4.6 - 5.0	160	160	0
over 5.0	160	160	0
B. If P is 70-150 kg/ha			
0 - 2.0%	160	130	240
2.1 - 3.0	160	130	220
3.1 - 3.5	160	130	120
3.6 - 4.0	160	130	120
4.1 - 4.5	160	130	60
4.6 - 5.0	160	130	0
over 5.0	160	130	0

[1]Bergman, E.L. and R.F. Fletcher. 1973. Soil Analyses and Computerized Fertilizer Recommendations for Vegetables in Pennsylvania. Acta Horticulturae (Tech. Comm. ISHS) 29:155-165.

[2]Amounts might have to be adjusted slightly to fit available fertilizer ratios.

Calcium (Ca) deficiency. Limestone application in advance of planting is the cheapest way of maintaining soil pH and Ca content. Applying high analyses fertilizers and triple superphosphate which contain little Ca contributes increasingly to Ca problems. Both cabbage and cauliflower do best in near neutral soil where sufficient Ca should be available. The immobility of Ca in the plant makes Ca deficiency symptoms appear first in the youngest plant parts, which in cabbage and cauliflower is the center or the heart. Deficiency symptoms include tip burn, chlorotic young leaves, and in severe cases breakdown of the heart and dieback of rootlets. However, these symptoms should not be confused with those of boron (B) deficiency, which are very similar. Nevertheless, Ca deficiency symptoms on cabbage and cauliflower are much more infrequent than those of B.

Earlier it was stated that a good soil has a Ca content of 60-80% of the cation exchange capacity. These values are very helpful in interpreting Ca problems, since if Ca falls below 30%, Ca deficiency can be predicted for cabbage and cauliflower--especially in a dry year--when N was applied in ammonium form, or large amounts of animal manure were used. Low pH (strong acidity) will increase B availability and reduce molybdenum availability.

There are many different kinds of limestone available. Some are very high in Ca and contain no Mg, while others, such as dolomitic limestone, contain as much as 9% Mg. Therefore, what to apply depends on availability, price and possible need for Mg. Gypsum (calcium sulfate) will not affect pH, is a good source for both Ca and S but not Mg, and can be used where Ca is needed but the pH situation does not allow liming.

Clubroot (Plasmodiophora brassicae) is a fungus disease infecting cole crops, with cabbage being most susceptible and cauliflower somewhat less. While not a direct nutritional problem, clubroot is closely connected since the fungus thrives on acid soils. A standard recommendation has been to apply hydrated lime to produce a soil pH of 7.2 several weeks in advance of planting, but more positive prevention is achieved by use of clean and healthy plants, and by crop rotation. In the past, calcium cyanamide was used successfully to combat clubroot, but is no longer available in many places and soil disinfecting chemicals are now used.

Magnesium (Mg) deficiency. Cabbage and cauliflower have a strong need for Mg and can be used as indicator plants for Mg deficiency. The deficiency is widespread, especially in soils developed from low-Mg parent materials, soils continuously cropped without addition of Mg, or soils high in K. As with N, P, and K deficiencies, oldest leaves are affected first. Both cabbage and cauliflower exhibit the classical Mg deficiency symptoms, interveinal chlorosis (loss of color) with green veins. Leaf margins stay green at first, but ultimately dry up, and might even fall out. On cauliflower, small chlorotic leaflets can protrude through the curd. Low Mg in plants brings about poor root development and shallow soil penetration, which prevents water uptake especially during dry seasons.

Magnesium deficiency should not be confused with virus infection which can exhibit similar symptoms. Virus infection would be spotty in a field rather than general on all plants, and Mg applications are of no value against virus. A ratio of one to two between soil K and Mg, based on percent of the cation exchange capacity, will meet nutrient needs of cabbage and cauliflower. To add Mg to complete fertilizer, units of Mg oxide available in fertilizer of feed grade form, can be added. Also Mg oxide can be applied directly, or water soluble Mg sulfate can be sprayed on the growing crop, but Mg-containing lime is the cheapest and most efficient way of application if lime is also needed. The strong antagonistic effect of K and Mg makes the simultaneous application of both elements questionable unless both elements are needed, since the effect of K on Mg is much stronger than Mg on K. Mg deficiency in cabbage and cauliflower can be induced quicker with high K than with low Mg.

Sulfur (S) deficiency. In the past, there have been no reports of S deficiency symptoms on cabbage or cauliflower grown in the field. However, more and more S deficiencies on other crops have been reported, including the closely related broccoli. These observations were made in areas where there is little air pollution or acid rain. Also, with less and less S being used in complete fertilizers or spray material, a need for S is possible. An upright growth habit in cabbage has been observed in plants in S-deficient nutrient solution. Youngest leaves turned purplish, cupped upward, and leaf edges were rolled in. Application of S as part of any one of the many available fertilizers will take care of this problem.

Iron (Fe) deficiency. In acid soils or those below pH 7, there is generally sufficient Fe available for cabbage and cauliflower absorption. Under high pH (alkaline) conditions, Fe deficiency symptoms can occur as interveinal chlorosis of youngest leaves, and all veins stay green (fishnet pattern). If severe, even the veins will lose their green color, and the entire leaf turns yellow and then light brown. Under such conditions if lowering soil pH is not possible, soil application of iron chelates or foliar spraying may help.

Manganese (Mn) deficiency. Cabbage and cauliflower prefer a near neutral soil pH range partly because in acid soils Mn is highly available. Both cabbage and cauliflower are sensitive to high Mn in the soil. Manganese toxicity is a big problem in both crops, and poor growth is the only symptom. Liming to increase pH will make Mn less available to plants. Some peat soils have a near neutral pH, are high in Ca, deficient in Mn, but surprisingly sufficient in copper.

Cabbage and cauliflower will show the typical Mn deficiency interveinal chlorosis, with dark green main veins (fishnet) of youngest leaves, but unlike Mg deficiency, Mn chlorosis extends into the leaf margins. In severe cases, the whole leaf turns yellowish and at this moment it is hard to distinguish between any one of the micronutrient deficiencies or even virus disease. Soil applications of Mn can be helpful.

Boron (B) deficiency. Cauliflower especially, and to a lesser degree cabbage, are severely affected by B deficiency which occurs more often than any other micronutrient deficiency. Practically it can be found on all soils but first of all on those derived from soil parent material low in B, on sandy and/or shaly soils, or acid soils. On the other hand, saline and alkali soils can be high in B. Dry weather has a strong negative influence on B availability. Under stress, B is not translocated from one plant part to another and therefore deficiency symptoms occur first on the youngest parts. In cauliflower, B deficiency is very distinct. Initially, youngest leaves turn chlorotic and their tips turn brown later on. Plants grow very unevenly with larger and smaller plants side by side. Under very severe conditions, some plants even die. The curd does not develop as a compact unit, is rather loose, and instead of staying clear white it turns purplish-brown. The whole plant appears as if it is wilting and leaves may curl downward. The trunk (stem) of these plants is hollow and when the top is cut off, a brownish-colored cavity is found, which is sometimes large enough to fit a man's thumb. This cavity will extend as much as 30 cm downward inside the trunk. Care must be taken that this hollowness is not confused with that sometimes found in well-grown plants of some cauliflower cultivars and where the hollowness is more like cracking and tissues are healthy greenish-white instead of brown. Also, plants grow evenly. If B deficiency is combined with that of Mg, little leaflets protrude through the loose curd.

In B deficient cabbage the cavity is extended into the head, the youngest leaves are chlorotic, leaf edges are scorched, blades turn downward and develop poorly.

Soil testing for B will give to some extent an indication of B availability; 1 ppm B in the soil is a safe amount for normal growth. Depending on severity of symptoms or soil content, applications of 1-5 kg B/ha are needed. Cauliflower can also be sprayed with B. Boron can be mixed with complete fertilizers. Care must be taken that the crop to follow cabbage or cauliflower receiving heavy B application will thrive under high B conditions, otherwise severe damage may result (beans are an example). Generally a heavy annual B application is not necessary and 1 kg B/ha mixed with a complete fertilizer will suffice.

Copper (Cu) deficiency. Deficiency of this element can be found on plants growing in mostly strongly acid soils high in organic matter. Cabbage and cauliflower have a medium need for Cu. Nevertheless, with Cu deficiency youngest leaves become chlorotic, growth is stunted, heads are poorly formed and sometimes the plant gives a wilted appearance. In most cases an application of 20-50 kg/ha of copper sulfate will alleviate the problem for several years. For emergency treatment of growing crops 1-4 kg/ha of copper chelate would be helpful.

Zinc (Zn) deficiency. While cabbage and cauliflower can be affected by zinc deficiency which occurs mostly in light sandy or alkaline soils, little has been reported on these crops. If there is a known deficiency, a green manure will increase available zinc. On acid soils as little as 1.5 kg/ha Zn sulfate can correct the problem. On alkaline soils, 4-12 kg/ha zinc chelate broadcast prior to planting should be adequate.

Molybdenum (Mo) deficiency. Occurrence of Mo deficiency is most pronounced on cauliflower and is called whiptail. It can occur on most any soil, especially acid soils. Symptoms start with cupping of the leaves and interveinal chlorosis, followed by twisting and narrow elongation of leaves. There is poor curd development, if any. Under severe conditions, the whip-like, long, slender leaves are an unmistakable identification. Application of Mo will remove the problem, but plants affected will not recuperate. One soil application before planting may last for several years.

PLANT ANALYSIS AS AN INDICATOR

There is a relatively short time span between seeding, transplanting (if necessary), and harvesting of cabbage and cauliflower, 90-120 days for cabbage and 70-100 days for cauliflower depending on cultivar and climatic conditions. It is therefore not possible to give days as a target date. The youngest mature leaves prior to heading are most suitable for plant tissue analysis.

The following ranges based on good growth and yield can be expected for cabbage, as based on percent or parts per million of dry weight:

N:	3.75 - 4.50%		Fe:	35^+ ppm
P:	0.30 - 0.50%		Mn:	30^+ ppm
K:	3.50 - 5.00%		B:	25^+ ppm
Ca:	3.00 - 4.00%		Cu:	4^+ ppm
Mg:	0.25 - 0.40%		Zn:	20^+ ppm

The $^+$ values mean that below them deficiency symptoms would appear, and values above these in most cases represent luxury consumption. Depending on cultivar, the major elements fall within the given values, which are not much different from those for cauliflower, which also holds true for microelements.

CULTURAL PRACTICES

Fertilizers. Fertilizer placement depends on available equipment. Plowing under of lime and complete fertilizer before planting or direct seeding is recommended. Fertilizer applications can be split with part applied before planting. A part can be applied as side-dressing, or by application through irrigation. When irrigation is used, the fertilizer must be completely water-soluble or in liquid form. If sprinkler irrigation is employed, plants should be wetted before the fertilizer is added, and after completion irrigation should continue to wash fertilizer off the plants, otherwise leaf burns can occur.

Nutritional sprays can be applied to both cabbage and cauliflower but a spreader sticker must be used otherwise the liquid will just roll off the plants. The cost of such a spray (labor, equipment, weather conditions, and the compounds) may not warrant application; however, to save a crop it is justified. Sometimes especially when it is dry and irrigation is available, plain water, equivalent to 5 cm rainfall, is sufficient to overcome Ca or B deficiency problems. Spraying for K and P is of little value. The practice of applying sprays containing many elements to catch the one unidentified and responsible for the problem is very much discouraged since it may antagonize rather than arrest it. There are many water-soluble compounds available that are manufactured specifically for spraying. Others such as borates, chelates, and molybdates are excellent if applied at the right time.

SUPPLEMENTAL READING

De GUES, J.G. 1967. Fertilizer Guide for Tropical and Subtropical Farming. Centre d'Etude de l'Azote, Zurich, Switzerland. 727 pp.

LORENZ, O.A. and D.N. MAYNARD. 1980. Knott's Handbook for Vegetable Growers. John Wiley & Sons, Inc., New York. 390 pp.

MAYNARD, D.N. 1979. Nutritional disorders of vegetable crops: A review. Journal of Plant Nutrition. 1(1): 1-23.

24
CUCUMBERS

Leonard M. Pike and Richard W. Jones

The cucumber (<u>Cucumis</u> <u>sativus</u>), a crop that is thought to have originated in India or Southeast Asia, is now grown throughout the world. It is grown in most countries in family gardens and in some areas as large-scale commercial production. The fruit are used primarily as fresh salad vegetables or as a processed crop for pickles. Some cultures eat it as a cooked vegetable. Cucumber fruit is low in calories and also relatively low in nutrients but has an excellent flavor as a salad or processed product.

Cucumbers can be grown widely due to their short growing season. Fruit are produced by many cultivars in 40 to 60 days after seeding. This short season provides great flexibility allowing the production of two or more crops per season in many areas, and also allows commercial growers a relatively short time between seeding and market, thereby decreasing the period of investment in the crop.

Optimum temperature for production is 70 to 75°F (20 to 29°C). Night temperatures should not fall below 60°F. Fruit quality also declines when temperatures rise to 95 to 100°F (31 to 39°C). Optimum soil temperature for germination is 60°F while maximum soil temperature for seed germination is 105°F.

Cucumbers grow well in most well-drained, non-saline soils. Optimum pH level is between 6 and 7 and the crop is moderately susceptible to the detrimental effects of acidic soils.

Fertilizer application and placement depend on the morphology of the root system of a given crop. The root system of cucumbers is extensive but shallow. The plants have a strong taproot, which may penetrate soils to a depth of 1-1.3 m. The taproot is not extensively

[1]Professor of Horticulture and Research Horticulturist, respectively, Texas A&M University, College Station, Texas 77843, USA.

branched below 0.6 m; however, from the soil surface to a depth of about 0.6 m, the cucumber produces a complex system of branched roots which completely occupies the soil volume. This lateral root development often exceeds the growth of the shoots. Therefore, in soils with good tilth, fertilizers should be placed near enough to the seed for this efficient root system to utilize during early growth and yet far enough away to avoid high salt levels. Banding fertilizers 10-15 cm away on one or both sides of the seed has proven to be an efficient placement method.

When environment and soil fertility are conducive, cucumbers are capable of producing large yields. Yields in excess of 30 mt/ha of fruit are not uncommon. However, the USA average yield is only 10 mt/ha. In most countries yields range from 3.7 to 7.3 mt/ha. Yield information can be very misleading due to the different methods of harvesting and growing cucumbers. In the USA, a destructive machine harvest is often used. The timing of this harvest coincides with the proper size of the early fruits on each plant so that about one to two fruits per plant are harvested. If cucumber plants are allowed to grow and the fruit is handpicked regularly, one plant may yield 10 fruits or more. In most countries the cucumber fruit are hand-harvested, and the crop is picked several times. The difference in average yield between the USA and other countries would be greater if the total USA crop was harvested in a multiple-pick system.

Much of the difference in yield between areas can be attributed to proper soil fertility and the correct use of fertilizers. The tropical soils of many countries are highly weathered, acidic, and low in fertility. Fertilization practice consists mainly of using inadequate amounts of animal manure as a fertilizer source, and many times no fertilizers are used at all if the intended fields have lain fallow for a season between crops. Very little inorganic fertilizer or liming material are used in those countries.

In many countries, manure is recommended as a fertilizer source at rates of 30-40 mt/ha. In addition to this, top-dressings of 80 kg/ha of ammonium sulfate when the vines begin running and again when fruiting begins are recommended. This recommendation probably provides the crop all of the N and K that is needed but is possibly deficient in P. Based on findings that the manure of most ruminants contains less than 1 kg P_2O_5 Per mt, this fertility program would only provide 60-80 kg P_2O_5 per ha. Current recommendations in the USA specify the application of 60-200 kg P_2O_5 per ha, depending on the soil.

Proper fertility maintenance programs could benefit cucumber producers in less developed countries. Such programs should include liming most soils (especially tropical soils) to increase pH to the 6 to 7 level (mildly acid), provide Ca and Mg as nutrients, while maintaining adequate levels of N, P, K, and S, and detecting and correcting deficiencies of any micronutrients.

Fertilizer recommendations should be based on soil tests. Estimates of nutrient removal by cucumber are shown in Table 24.1. In all cases the lower estimate is based on a short-season crop in which the vines are destroyed during harvest, and the upper estimate is based on multiple pickings in which 10 or more fruit may be harvested from a single vine. The dichotomy between the two cultural systems means that in the destructively harvested crop the nutrients should be readily available to the seedlings to ensure rapid early growth.

TABLE 24.1
Yields of cucumber fruit and nutrient absorption estimates based on various fertility rates

Fertilizer Application (kg/ha)					Fruit Yield (mt/ha)		Total Nutrient Absorption (kg/ha)				
N	P	K	Ca	Mg			N	P	K	Ca	Mg
854	171	394	41	60	366	x	583	184	86	683	155
90	47	90	--	--	20	y	100	13	162	--	--
1408	296	439	188	109	238	x	408	92	550	236	57
67	293	558	--	--	20	y	--	--	--	--	--
134	112	134	--	--	30	z	100	40	134	--	--

x   Greenhouse, multiple harvest
y   Pickle, once over harvest
z   Pickle, multiple harvest, 8 pickings

Nitrogen = N   Phosphorus = P   Potassium = K
Calcium = Ca   Magnesium = Mg

In the multiple-picked crop the vines must be maintained for a long time. Some researchers have advocated tissue sampling for the assessment of crop nutrition status. However, studies of cucumbers indicate that the growing season of the once-over harvest crop is too short for nutrients added after planting (i.e., side dressing) to be of any benefit in increasing yield. Results from tissue analysis could probably be used to increase yield and quality in multiple-picked crops. The petiole of the sixth leaf has been set as the standard for tissue analysis of nutrient levels (Table 24.2).

Cucumbers have a remarkable capacity to set fruit, even under severe nutritional stress. However, the nutritional status of the plant largely affects fruit quality. Too little or too much fertilizer will result in a high percentage of "culls" or nonmarketable fruit. Plants grown at marginal fertility levels also have a slower growth rate of fruit than plants at optimum fertility levels. The developing fruit inhibit further fruit set to a greater degree when plants are grown at suboptimal fertility levels.

TABLE 24.2
Plant tissue analysis guide for sampling time, plant part and nutrient levels of cucumber plants

Time of Sampling	Plant Part Tested	Nutrient	Defi-cient	Suffi-cient
Early fruit set	Petiole of 6th leaf	NO_3-(ppm)	5000	9000
		PO_4-(ppm)	1500	2500
		K (%)	3	5
First harvest	Petiole of 4th fully expanded leaf from terminal bud	$NO_3 - N$(%)	- [1]	.82
		P(%)	-	9.70
		K(%)	-	2.30
		Ca(%)	-	.91
		Mg(%)	-	1.61
		Na(%)	-	.43
		Cu(ppm)[2]	-	48
		Fe(ppm)	-	149
		Mn(ppm)	-	82
		Zn(ppm)	-	71

[1]Deficient levels not established for Minor Nutrient Elements, copper, iron, manganese, zinc, and sodium.
[2]ppm = parts per million.

EFFECTS OF MINERAL NUTRIENTS

Nitrogen (N). Vegetative vigor of cucumber plants and good fruit development are largely dependent on an adequate N supply, which may be provided through organic (animal manures) or inorganic (fertilizer) sources. The N should be available to seedlings immediately after germination. Caution should be exercised in the placement of inorganic N salts, which are highly soluble, as cucumbers are injured by high salt concentrations. Inorganic N fertilizers may either be banded on one side of the bed below the seeds, or broadcast and thoroughly incorporated in the upper soil layers. Growers should also take care when using fresh manure, urea, anhydrous ammonia, or other ammonia sources because ammonia is highly toxic to cucumbers. Under good growing conditions, however, the ammonia released from these materials is rapidly converted to nitrate by soil bacteria. In cold, wet, or acidic soils these bacteria are inactive and toxic levels of ammonia may accumulate. Very small amounts of ammonia in

the soil decrease the uptake of K and P by cucumbers, resulting in slow root and shoot growth.

Nitrogen-deficient plants have small leaves and fibrous, spindly stems. The foliage lightens in color from
green to yellow beginning with the older leaves. The
fruit are pale and do not fill out at the blossom end.
Over application of N also causes an increase in misshap-
en, unmarketable fruit. Highly soluble salts will lead
to stunted, dark green plants which resemble P deficient
plants. Ammonia toxicity produces stunted plants that
may have some of the chlorophyll destroyed, leaving white
areas on the leaves.

Applications of 85-170 kg N/ha are recommended and
should be based on soil tests. Any N source may be used
if caution is exercised regarding ammonia toxicity problems on cold, wet, or acidic soils. The N should be applied immediately before planting, or at planting time,
to ensure maximum availability to the young seedlings and
to minimize leaching losses. Nitrogen applications during crop growth are of little economic benefit in the
single picking, destructively harvested crop, but do
benefit the multiple-picked crop.

Soils used to grow cucumbers should contain some
organic matter. Organic matter contains N, P, and K that
is made available slowly to the plant through microbial
degradation. These nutrients are held against leaching
and soil fixation. In addition to this, research has
shown that the humic acids that are associated with soil
organic matter hasten early flowering and increase the
number of female flowers produced. Green manure crops,
crop residues, and manure all contribute to the soil
organic matter content.

Phosphorus (P). Soil P availability is affected by
soil pH. At a pH of 6.7, available P is at a maximum.
As the pH deviates from this level, insoluble or only
slightly soluble phosphate compounds are formed. Since
the pH of most soils is not at this optimum value, P fertilizers should be banded close to the seeds at planting
time. This ensures maximum P availability for early crop
growth.

Phosphorus applications should be based on soil
test. Some soils have high levels of P, and therefore no
response of cucumbers to P fertilizers can be expected.
Rates of 67-225 kg P_2O_5/ha are recommended for most
soils. Superphosphate fertilizers are good sources of P.

Phosphorus deficiency produces stunted plants with
small, dark green leaves. Some leaves may have a purple
tinge on the underside. Roots will be poorly developed,
and fruit setting and ripening delayed.

Potassium (K). Potassium enhances disease resistance in cucumbers. Light, sandy soils and organic soils
(peats) tend to be low in K. Also, old highly leached

soils may be K deficient. Even on soil where there is an abundance of K bearing minerals, K may be deficient due to fixation in the soil or because of K depletion through cropping. In soils with sufficient K and if the K is in available form, cucumbers will absorb more than the required amount for optimum growth. In greenhouse studies, K concentrations in the petiole of cucumber leaves have been reported to be as high as 15% of the petiole dry weight. This high level seems to have no detrimental effect on the plant.

Potassium deficiency does not immediately result in visible symptoms. The early plant response will be a reduction in growth rate. For this reason it is important to have the soil tested and to maintain adequate K levels in the soil solution through a good fertility program. Recommended rates of K fertilization range from 67-270 kg K_2O/ha. Potassium chloride is a good source of K, as are potassium nitrate and potassium sulfate. Symptoms of severe K deficiency are plants with low vigor and reduced disease resistance. Leaves may be yellowed or bronzed, at first between the veins, and then over the whole surface. The leaf margins dry out and die. These symptoms occur first in the older leaves. Fruit may be distinctly tapered at the stem end and enlarged at the blossom end.

Sulfur (S). The main reservoir of S in most humid region soils is organic matter. This organic S is converted by microbial activity to SO_4^{-2}, a form that is available to growing plants. In the soils of arid regions high amounts of sulfate salts of Ca, Mg, and Na may exist. If the total salt concentration in these soils is above about 1600 ppm (parts per million), yield decreases are expected because of high soluble salt levels, but not toxicity.

Sulfur deficiency in the field resembles N deficiency in its early stages, therefore a tissue analysis may be helpful in determining which element is needed. Reduced growth rate, weak stems, and brittle tissues characterize both deficiencies, however, chlorotic symptoms (reduced green color) occur first in the youngest leaves as a result of S deficiency.

The atmosphere is a major contributor to the S content of many soils. This is due to air-borne SO_2 being dissolved in raindrops. Industrial pollutants are the major source of atmospheric SO_2. Soils near the ocean have also been found to be high in S content. In areas that are far from industrial or maritime regions, S must be added to deficient soils. Gypsum, superphosphate, ammonium sulfate, potassium sulfate, and sulfate of potash magnesia are all good sources of S. In any soil with a high pH, elemental S can be used as an amendment to lower the pH. Microbial activity will also lower the soil pH and provide available S for the crop. Sulfur application rates in deficient soils range from 12-56 kg S/ha.

Calcium (Ca). Calcium is a major constituent of the parent minerals of many soils and is therefore not always limiting. However, when the pH is low (strongly acid), or when soils are sandy and highly leached, Ca deficiency may occur. Calcium deficiency may also be induced in soils that have adequate Ca through addition of ammonium fertilizers, soil water stress, or high soil salt concentrations, since these factors decrease Ca translocation in the plant stem xylem. The xylem is the only tissue through which Ca moves. Because Ca is excluded almost entirely from the phloem (food transport tissue), it is very immobile within the plant. Also, cucumbers are native to high Ca soils. Plants adapted to calcareous soils have developed a mechanism for depositing excess Ca in the lower leaves. When this mechanism operates in low Ca soils, it leads to Ca deficiency in the young plant parts. Cucumbers are susceptible to Ca deficiency on low Ca soils.

Symptoms of Ca deficiency in cucumbers are pale young leaves and a reduced growth rate. Young leaves may be cupped downward with a necrotic (dead) leaf margin. The blossom end of the fruit may be poorly developed. This phenomenon also is typical of Ca deficiency in the watermelon, causing blossom end rot. Ground limestone will correct Ca deficiency in cucumbers. Superphosphate and gypsum also are satisfactory sources of Ca. Any additions of Ca-containing amendments should be based on soil test.

Magnesium (Mg). Magnesium deficiency is relatively common on acidic, highly leached sandy soils. Other cations such as NH_4^+, K^+, Ca^{+2}, or Na^+ compete with Mg^{+2} uptake by cucumbers. Amendments containing Mg are needed where soil deficiency occurs.

Cucumbers are susceptible to Mg deficiency. Symptoms appear first on the older leaves which show an interveinal chlorosis, and the chlorotic areas eventually die. Symptoms usually occur late in the growing season. Magnesium sulfate at the rate of 12-17 kg/260 liters of water may be used as a foliar spray to correct deficiency symptoms in greenhouse and long season (multiple-pick) cucumber crops. Foliar sprays would probably be of little value with the once-over (destructive) harvest cucumbers due to yield loss already suffered by the crop when visual symptoms are seen. If soil tests indicate Mg deficiency, apply dolomitic limestone at a rate of 330 kg/ha or magnesium sulfate at a rate of 220 kg/ha. A foliar spray containing 12 kg of $MgSO_4$ per 650 liters of water per ha may be used. In addition to this, avoiding over-fertilization with antagonistic cations (NH_4^+, K^+, Ca^{+2}, Na^+) will reduce Mg deficiency.

Magnesium stress has been shown to increase the severity of powdery mildew disease on cucumbers.

ESSENTIAL MINOR ELEMENTS

Application rates of essential minor elements for cucumber are given in Table 24.3.

Iron (Fe). While Fe is present in almost all soils the availability to plants of inorganic Fe in most agricultural soils is governed primarily by pH values. At pH levels of 6.5-8.0 (mildly acid to mildly alkaline) the concentration of soluble Fe is generally low. Acid soils are therefore relatively higher in soluble inorganic Fe than calcareous soils. Iron in a readily available form occurs naturally as organic chelates in soils; such organic complexes may be a significant source of Fe for plant nutrition in some soils.

Iron deficiency symptoms occur first on the youngest leaves. Characteristic signs of Fe deficiency are distinct interveinal chlorosis, with the chlorotic areas eventually becoming white. The veins will become chlorotic if the deficiency is not corrected. Necrosis is usually absent. Fe uptake and utilization by the cucumber may be depressed not only by high pH (alkalinity), but also by high phosphate, K^+, Mg^{+2}, and Ca^{+2} concentrations in the soil solution. Heavy metal ions such as Mn, Cu, nickel, cobalt, and Zn also will compete with Fe absorption by plants.

Iron deficiency can be alleviated by several methods. In alkaline soils, reducing soil pH (making the soil more acid) will often increase the concentration of available Fe. For rapid correction of Fe chlorosis, foliar applications of Fe chelates (e.g., Fe EDTA, Fe citric acid, Fe EDDHA) are an effective remedy. Iron chelates are also excellent soil amendments because the Fe contained in them cannot be rendered insoluble in the soil. Applications of organic fertilizers and manures may also aid in alleviating Fe deficiency.

Manganese (Mn). Manganese is present to some degree in the parent material of most soils. Its availability is dependent on soil acidity, therefore deficiencies are most often encountered on calcareous (high lime) soils. Toxicity symptoms may be seen on some very acid soils.

Cucumbers are moderately susceptible to Mn deficiency. Manganese is relatively immobile within the plant, thus deficiency symptoms are seen first in the youngest leaves. A major symptom is chlorotic mottling between the veins. This chlorosis is less marked near the veins. These areas may eventually become necrotic (dead). Small leaves and weak, slender stems are also characteristic. Manganese toxicity is characterized by brown spots on older leaves.

Most soils contain adequate amounts of Mn for plant growth. On calcareous soils Mn deficiency may be corrected by lowering the pH by the addition of elemental S.

TABLE 24.3
Soil and foliar application of trace nutrients for cucumber

Nutrient	Material and Approximate Analysis	Soil Application (kg/ha)	Foliar Application (kg/378 liters of water)
Boron (B)	Borax 10.6% B	11–28 (mineral) 28–56 (muck)	1.0–2.0
	Solubor 20.5% B	6–13 (mineral) 17–28 (muck)	0.4–0.7
	Sodium pentaborate 18.1% B	7–18 (mineral) 18–36 (muck)	0.4–1.4
	Sodium tetraborate pentahydrate 13.7% B	9–22 (mineral) 22–44 (muck)	0.4–1.8
Copper (Cu)	Copper sulfate 25.5% Cu	28–56 (mineral) 112–336 (muck)	1.0–2.0
	Copper oxide 79.6% Cu	9–18 (mineral) 34–102 (muck)	
Iron (Fe)	Ferrous sulfate 20% Fe	11–22	1.0–1.4
	Chelated iron 9–12% Fe	20–40	0.3–0.4
Manganese (Mn)	Manganese(ous) sulfate 24.6% Mn	22–33 (mineral)	1.0–1.8
Molybdenum (Mo)	Ammonium molybdate 48.9% Mo	2 in bands, 4 broadcast	–
Zinc (Zn)	Sodium molybdate 39.7% Mo	1/2–1	0.1–0.2
	Zinc sulfate 22.7% Zn	11–44	1.0–1.8
	Chelated zinc 14% Zn	17–44	0.3–0.4

Taken from Lorenz, O.A. and D.N. Maynard. Knotts Handbook for Vegetable Growers. Wiley-Interscience, New York, 1980.

Some highly leached, sandy soils and calcareous organic soils have inherently low Mn levels. It is necessary to apply Mn to these soils. Manganese sulfate placed in a band near the plants is the most efficient way to alleviate Mn deficiency. Foliar application of Mn will also correct deficiency symptoms. Manganese sulfate is the most desirable Mn carrier for foliar sprays.

Zinc (Zn). Little research has been done on the tolerance of cucumbers to low levels of Zn. Deficiency symptoms occur on the youngest leaves as interveinal chlorotic mottling, shortened internodes, and small leaves. Chlorotic areas may become necrotic as the deficiency progresses.

In most cases the amount of Zn contained in a soil exceeds the crop requirement. However, Zn availability is dependent primarily on soil pH. Zinc deficiency is observed most often on calcareous or highly limed soils. Deficiency symptoms may also occur on soils with inherently low Zn levels such as acidic, highly leached, sandy soils and organic soils. High P levels in the soil solution may contribute to Zn deficiency in plants due to physiological Zn/P interactions. Zinc sulfate is the most commonly used soil amendment or ingredient of foliar spray. Acidifying calcareous soils with elemental S may correct Zn deficiency on many soils.

Copper (Cu). Cucumbers cannot tolerate Cu deficiency. Cu is required by the plant in very small quantities, and therefore Cu deficiency is relatively rare on most soils. Cu is bound very tightly (fixed) by organic matter in the soil. Because of this Cu deficiency symptoms are seen most often in humus-rich soils or organic soils. Deficiencies may also be seen on highly leached, sandy soils.

Deficiency symptoms occur first on the youngest leaves which may wilt first, become chlorotic, and then become necrotic. On peat soils, applications of high rates of N or P fertilizers have provoked Cu deficiency symptoms. Zinc fertilizers may also aggravate Cu deficiency on soils with marginal Cu levels.

Copper sulfate is the most frequently used amendment for Cu application and may be applied as a soil amendment or a foliar spray. There are, however, problems associated with its use such as toxicity and rapid immobilization of Cu on organic soils. Copper chelates are most desirable for supplying Cu in available form to plants.

Molybdenum (Mo). Molybdenum is present in the cucumber in lower concentrations than any other essential nutrient element. Deficiencies are usually seen on acidic, highly leached, sandy soils.

A deficiency of Mo results in early symptoms that resemble N deficiency. As the deficiency progresses, the older leaves develop interveinal mottling and puffy, necrotic areas. The leaves may cup, and the younger leaves become twisted and distorted. Deficiency can be corrected on many acid soils by raising the pH with lime, thereby making Mo more available. Sodium or ammonium molybdate may be used as soil amendments on soils with inherently low Mo contents. Foliar applications of ammonium molybdate can also be used.

Boron (B). Boron occurs in the soil as boric acid or borate. Although cucumbers are tolerant of low soil B, deficiencies are most likely to show up on highly leached soils or on alkaline organic soils. Characteristic deficiency symptoms are shortened internodes in young tissue, brittle leaves and stems, and distorted or necrotic young leaves followed by death of the terminal bud. Cucumber fruit may be cracked, rough, or spotted. Flowering and fruit set will be greatly reduced. The most common way to correct B deficiency is soil application of borax or foliar application of boric acid. Care is needed when applying B due to the possibility of toxicity if excess B is applied.

Chlorine (Cl). Chlorine deficiency is very rare in the field because of the presence of Cl in the atmosphere or in rainwater, which is generally more than enough to meet the requirements of cucumbers. Chlorine is also present in adequate concentrations in most soils. Deficiency symptoms appear as chlorosis in the younger leaves and an overall wilting of the plant. Potassium chloride is a good source of Cl if a soil application is needed.

SUMMARY

Proper fertility is essential for cucumber production. The crop requires large inputs of money and labor, as compared to agronomic crops such as maize, wheat, or sorghum. If a high quality crop is produced, large profits can be realized by the producer. Fertility is a major factor determining quality and yield. Good fruit color, shape, and texture all depend on fertility. Cucumber may produce acceptable yields but poor quality fruit under nutrient stress. Therefore, fertility is probably as important in determining quality as yield. Levels of fertilizer are critical as too high or too low rates can lower yields or quality. Excess N is a good example; N may reduce total yield, fruit quality and promote development of diseases. Since fertilizers compose a small percentage (less than 10%) of the total cost of production, cucumber growers should strive to maintain optimum fertility levels. Therefore, based on many observations of cucumber production trials and commercial operations, multiple-harvest crops should be provided 112 to 134 kg N/ha, 90 to 112 kg P/ha, 134 to 168 kg K/ha, and sufficient levels of other nutrients to produce a crop of 30 mt, all based on soil analysis. Once-over-harvest crops should be provided 90 to 112 kg N/ha, 45 to 67 kg P/ha, 112 to 123 kg K/ha, and sufficient levels of other nutrients based on soil analysis. Greenhouse production, where cucumbers are grown over a 4-5 month season and yields of up to 400 mt of fruit are expected, might require up to 680 kg N/ha, 200 kg P/ha, 900 kg K/ha, applied over a long period of time. Tissue analysis as indicated in Table 24.2 is recommended to maintain sufficient levels.

25
EGGPLANTS

James W. Paterson

Eggplants belong to the Solanaceae family which comprises about 75 genera with over 2,000 species. Most of the plants in this family are herbaceous, however, some are shrubs or small trees. Eggplant or aubergine, <u>Solanum melongena</u>, is a native plant of India and has been grown in the USA for many years. The major states producing eggplants are Florida, New Jersey and California. The fruit is usually baked, sauteed or fried before eating. In 1976, the average USA per capita consumption of fresh eggplants was 0.3 kg.

A raw eggplant which is about 92.5% water contains 24-25 calories, 1.1-1.2 g protein, 0.2 g fat, 5.5-5.6 g total carbohydrate of which 0.9 is crude fiber, 0.5-0.6 g mineral ash, 12-15 mg calcium, 26-37 mg phosphorus, 0.7 mg iron, 2 mg sodium, 214 mg potassium, 10 I. U. Vitamin A, 0.05 mg thiamine, 0.05 mg riboflavin, 0.6 mg niacin and 5 mg ascorbic acid in 100 g of edible portion.

ECOLOGICAL AND PRODUCTION REQUIREMENTS

Eggplants are well adapted to the tropics and the warm regions of the world. This warm season crop is very sensitive to frost damage. When temperatures become very high, the blossoms tend to abort. The optimum temperatures for seed germination are 21 to 23.9°C (70 to 75°F) and the best temperatures for growing the crop are 21 to 29.4°C (70 to 85°F). Temperatures below 18.3° C (65°F) result in poor growth and poor fruit set.

In many warm areas of the world, eggplants are sown directly into the field; however, other areas seed in the greenhouse and then field transplant after the last frost-free date. Seedlings can be produced in flats and then transferred to 5 cm or larger pots or containers

---

[1]Extension Specialist in Soils and Crops, Rutgers-The State University of New Jersey, Rutgers Research and Development Center, Bridgeton, NJ 08302, USA.

anytime after the first true leaves appear, or seeds may
be sown directly into the pots and then thinned back to a
single plant per pot.

A popular growing medium for producing eggplant
transplants in New Jersey is half peat and half vermicu-
lite supplemented with slow release fertilizers, limestone, superphosphate, potash and trace elements. If the
grower does not wish to use a slow release fertilizer, he
may use a farm-type, complete (NPK) fertilizer in the
basic growing media and then drench the plants after they
emerge with a soluble complete fertilizer, such as a 20-
20-20, at the rate of 1 to 1.4 kg per 370 liters of water
on at least a weekly basis until the plants are ready for
field transplanting. Using the amended peat-vermiculite
soilless mixes enables the grower to produce an excellent
plant for field planting in 8 to 10 weeks from seed.

Eggplant has a strong taproot with an extensive
branch root system which does not spread widely and develops very actively in well-drained loam and sandy loam
soils that are fairly deep. For good production on the
lighter, sandier soils, frequent irrigations are necessary if rainfall is short.

PRODUCTION AND YIELDS

Average eggplant yields for New Jersey are approximately 18,500 kg/ha (500 bushels per acre). However, in
experimental plots more than 90,000 kg/ha of quality eggplants have been produced, while some commercial growers
are producing as much as 74,000 kg/ha using the latest
cultural techniques. Eggplant production in various
areas of the world is shown in Table 25.1.

SOIL ANALYSIS

Considerable amounts of eggplant are grown in the
Coastal Plain soils along the east coast of the USA.
These soils, in general, are low in organic matter, quite
deficient in N, high in available P, medium to low in
available soil K and naturally acidic. These soils are
very responsive to soil amendments.

The most widely used soil extractant for these
coastal soils is a weak, double acid solution developed
by North Carolina State University. The values presently
employed and their relative soil test level designations
are given in Table 25.2.

TABLE 25.1
Eggplant production in 1980

Area	Area Harvested (1,000 ha)	Yield (kg/ha)
Africa	31	14,498
Egypt	15*	21,215
Ghana	10*	2,786
North and Central America	4	21,060
Mexico	1*	20,143
USA	2*	21,127
South America	1	14,300
Asia	267	12,816
China	155*	9,865
Japan	22*	30,648
Philippines	17*	5,882
Thailand	10*	5,490
Turkey	35	17,286
Europe	24	23,926
Italy	13	25,253

*FAO estimates.

TABLE 25.2
Interpretation of soil test results by the North Carolina
Double Acid Extraction Method

Phosphorus Lb/A	Potassium Lb/A	Magnesium Lb/A	Calcium* Lb/A	Relative Level in Soil
0-13	0-29	0-35	0-400	Very low
14-27	30-70	36-70	401-800	Low
28-45	71-134	71-125	801-1200	Medium
46-89	135-267	126-265	1201-1600	High
90+	268+	266+	1601+	Very high

*Values apply only to sandy loam soils.

+ Lb/A = pounds per acre, multiplied by a factor of 1.12 = kg/ha.

PLANT TISSUE ANALYSIS

Liming and fertilizing can influence the concentration of the essential plant nutrients in vegetables. Table 25.3 shows the approximate concentrations of essential plant nutrients that may be found in the most recently matured leaf at full bloom of unmulched and mulched eggplants grown on well limed and fertilized soils in southern New Jersey.

TABLE 25.3
Plant nutrient concentration in eggplants*

Material	Unmulched	Mulched
Nitrogen (N), %	4.8-5.0	5.0-5.3
Phosphorus (P), %	0.3-0.6	0.5-0.7
Potassium (K), %	4.5-4.7	4.8-5.1
Calcium (Ca), %	1.3-1.8	1.0-1.6
Magnesium (Mg), %	0.6-0.7	0.5-0.6
Boron (B), ppm	40-50	70-80
Copper (Cu), ppm	10-12	12-16
Iron (Fe), ppm	100-400	125-450
Manganese (Mn), ppm	40-100	35-200

*On dry weight basis.

DEFICIENCY SYMPTOMS

Visual observation of the growing plant can be helpful in determining whether the plant is receiving adequate nutrients or not. Some general plant nutrient deficiency symptoms which may be observed in various vegetables follow, some of these may apply to eggplant.

Nitrogen (N). Light green to a yellowing of older, lower leaves first occurs, progressing to the newer leaves as deficiency continues. Firing of the older, lower leaves. Plants usually stunted with hard, fibrous, slender stems.

Phosphorus (P). Leaves are usually small and darker green than normal. Some leaves turn reddish purple on the underside, beginning on the older leaves. Thin stems and delayed maturity.

Potassium (K). Older leaves affected first. Leaf tips and margins turn yellow and then become scorched, continuing inward to the leaf center. Leaf margins sometimes cup downward. Interveinal leaf necrosis and sometimes patches of dead tissue. Restricted growth.

Calcium (Ca). Necrosis at tip and margins of newer, immature leaves nearest the terminal growth with a distorted appearance. Stems become thick and fibrous with retarded vegetative growth. New growth often lacks turgidity. Terminal buds will eventually die.

Magnesium (Mg). Interveinal chlorosis of older, larger leaves, while veins remain green. Chlorosis can develop into necrotic areas with time and die. Leaf margins sometimes curl upward. Some may develop purple tinting on older leaves rather than chlorosis. Smaller fruit on eggplants.

Sulfur (S). Newer leaves become generally light green to yellowish. Sometimes the leaf veins appear lighter in color than the adjacent interveinal tissue, with leaf tips cupping downward. Stems may be hard, fibrous and spindly.

Boron (B). Leaves become somewhat chlorotic, as well as small, thick, brittle and misshapened. Sometimes the base of the new leaves will be wrinkled or deformed. Fruit may be rough and cracked. Internodes may be short with a rosetting appearance of the terminal growth. The terminal bud eventually dies.

Copper (Cu). Retarded growth with some leaf necrosis and a bleached appearance. Sometimes leaf veins appear lighter in color than the adjoining interveinal tissue. New leaves lack turgidity. Stems are small and leaves sometimes curl inward.

Iron (Fe). Main, larger leaf veins remain a pronounced green, while the interveinal portion turns yellow on the newer leaves. Chlorotic tissue usually does not die and young leaves may be small but not deformed.

Manganese (Mn). Upper, newer leaves or middle leaves affected first. Interveinal chlorosis with main and smaller veins remaining green. With time the symptoms progress to older tissue. Chlorotic spots become necrotic and brown with time. Plants may be stunted.

Molybdenum (Mo). Interveinal mottling on older leaves, while veins remain light green. Mottling also occurs on the newer leaves as they mature. As deficiency continues the leaf tip and margin become necrotic with the leaf curling inward.

Zinc (Zn). Young leaves are abnormally small and narrow with interveinal yellow to white coloration. Necrotic spotting of older leaves. If deficiency is severe, defoliation may occur.

MANAGING THE NUTRITION OF EGGPLANT

There are some excellent varieties of eggplants being developed. The newer varieties, as do some older varieties, appear to respond to good fertilization practices; however, the newer varieties produce a much higher percentage of quality fruit than the older selections.

Animal manures, as well as green manures and crop residues, can be very beneficial for eggplant not only because of their nutritive value but also for the improved physical condition of the soil. Applying 20 mt of well handled cattle manure or turning under a good stand of alfalfa can supply the following crop with as much as 110 kg N/ha. Legume crop residues or animal manures can substitute for commercial fertilizers.

There are many important management decisions involved in profitable eggplant production. One of the most important of these is the proper selection and use of lime and fertilizer materials. These two important inputs often account for 50% or more of the crop production costs.

The most economical means of determining lime and fertilizer needs for eggplants is to test the soil. The desired soil pH for eggplant is between pH 6 and 7. If soil acidity is too high (pH below 6), apply ground or pulverized agricultural limestone. If the soil also tests too low in magnesium, use a dolomitic-type limestone which carries high levels of magnesium, as well as calcium. If the soil test pH is too high, elemental sulfur or other sulfur compounds can be used to reduce pH.

The plant nutrient recommendations for eggplants in Table 25.4 were developed on Coastal Plain soils of the east coast of the USA. When the plant nutrients are applied according to the soil test recommendations, a slow but steady increase of P and K will result in soils testing low in these elements. Soils testing medium in P and K will remain essentially constant and those testing high and very high will slowly decrease to a medium level.

Eggplants should be fertilized according to soil test levels as shown in Table 25.4. Also shown are the timing and method of fertilizer applications.

TABLE 25.4
Eggplant plant nutrient recommendations

Method and Timing of Application	kg N/ha	Recommended Nutrients Based on Soil Tests			
		Low	Medium	High	Very High
	Nitrogen	kg P_2O_5 and K_2O per ha			
Total recommended	140–168	280	168	112	56
Broadcast and disk in	56–112	224	112	56	0
Sidedress 3–4 weeks after planting	28–56	56	56	56	56
Sidedress 6–8 weeks after planting	28–56	0	0	0	0

At planting some growers may apply a starter fertilizer solution in the root area of the transplant. The rate used is usually 2 kg per 370 liters of water of a fertilizer high in P such as 1-3-2 or 1-4-2 ($N-P_2O_5-K_2O$) ratio material; about 150-200 ml of solution is applied to each plant. The high P encourages root branching and early establishment of the plants.

Mulching eggplants encourages early fruit production, and applying N under mulches increases total yields. If plastic mulching is employed, all the N, P and K should be applied before the plastic is laid. It is inefficient to sidedress plastic-mulched crops. If low volume trickle irrigation is used under the mulch, the plant nutrient can be delivered through the irrigation system according to application of micronutrients through this system, provided that all plant nutrients applied are soluble in water.

On unmulched or bare soils, at least half of the total N and most of the P and K should be applied preplant and worked in deeply, particularly on soils which have tested low. Another 25 to 30 percent of the N should be sidedressed about 3 to 4 weeks after transplanting and the remainder of the N applied 6 to 8 weeks after transplanting, particularly on lighter, sandier soils. On the heavier loam type soils, most of the N can be applied at planting with little or no sidedressing.

Many soils are deficient in available soil boron. In New Jersey it has been shown that boron applications contributed to a 15% increase in eggplant production on soils testing low in that element. Depending on the soil test level, 0.4 to 0.8 kg/ha of boron are recommended.

Copper is not usually a problem in mineral soils, however, the element may become limiting in organic soils, particularly newly drained soils. To correct a deficiency of copper, apply 17 to 34 kg/ha of the element. Apply lower rates if a chelate is used.

Iron deficiencies occur most often in coarse textured soils high in pH and low in organic matter. An iron deficiency may be temporarily rectified by foliar application of iron as a chelate or by reducing the soil pH.

Manganese deficiency occurs mainly in coarse textured soils which are heavily limed. More manganese will become available to the plant by lowering the soil pH or applying 12 to 24 kg/ha of the element to the soil. Banding near the roots is a more efficient way to apply manganese but the rate should be reduced.

Molybdenum deficiencies also occur most often in the coarser textured soils, particularly in acid soil with pH below 5.5. The easiest and most economical way to correct this deficiency is by liming the soil. If a deficiency occurs on an adequately limed soil, 0.2 to 0.4 kg/ha molybdenum may be applied.

Zinc deficiencies have been noted mainly in highly limed, coarse textured soils which are cold and wet. Deficiencies may be reduced by applying 5 to 10 kg/ha Zn. Reduced rates should be used if Zn is to be banded or applied as a chelate.

Eggplants in poor nutrition are more susceptible to pest damage. Some common insects which attack eggplants are the Colorado potato beetle, aphids, flea beetles and mites. Common diseases associated with eggplants are Verticillium and bacterial wilts and fruit rots. The type of weed problems occurring would be peculiar to the area in which the eggplant are grown. High populations can reduce crop yields through competition for space and nutrition.

SELECTED REFERENCES

KITCHEN, H.B. (ed.) 1948. Diagnostic Techniques for Soils and Crops. The American Potash Institute, Washington. 259 pp.

PATERSON, J.W. 1982. Proceedings Ninth International Plant Nutrition Colloquium, Warwick University, England. p. 461-466.

POLLACK, B.L. (ed.) 1982. Commercial Vegetable Production Recommendations. Rutgers-The State University of New Jersey. Ext. Bull. 406-I. p. 3-11 and 30-31.

SEELING, R.A. and C. Magoon (ed.) 1978. Fruit and Vegetable Facts and Pointers. 4th Edition. United Fresh Fruit and Vegetable Association, Virginia. p. 1-8.

26
MELONS

Stanley F. Gorski

Melons belong to the Cucurbitaceae (or gourd) family which contains many members. This family includes 90 genera and approximately 700 species. Included in the family are watermelons, muskmelons, honey dews, cucumbers, pumpkins, and squash.

Watermelon (Citrullus lanatus) is a trailing annual which originated in Africa. Cultivation of watermelons dates back 4,000 years to the Egyptians. The fleshy part of the fruit is most often eaten raw, primarily as a dessert. In other parts of the world the watermelon is used in different ways. In Africa, Egypt and Iraq the flesh is used as a staple food and animal feed. The flesh is over 92% water and therefore is used as a source of water in dry areas. In Russia, beer is made from the juice of melon, or the juice can be boiled down into a thick molasses type syrup for its sugar. The rind (ovary wall) is often used for pickling. In some areas the seeds are eaten. They can either be roasted and salted or eaten raw. The seeds can also be ground and used to make bread. Watermelon seeds are considered to be a good dietary supplement because of their high protein content (30-40%). A valuable cooking oil can also be extracted from the seeds. Table 26.1 presents information on the nutritional value of melons.

GROWTH REQUIREMENTS

Melons are grown in both tropical and temperate regions of the world. They are considered to be warm season crops, are quite sensitive to frost, and should not be planted until all danger of frost is past, unless some type of plant protection or covers are used. Melons grow best in areas with high temperatures (Table 26.2) and a dry climate since high humidity favors foliar diseases.

---

[1]Assistant Professor, The Ohio State University and the Ohio Agricultural Research and Development Center. Columbus, Ohio 43210, USA.

TABLE 26.1
Food composition of 100 grams of edible melon fruit

	Musk-melons	Honey Dew	Water-melon
Calories	20.0	32.0	28.0
Water (g)	94.0	90.5	92.1
Protein (g)	0.6	0.5	0.5
Total Carbohydrates (g)	4.6	8.5	6.9
Crude Fiber (g)	0.6	0.4	0.6
Vitamins			
A (IU)	3,420.0	40.0	590.0
B_1(mg)	0.05	0.05	0.05
C (mg)	33.0	23.0	6.0
Minerals			
Calcium (mg)	17	17	7
Phosphorus (mg)	16	16	12
Potassium (mg)	230	--	110

TABLE 26.2
Temperature range for germination and growth of melons

	Germination Temperature (°C)			Growth Temperature (°C)		
	Min.	Opt.	Max	Min.	Opt.	Max.
Muskmelons	15	24-35	38	15	18-24	32
Watermelons	15	21-35	40	18	21-29	35

The root system of melons is quite extensive and deep. For this reason vine crops do not respond as well to irrigation as other crops. Melons grow best on a deep, well-drained soil. If early maturity is desired, a light textured sandy loam or silty loam soil is preferred since such soils warm up earlier in the season. However, heavier soils with a better water holding capacity generally give the highest yields. In areas where rainfall is limiting these heavier soils are recommended.

YIELDS

Varying amounts of melons are grown in numerous countries around the world (Table 26.3).
Yields vary from country to country and within a single country. Yield is highly dependent on numerous factors including climate, soil, moisture, nutrition, pest control, variety and many more. This makes it difficult to determine what is a "good" or "acceptable" yield. Such values can only be established for a parti-

cular region. In the USA a "good" yield for watermelons and muskmelons is considered to be 22,000 kg/ha, for honey dew melons 27,500 kg/ha, and 16,500 kg/ha for persian melons.

TABLE 26.3
Melon production for 1979 (FAO)

| | Watermelons | | Muskmelons and Other Melons | |
	Area Harvested (1000 ha)	Yield (kg/ha)	Area Harvested (1000 ha)	Yield (kg/ha)
Africa	121	17,186	34	16,178
Egypt	53*	25,592	13*	20,643
USA	88*	12,393	46*	15,804
Central America	122	11,582	77	13,235
South America	131	7,897	23	12,074
Asia	819	16,084	228	13,069
China	222*	18,333	83*	17,212
Europe	158	18,667	119	12,987
E. Europe and USSR	639	5,873	14	10,000
World	1,965	12,260	481	13,245

*Indicates FAO estimate.

NUTRIENT REQUIREMENTS

A muskmelon crop removes considerable quantities of N, K and Ca from the soil. A 1 ha field will produce approximately 15,000 kg of fruit. At maturity this fruit will have removed 56 kg N, 17 kg P_2O_5, 100 kg K_2O, 70 kg Ca and 15 kg Mg. The vines will also remove considerable quantities of these elements, especially calcium, which would normally be returned to the soil after the crop is harvested. A general nutritional guideline is that a melon crop should receive 67-110 kg N plus 110 kg each of P and K. Recommendations of this type will be discussed later.

Plant Tissue Analysis

Leaf petioles reflect with greater sensitivity the N and P content of the plant and are therefore preferred over the leaf blades for analysis. The petioles are higher in NO_3-N, PO_4-P and total K than the leaf blades but are lower in Ca and Mg. Early season samples are preferred over mid or late season plant samples because the N, P and K content increases with age. Table 26.4 presents some values that help to interpret tissue analysis results.

TABLE 26.4
Interpretation of foliar analysis tests

Crop Growth Stage[a]	Nutrient		Deficient	Sufficient
Muskmelon				
Early growth	NO_3-N	(ppm)	8,000	12,000
(short runners)	PO_4-P	(ppm)	2,000	4,000
	K	(%)	4	6
Early fruit set	NO_3-N	(ppm)	5,000	9,000
	PO_4-P	(ppm)	1,500	2,500
	K	(%)	3	5
First mature fruit	NO_3-N	(ppm)	2,000	4,000
	PO_4-P	(ppm)	1,000	2,000
	K	(%)	2	4
Watermelon				
Early fruit set	NO_3-N	(ppm)	5,000	9,000
	PO_4-P	(ppm)	1,500	2,500
	K	(%)	3	5

[a]Tissue samples used were petioles of the 6th leaf from the growing tip.

Visual Symptoms

An alternative to foliar analysis is recognizing nutrient deficiency symptoms exhibited by the crop. This method will not allow us to see a deficiency problem developing as early as a foliar analysis but is still a very valuable diagnostic tool. Careful inspection of the crop is necessary to distinguish nutrient deficiencies from other problems such as disease and herbicide injury. The following descriptions were developed to aid in the identification of nutrient deficiencies in melons from foliar symptoms.

Nitrogen (N). Leaves are yellow green to yellow in more severe cases. Stems are slender, hard and fibrous. Roots may be stunted, brown and dead. Fruit may be poorly netted and may be sunburned due to lack of foliar cover.

Phosphorus (P). Leaf color will change from a dark green color to a much duller green. The stems may also be slender.

Potassium (K). Leaf margins of the older leaves become yellow. In more severe cases browning and death of this tissue occur. Stems will split longitudinally. The fruit will also split at the blossom end and maturity will be delayed.

Calcium (Ca). Interveinal chlorosis and marginal leaf necrosis will occur along with a downward curling of the leaf margins. Leaves may have a mottled appearance. The internodes will be shortened, giving the plant a stunted appearance.

Magnesium (Mg). Older leaves are affected and develop small, light tan areas that become necrotic. These areas enlarge and coalesce.

Sulfur (S). Leaves become yellow in color and growth is slowed.

Boron (B). Terminal buds fail to develop.

Manganese (Mn). Interveinal leaf tissue turns a yellowish white color. The leaves will generally be small. Stems will be slender and weak. Flower buds will frequently be yellow.

Copper (Cu). Upward cupping and crinkling of the leaves. Internodes may be shortened, and growing tips may die.

FERTILIZING MELONS

Manure is a valuable addition to the soil if it is available at reasonable cost. The manure should be free from organisms that cause disease in melons. Avoid using manure where the hay or bedding came from fields where diseased melons have been grown. Twenty to thirty mt/ha of manure per hectare can be broadcast. Lesser amounts can be banded in the crop rows. If manure is not available, green manure crops such as alfalfa, clover and cowpeas can be grown to increase soil organic matter and reduce soil compaction. Manure and cover crops should be plowed under and allowed to decompose before melons are planted. Even with the addition of manure the use of a commercial fertilizer is necessary to achieve maximum melon yields.

Lime should be applied if necessary to provide the proper pH level for good crop growth. The optimum pH level for melons is 6.0-6.8. Watermelons will respond well even when the pH falls to 5.0.

General fertilizer recommendations vary from region to region depending on the soil type. Thus, a soil analysis is always recommended before applying fertilizer. Table 26.5 presents information on how soil test results may be used to arrive at fertilizer recommendations. In most areas of the world a complete NPK fertilizer is required.

TABLE 26.5
Melon nutrient recommendations

	Nitrogen[a] kg/ha	Soil Phosphorus Level[b] kg P_2O_5/ha				Soil Potassium Level[b] kg K_2O/ha			
		40-69	70-115	115-228	229+	40-94	95-180	181-358	359
Total Recommended	95-112	168	112	56	28	224	168	112	56
Broadcast and Disk in	28-56	112	56	0	0	168	112	56	0
Band Place With Planter	28	28	28	28	0	28	28	28	28
Sidedress When Vines Start to Run	28-56	28	28	28	28	28	28	28	28

[a]Where plastic mulches are being used, broadcast 168 kg N/ha with recommended P_2O_5 and K_2O and disk in prior to laying mulch.

[b]Kilograms of phosphorus (P_2O_5) or potassium (K_2O) per ha as determined by the double acid, soil extraction method.

Nitrogen fertilizer increases the early yield of melons. Plants that receive low N levels will be smaller and show N deficiency symptoms. Maximum fruit set can be delayed up to 11 days if N is limiting. The application of 34, 68 or 100 kg N/ha increases flower production, fruit set and the soluble acid content of the fruit. Total yields increase due to increased fruit number rather than increased fruit weight. Higher N rates reduce these factors and can delay flowering as much as 1 week. Nitrogen fertilizer by itself will increase yields up to 20%.

Only a small amount of the total nutrient uptake by melons occurs in the first half of the growing period. During the later stages of growth, as much as two-thirds of the nutrients are absorbed. Therefore, a grower should consider applying much of the N, and possibly part of the K and P, as a sidedressing. This is especially true on lighter textured, sandy soils that are subject to leaching.

Most authorities agree that N, and possibly P and K, should be applied at two or three different times during the season. Apply and disk in one-half of the N just prior to planting. This application could be further split by applying one-fourth of the N and disking it in prior to planting and banding the second quarter at planting. The N should be banded a few inches to the side and below the seed. If N is placed directly beneath the seed, injury may occur. As much as half of the total amount of N can be sidedressed when the vines begin to tip and form runners.

Phosphorus is considered to be immobile in cold soils, therefore, the P must be placed close to the germinating seed or transplant for uptake early in the season. Split P applications are recommended as they are for N. Research has shown that banding P is more effective than broadcasting. The addition of P is responsible for increasing yields by approximately 15%.

The use of starter fertilizers is a common practice with transplanted melons and is becoming very popular with direct-seeded melons as well. A typical analysis for a starter fertilizer is 21-53-0, 10-52-17, or 10-55-10. Add 1.4 kg of fertilizer to 180 liters of water. Apply 300 ml of this solution to each transplant. The high amount of P is important due to its inability to move in cool soils and the plant's need for it during early growth. This supplies the plant with enough nutrients at this critical time.

Some soils are high in available K and do not require additional K. If needed a split application is recommended.

Many soils require the addition of boron for good plant growth and maximum yields. This is particularly true for the lighter textured, sandy soils which are low in organic matter. Boron can also be limiting in areas of moderate to heavy rainfall or under extremely dry conditions. Soils that are neutral or alkaline are often low in B. Boron is limiting in soils when a soil test shows about 0.35 ppm B and application of 1-2 kg B/ha is then recommended. Soil levels above 0.7 ppm are considered high and do not require additional B.

When fertility rates are optimum melon yields will be maximized. A reduction in plant population will increase mean fruit weight but will reduce the number of melons per ha, as well as the total yield per ha. Optimum plant spacing will vary depending on the cultivar planted, soil type and climate.

A significant yield increase will normally occur when melons are grown on plastic mulch. Besides an increase in total production, earlier yields (up to 2 weeks) can be expected. Cleaner fruit with less fruit rotting is common since the fruit does not rest on the soil. In some areas higher yields are observed with clear plastic as compared to black or brown plastic. Clear plastic allows the sun's rays to penetrate, warming the soil faster and higher than opaque mulches. When plastic mulches are used, all of the fertilizer is applied before the mulch is laid. Researchers are currently studying methods to continuously feed melons grown on plastic mulches. This is done by injecting fertilizer through a trickle irrigation line placed under the mulch and next to the plant. Some very promising results are being observed.

Weed control in melons is very important since weeds will compete with the crop for available nutrients, moisture, light and space. Yield reductions will be most severe if weeds are allowed to compete with the crop during the first several weeks of growth. Weed competition late in the season generally does not reduce yields but may hinder the harvest operation and promote disease problems. The addition of more fertilizer will not overcome this competitive effect. The opaque plastics (black and brown) and paper, as well as organic mulches (hay, straw, etc.), are quite effective in reducing weed populations.

The level of N may influence melon tolerance to certain diseases. Generally, if liberal amounts of N are applied and vine growth is vigorous and succulent, melons will be more susceptible to disease. A normal growth rate with a proper nutrient balance will produce the greatest degree of disease resistance. However, this is not true under dry conditions. Boron deficiencies have been associated with increased disease susceptibility in melons.

FUTURE PROSPECTS

Muskmelons are relatively important in the USA, ranking twelfth in hectarage and ninth in value in 1981

among the 22 principal vegetable crops. Muskmelon hectarage has fluctuated downward for many years; however,
yields per ha have increased substantially and total production has slightly increased. The 1981 production of
some 6 million kg of melons was valued at $188,070,000.

Honey dew melons have increased in hectarage from
approximately 4000 ha in 1949 to 7700 ha in 1981. The
value of the 1981 USA honey dew crop was $56,187,000.

Watermelons ranked seventh in hectarage and thirteenth in value among the 22 major USA vegetable crops
during 1981. Like muskmelons, the crop area has declined
from 155,000 ha in 1949 to 82,000 ha in 1981, but yields
have nearly doubled from 7500 kg/ha in 1949 to 14,400
kg/ha in 1981. The 1981 crop was valued at $158,923,000.

Melons are not as perishable as many other vegetable
crops; however, they are not adapted to long periods of
storage. Watermelons will store slightly longer than
muskmelons. After 4 to 5 weeks in storage, melon quality
deteriorates significantly. Under optimum storage conditions of 4 to 7°C and a relative humidity of 85-90%, melons should be stored for no longer than 2 weeks.

SUPPLEMENTAL READINGS

Anon. 1981. Commercial Vegetable Production Recommendations. Rutgers, The State University of New Jersey. Ext. Bull. 406-H. 56 pp.

Anon. 1981. Vegetables 1981 Annual Summary Acreage, Yield, Production and Value. United States Department of Agriculture. Washington, D.C. 56 pp.

Anon. 1979. FAO Production Yearbook. Vol. 33 p. 162-163.

BRANTLEY, B.B. and G.F. WARREN, 1961. Effect of nitrogen nutrition on flowering, fruiting, and quality in the muskmelon. Proceedings American Society of Horticultural Science, 77:424-431.

California Fertilizer Association Soil Improvement Committee. 1980. Western Fertilizer Handbook. 6th edition. The Interstate Printers and Publishers Inc., Danville, IL. 264 pp.

CHOUDHURY, B. 1967. Vegetables. National Book Trust. New Delhi, India. 198 pp.

DE GEUS, JAN G. 1973. Fertilizer Guide for the Tropics and Subtropics. 2nd Edition. Centre D'Etude De L'Azote, Zurich. 774 pp.

LINGLE, J.C. and J.R. WIGHT. 1964. Fertilizer Experiments with Cantaloupes. California Agricultural Experiment Station Bulletin 807. 22 pp.

LOCASCIO, S.J. and J.G. FISKELL. 1966. Copper requirements of watermelons. Proceedings American Society Horticultural Science, 88:568-575.

LORENZ, OSCAR A. and DONALD N. MAYNARD. 1980. Knott's Handbook for Vegetable Growers. 2nd Edition. John Wiley and Sons, New York. 390 pp.

MCMURTREY, J.E. JR. 1948. Visual Symptoms of Malnutrition in Plants. In: H.B. Kitchen (ed.). Diagnostic Techniques for Soils and Crops. American Potash Institute, Washington, D.C. 308 pp.

PURVIS, E.R. and R.L. CAROLUS. 1964. Nutrient Deficiencies in Vegetable Crops. In: Howard B. Sprague (ed.). Hunger Signs in Crops. David McKay Co. New York, NY 461 pp.

SHARPLES, G.C. and R.E. FOSTER. 1958. Growth and composition of cantaloupe plants in relation to the calcium saturation percentage and nitrogen level of the soil. Proceedings American Society Horticultural Science, 72: 417-425.

THOMPSON, H.C. and W.C. KELLY. 1957. Vegetable Crops. 5th Edition. McGraw-Hill Book Co., New York. 611 pp.

TYLER, K.B. and O.A. LORENZ. 1964. Diagnosing nutrient needs of melons through plant tissue analysis. Proceedings American Society Horticultural Science, 85: 393-398.

27
PEPPERS

B. A. Kratky and John E. Bowen

There are about 20 species of peppers in the Sola-naceae family. Pepper is a much branched, shrubby pe-rennial herb which is native to tropical America. Cap-sicum annuum and Capsicum frutescens are the primary cultivated species.

Capsicum annuum includes most of the cultivated peppers of North and South America. There are five ma-jor groups of peppers; each group has many cultivars. Some cultivars of the Cerasiforme, Conoides and Fascicu-latum groups are grown as ornamentals. The Grossum group contains the widely cultivated bell, sweet, green and pimento peppers. Fruits are usually large and thick fleshed, broadly oblong and bell or apple shaped. Fruits may be red or yellow when mature, and the flavor is mild and not pungent. The Longum group contains the capsicum, cayenne, chili, long and red peppers. The fruits are mostly drooping, elongated, tapering to the apex, and very pungent. This group supplies the condi-ment pepper, chili powder, paprika and medicinal capsicum.

Capsicum frutescens is the tabasco pepper which is pungent and used in the commercial production of hot sauces. These are long season varieties and do not ma-ture well in temperate regions. In the USA, they are principally cultivated in Louisiana, Mississippi, New Mexico and Texas.

Peppers have a pungent quality that may be de-scribed as a hot type of sharpness and a biting proper-ty. This is caused by an acrid substance, capsaicin, which is a volatile alkaloid. The percentage of capsaicin is commonly very high in small-fruited forms such as the tabasco; the large bell pepper cultivars usually contain only traces of capsaicin.

[1]Horticulturist and Plant Physiologist, respectively, Hawaii Agricultural Experiment Station, Hilo, Hawaii, 96720 USA.

Sweet and mildly pungent peppers are used as fresh
or cooked vegetables, or they may be canned or pickled
in brine. Paprika peppers are used to make a red fla-
voring powder. Pungent peppers are used for flavorings
in dry forms and also as sauces.

ECOLOGICAL ADAPTATION

Peppers are very sensitive to temperature and are
considered to be a warm weather crop, even more so than
tomatoes. Peppers grow very poorly at temperatures below 13°C, and a light frost will kill the plants.
Many cultivars require 65-80 days after transplanting to reach market maturity, but an additional 2-4
weeks are needed to obtain substantial yields of ripe
fruits. Therefore, cool springs and early frosts in
temperate regions will cause reduced yields. Some peppers require warmer temperatures and a longer maturation
time; these types must be raised in regions with long
growing seasons. In the USA, this is the reason tabasco
peppers are produced in the Gulf States and are not a
feasible crop for the temperate Northern States.
Optimum temperature for fruit set is 18-27°C.
Blossoms suffer heat injury at temperatures above 27°C,
causing small or poorly shaped fruits. Blossom dropping
becomes excessive at temperatures above 32°C and there
is little fruit set at temperatures above 35°C. Dry air
and drying winds tend to aggravate the injurious effects
caused by high temperatures. Large-fruited cultivars
are generally less tolerant of high temperatures than
small-fruited cultivars.
Temperature affects the amount of the red pigment
of bell pepper fruit. After the fruit has reached a
mature green stage, optimum development of red color
occurs at 18-24°C. Color development stops at temperatures below 13°C. Fruits develop a yellowish cast if
the bell pepper fruits are exposed to temperatures above
27°C during most of the coloring period.
Plastic mulches have helped to increase crop yields
in cool areas by raising soil temperatures. Plastic row
covers help to increase air temperatures and provide a
windbreak effect; this can also cause higher pepper
yields. For example, in an upper elevation trial in Hawaii during February through August, mean monthly minimum and maximum air temperatures ranged from 12-15°C and
19-22°C, respectively. Plastic row covers increased air
temperatures by 4-7°C at midday, and plastic mulch increased soil temperatures by 5-8°C; the resulting pepper
seed germination is also affected by temperature. Germination occurred in 12 days at a soil temperature of 20°
C and within 8 days when the soil temperature was 25°C.
Peppers may be grown successfully on a wide range
of soil types. Sandy loams and loams are preferred.
Good drainage is important.

The optimum pH range for pepper production is 6.0-6.8. Acidic conditions should be corrected with lime. If magnesium levels are also low, dolomitic lime should be used.

Peppers are commonly grown in intensive culture. Gross incomes of above US$100,000/ha/year in intensive greenhouse plantings have been reported. Even average US fresh market bell pepper yields of 12,000 kg/ha generate a large gross income. Since fertilizers often cost less than 10% of the gross income of peppers, growers are prone to use more expensive fertilizers (e.g. homogenous, highly soluble or slow release) if they can be proven to increase yields or reduce labor.

MINERAL NUTRIENT REQUIREMENTS OF PEPPERS

The world production of chili plus green peppers in 1979 was 6.67 million metric tons produced on 890,000 ha. Florida, California and Texas have the largest plantings of peppers in the USA.

The average world pepper yield in the decade from 1969-79 was 7440 kg of marketable fruit per hectare, with yields ranging from a low of 780 kg/ha, to more than 42,000 kg/ha in Japan. Recent research indicates that yields of 25,000 to 35,000 kg/ha should be readily attainable on a highly efficient and intensively managed farm in those regions of the world that are optimal for this crop.

A major, although obviously not the only, determinant of yield is the fertilization regime; i.e., what nutrients should be applied and in what chemical forms, as well as how, where and when to apply them.

Mature pepper fruit contain about 92% water and, on a fresh weight basis, 0.21% K, 0.027% P and 0.2% N. The total amount of N, P and K in pepper plants (excluding fruit and roots) yielding 30,000 kg/ha will be approximately 128 kg N/ha, 8.5 kg P/ha and 121 kg K/ha. The total crop removal for these plants per hectare by the mature fruit and above ground plant parts will be 190 kg N/ha, 17 kg P/ha and 180 kg K/ha.

These values provide insight into the absolute minimum amount of N, P and K that must be available to the plants. These nutrients may either be added as fertilizers or be already available in the soil. Some soils contain large amounts of K that will be released for plant uptake so the amount of K fertilizer can be reduced in these cases. A previous crop and fertilization history will provide insight into the residue of available nutrients. The amounts of nutrients that must be available per hectare will be higher than the minimum amounts expressed above because it is not possible to attain a 100% utilization efficiency. Some N and K will be lost by leaching; the extent of leaching is dependent upon the amount of rainfall. For nitrogen, nitrates leach more readily than the ammonium form. In some soils large amounts of P will be rendered unavailable to the plants by chemical fixation. Soil tests should be made to verify the P status of local soils.

Specifically how much fertilizer should be applied to a pepper crop? This becomes an impossible question to answer in a general context because there are so many local factors that must be considered. Therefore, we will provide some examples of fertilization regimes used for peppers in different areas and then some guidelines on how to use soil and plant tissue analyses to modify these regimes for specific conditions.

For open field culture on the sandy soils of Texas in the southern USA, a balanced fertilizer such as 12-24-12 is broadcast at the rate of 330 to 560 kg/ha and disked into the soil before planting. Two or three sidedressings of N are applied during the growing season to maintain vigorous growth and to promote continuous blooming and fruit set. The total amount of fertilizer applied per crop may be as high as 280 kg N/ha, 74 kg P/ha and 55 kg K/ha. These soils already contain high levels of K, so additional K fertilization is not necessary or certainly may be much less than the total plant K requirement.

Pepper growers on Florida's irrigated mineral soils fertilize with a basic application of 120 kg N/ha, 70 kg P/ha and 131 kg K/ha followed by one to six sidedress applications of 34 kg N/ha and 28 kg K/ha. For the high pH marl soils of Florida, the regime is 60 kg N/ha, 35 kg P/ha and 66 kg K/ha applied prior to planting followed by one to three supplemental applications of 34 kg N/ha and 28 kg K/ha.

It may be necessary to apply calcium and magnesium in some soils. Micronutrients in various combinations may also be needed and can be added to the basic fertilizer mix at the following rates: 0.3% MnO, 0.2% CuO, 0.5% Fe_2O_3, 0.2% ZnO and 0.2% B_2O_3. Micronutrients are best added to fertilizer mixes as oxides, sulfates or fritted materials. Sodium tetraborate (borax) is often used as a source of boron.

The other alternative is to apply micronutrients as foliar sprays to the growing crop. Foliar sprays should be applied at 7 to 10 day intervals for 4 to 6 weeks during bloom and early fruit set. These are commonly applied in combination with an insecticide or fungicide. Chemicals in the contemplated formulation must be compatible, however. Also, micronutrient sprays must be applied in addition to the basic N, P and K fertilizer because the sprays alone do not contain adequate N, P and K to support the crop.

Any chemical containing the desired micronutrient can be sprayed if it is sufficiently soluble in water. Sulfate salts meet this criterion and are often used. Chelates are excellent materials for foliar sprays and are immediately absorbed but are expensive. Chelates are usually economically feasible for peppers.

NUTRIENT MANAGEMENT AND NUTRIENT DEFICIENCY SYMPTOMS

High N levels are necessary early so that the plants will grow rapidly after germination or trans- planting. Otherwise they will start to bloom and set fruit while too small, stunting plant growth and lower- ing yield. Another reason for early heavy fertilization is that the greatest absorption of N, P and K by peppers occurs 8 to 14 weeks after transplanting. When the soil has been fumigated, most of the nitrogen (60-70%) should be supplied as nitrate nitrogen.

Pepper plants that have received inadequate nitro- gen are typified by stunted growth and moderately to severely chlorotic leaves. The appearance of visibly detectable N deficiency symptoms is associated with a whole plant (excluding fruit and roots) tissue N level of 1.25% or less (dry weight basis). Moderate N defi- ciency causes a mild chlorosis of the mature fruit but would not usually affect their marketability.

Peppers respond well to phosphorus placed under the seed or seedling at the time of planting or transplant- ing. Banding phosphorus 5 to 8 cm deep in the rows will optimize its availability and cut fertilizer costs as compared to broadcasting. Typical rooting depth is 90 to 120 cm if there is no barrier to root penetration. Thus, depth of fertilizer placement is not a critical factor although it must not be so deep as to render it unavailable to newly developing roots.

Symptoms of phosphorus deficiency are manifested in peppers when the whole plant (excluding fruit and roots) tissue P level drops to 0.09% or less. The P-deficient plants are severely stunted, and leaves become quite narrow and develop an abnormal grayish-green color. The typical purpling of plants due to P deficiency does not occur in pepper, but may indicate other stresses such as wind or cold weather.

Fruit produced by P-deficient plants are commonly misshapen. Fruit length may be reduced somewhat but most noticeable is the smaller circumference. Signifi- cant decreases in marketable fruit yield can be antici- pated if the P content of the fruit is permitted to fall to 0.13%.

Inadequate potassium is manifested initially on peppers as a non-specific reduction in growth. As the condition worsens, the leaves develop a characteristic "bronzing", followed by the appearance of necrotic le- sions along the veins. Foliar abscission follows in ex- treme cases of deficiency. The minimum level of suffi- ciency in pepper is approximately 1.5%; mild to moderate visually apparent deficiency symptoms appear when the tissue K concentration is 1.2% or less.

Both fruit size and total yield per plant are decreased by K insufficiency, which is defined as tissue K levels of 1.85% or below in the mature fruit.

Calcium concentrations in vegetative tissues must be maintained at 0.6% or above if pepper plants are to remain healthy and the maximum yields of marketable fruit are to be attained. Calcium deficiency on pepper is very difficult to identify visually without confirming tissue analysis. Calcium-deficient plants lack vigorous growth and the leaves develop an abnormally dark green color that would not likely be apparent unless it was compared directly with a healthy leaf.

The most striking symptom of Ca deficiency, as well as the most important economic effect, is the condition known as "blossom end rot". This condition is characterized by water-soaked spots on the lower half of the fruit near the blossom end. The spots enlarge, become light brown to brownish-black, and eventually dry to form a leathery texture. The fruit is rendered unsaleable. Fifty percent of the fruit can be affected if the Ca content of the fruit drops to 0.1%. Otherwise, Ca deficiency causes a moderate reduction in fruit size but there is no detrimental effect on fruit shape.

It should also be noted that a lack of calcium is not the only condition that can lead to Ca deficiency in pepper; e.g. very high amounts of NH_4 or K can induce Ca deficiency by interfering with the uptake and translocation of Ca by the plants.

Excessive Mg can also depress uptake of Ca and this can increase blossom end rot if the Ca supply is marginal. Calcium deficiency can also be caused by moisture stress. There are several options available to correct blossom end rot. Lime or gypsum may be added to increase soil Ca levels, and adequate soil moisture levels should be maintained to ensure uptake of Ca.

Solutions of calcium chloride or calcium nitrate (0.5 - 1.0 kg/100 liters of water) may be periodically sprayed on the foliage at a rate of 1000 liters/ha. Calcium chelates may also be used. Since Ca is only slightly mobile in plants, it should be sprayed on immature plant tissues, apical meristems, emerging and elongating leaves and developing fruits. Susceptibility to blossom end rot also varies with cultivar.

Magnesium deficiency is manifested as an interveinal chlorosis of immature and recently matured leaves. As the Mg insufficiency worsens, necrotic lesions will appear within the chlorotic areas. The tissue Mg concentration must be kept at 0.6% or higher to avoid Mg deficiency. Magnesium inadequacy causes a decrease in fruit number and size but does not affect the shape. The Mg content of the mature fruit should not be less than 0.25%.

FERTILIZATION PRACTICES FOR DIFFERENT CULTURAL SYSTEMS

Peppers can be grown in open field culture without mulching, in strip-mulch culture (about 30 cm wide), and in full-bed mulch. The recommended fertilization regime varies somewhat with each of these techniques because of the physical barrier of mulch and also because less leaching occurs in mulched soils.

For open culture on a mineral soil, about 60% of the projected total N P K fertilizer can be broadcast and disked into the soil prior to planting. Also, all necessary calcium and magnesium should be applied at this time.

As an alternate system, the calcium and magnesium may be broadcast and disked in, but the N P K fertilizer is banded 5-10 cm below the soil surface and to the side of each row. A broad band would perform as well as a narrow band and also reduce the risk of salinity injury.

The remaining 40% of the fertilizer is applied in one to six supplemental sidedressings of N and K to maintain healthy growth and uniform fruiting. Supplemental sidedressings of N are especially beneficial in high-density plantings when the plants are 15 to 20 cm tall.

For strip-mulch culture, the same fertilizer mix and amounts are used as for open culture. Eighty percent of the fertilizer is placed under a 20 to 30 cm strip of mulch to minimize nutrient leaching. Supplemental sidedressings may be applied whenever the need arises, such as after heavy rains.

In full-bed mulching, all of the fertilizer must be applied prior to mulching, since it is very difficult to correct a nutrient deficiency after mulching except through use of foliar sprays.

Lime, superphosphate, micronutrients and up to one-third of the mixed fertilizer (e.g. 6-12-12) are broadcast and incorporated into the soil along with any recommended insecticides and herbicides. Beds are shaped and possibly fumigated. A surface herbicide may be applied and a N K fertilizer is banded on each side of a one-row bed or in three bands on a two-row bed. The mulch is then applied and the seedlings are transplanted.

OTHER TYPES OF FERTILIZERS - SLOW RELEASE AND ORGANIC

Slow release fertilizers may be used more extensively in the future with all three planting systems. These materials will lessen or eliminate the need for sidedressing with open culture and strip mulch culture methods and provide for a more orderly release of available N and K for full-bed methods. Higher yields may also be possible because N and K will always be available, even after a heavy rain. Growers will have to decide if these advantages are worth the increased cost of slow release fertilizers.

Manures, composts and other organic fertilizers are a type of slow release fertilizer. They also provide organic matter to the soil increasing the cation exchange capacity, benefitting soil structure, and improving soil moisture holding capacity. Manures and composts vary greatly in their nutrient content, so the grower is never sure how much nutrient has been added nor how fast it is being released.

Manures and composts are low analysis fertilizers, therefore large quantities must be applied. They range from fresh cow manure with an approximate analysis of 0.5-0.2-0.5 to dried chicken manure (1.5-1.75-2.0) and hog manure (2.25-2.0-1.0). Manures are usually broadcast and incorporated into the soil when large quantities are applied. However, dried manures are often applied in smaller quantities (1-5 ton/ha) and may be banded, providing care is taken to prevent burning.

SALT TOLERANCE

Peppers are less salt tolerant than tomatoes. Yield decreases of 10, 25 and 50% occur at EC (electrical conductivity of the soil saturation extract) readings of 2, 3 and 5 mmho/cm, respectively.

There is a tendency for growers to over-fertilize peppers because fertilizer cost is but a small fraction of the value of the crop. Therefore, greenhouse growers and growers in dry climates are advised to take precautions not to allow salt build-up. High salt levels in soils can be reduced by sprinkler irrigating with 10-14 cm of water; the depth of water applied can be measured by placing open-top cans in the area being irrigated. The irrigation rate should be 1 cm/ha or less. Higher irrigation rates are less efficient because during rapid irrigation most of the water travels in the large soil pores, leaving the salt behind in small pores which can only be leached efficiently by a slow irrigation.

Proper water management is needed when saline irrigation water must be used for irrigation. Drip irrigation, when properly managed, minimizes salt injury. Salts follow the movement of water in the soil, therefore placement of drip irrigation lines immediately adjacent to the rows results in a high salt concentration midway between two rows where it causes the least damage to the peppers.

SUPPLEMENTAL READINGS

Anonymous, 1974. Pepper production guide. Florida Cooperative Extension Service Circular 102 D, University of Florida, Gainesville, Florida.

BOSWELL, V. R., S. P. DOOLITTLE, L. M. PULTZ, A. L. TAYLOR, L. L. DANIELSON and R. E. CAMPBELL 1964. Pepper production. USDA-ARS Agriculture Information Bulletin No. 276.

MILLER, C. H. 1961. Some effects of different levels of five nutrient elements on bell peppers. American Society of Horticultural Sciences, 77:440-448.

LONGBRAKE, T., S. COTNER, R. ROBERTS, J. PARSONS, and W. PEAVY 1976. Keys to profitable pepper production. Fact Sheet L-966, Texas Agricultural Extension Service, Texas A&M University, College Station, Texas.

LORENZ, O. A. and D. N. MAYNARD 1980. Knotts' Handbook for Vegetable Growers, 2nd ed.. John Wiley & Sons, New York.

SIMS, W. L. and P. G. SMITH 1976. Growing peppers in California. Cooperative Extension Service Leaflet 2676, University of California at Davis.

28
APPLES

G. H. Oberly

Apples (Malus domestica) are available for human consumption in most parts of the world throughout the year because of exceptional storage and shipping qualities, and production in both the Northern and Southern hemispheres. In the USA, apples comprise about 25% of all deciduous fruit and nut production, exceeded only by grapes.

Apples are not rich in calories, mineral elements, vitamins, or basic foods; however, they are beneficial in the digestive processes, appear to reduce respiratory problems, and are good for teeth. Apples are eaten fresh, cooked, added to bakery products, canned, frozen, dried, juiced and made into vinegar.

Annual production, within a region, fluctuates from year to year because of climatic conditions. Data are not available on the number of trees or acreage by country. However, there has been an increase in the number of bearing apple trees in most of the apple-producing countries of the world (see Table 28.1).

ECOLOGICAL ADAPTATION

Apple production is limited to areas where there is adequate chilling to break the rest period, minimum winter injury to the trees, adequate number of frost-free days from bloom to maturity, and well drained soils with adequate soil moisture. Most apples are grown between 30° to 60° North and South. Some are grown in isolated areas outside these boundaries where the climate is modified by large bodies of water or high elevation. The selection of cultivars can also extend the range by genetic differences in hardiness, date of bloom, length of rest period, or number of days to maturity.

1Pomology Department, Cornell University, Ithaca, New York 14853, USA.

TABLE 28.1

Production of fresh and cooked apples by region and major producing countries[a]

	1969-71	1977	1978	1979
World	18,276	30,637	32,778	35,707
Africa	285	362	446	455
South Africa	239	292	270	379
All other Countries	46	70	76	76
North and Central America	3,508	3,630	4,226	4,349
Canada	413	411	452	437
Mexico	128	187	342	390
USA	2,962	3,026	3,427	3,515
All other Countries	5	6	5	7
South America	718	1,135	1,164	1,339
Argentina	435	820	810	972
Chile	142	140	165	175
All other Countries	141	175	189	192
Asia	4,710	6,347	6,720	7,348
China	1,953F	2,111F	2,418F	2,718F
India	270	720F	740F	760F
Japan	1,037	959	837*	840F
Turkey	716	900	950	1,150
All other Countries	734	1,657	1,755	1,880

Europe	13,966	11,062	13,579	14,196
France	3,895	1,686	3,022	2,950
West Germany	2,110	1,175	1,783	1,951
Hungary	700	1,105	779	980F
Italy	1,923	1,828	1,874	1,800
Poland	636	912	1,030	1,030F
Spain	535	730	1,072	1,156
All other Countries	4,167	3,626	4,019	4,329
Oceania	556	447	444	521
Australia	430	301	259	335
New Zealand	126	146	185	186
USSR	4,533	7,653*	6,199*	7,550*

aTable does not include processed apples.

F – FAO Estimate.
* – Unofficial.

CLIMATE ADAPTATION

The limits of apple planting should be in areas where the minimum temperatures do not drop below -37°C for the hardiest cultivars and -32°C for most other cultivars. Winter injury may be in the form of bud killing, injury to the conducting tissues or complete killing. Not only is there a minimum killing temperature, but it is compounded by the general physiological condition of the trees and the rapidity at which the temperature drops. Severe injury can occur when the temperature suddenly drops to -9°C in the early winter before maximum hardiness has developed. The same trees may withstand -37°C in mid-winter.

In many areas, the frequency of spring frosts may be more of a hazard than low winter temperatures. Within a recognized apple producing area, the frequency and severity of spring frosts may be an area or a localized effect. Frost damage may be modified by the presence of large bodies of water, or in locations where there is good air drainage to allow the cold heavier air to settle to lower elevations.

Temperature effects at lower latitudes also limit apple production. Both vegetative and flower buds open sporadically if the trees do not receive adequate chilling to break the rest period.[1] Some limbs may be in leaf or flower while other limbs on the same tree are still in the resting stage.

SOIL ADAPTATION

For maximum production the physical characteristics of the soil are more important than the fertility level. Apples will produce over a wide range of fertility if the soil has good physical properties. The soil should be deep and well drained to permit deep and widespread rooting. About 50% of the soil volume of a good orchard soil is pore space, and there is an inverse relationship between the space occupied by air and that occupied by water. Where there is inadequate exchange of soil gases, rooting is limited and production is greatly reduced. Most of the nutritional problems can be corrected, but little can be done to correct the physical properties of the soil.

---

[1] In warmer regions, where apples are grown with irrigation, apple trees should be forced into a dormant state by withholding all water until leaf fall has occurred. Subsequent irrigation will stimulate a new vegetative cycle, comparable to that occurring in spring in temperate zones.

Place in Farming Systems

Because of the specialized equipment for production and storage and the critical timing of many cultural practices and intensive labor requirements, apple production does not lend itself to mixed farm enterprises. There are some successful fruit-vegetable and fruit-beef cattle operations, but for the most part the successful operations are those which concentrate on apple production.

MINERAL NUTRIENT REQUIREMENTS

The most important factor in obtaining maximum yields of apples is management. In surveys to determine why some growers obtained greater yields than their neighbors, physical properties of the soil, nutritional levels, rootstocks and many other production practices were evaluated. From these surveys it was found that maximum production was obtained when all of the known good orchard practices were utilized. Where any one practice was neglected, it became the limiting factor and production was reduced.

The maximum yield of marketable size fruit of fancy and extra fancy apples varies with areas of production. The difference in potential yields is dependent upon the available sunlight for photosynthesis and the control of available soil moisture. For example, with present technology in the Northeast USA where sunlight is limited and production depends on natural rainfall, yields of about 64 mt/ha (1200 bu/acre) can be obtained. In the Northwest USA, with more hours and greater intensity of sunlight, and where soil moisture is controlled by irrigation, yields of 106 mt/ha (2000 bu/acre) may be realized. Higher yields have been reported in both areas, but generally those are comprised of a high percentage of small size fruit not marketable as fresh fruit. The lack of data on tree numbers or hectares planted makes it impossible to compare yields from other apple-producing countries.

Average yields are approximately 50% of now attainable yields. This figure reflects the differences between trees receiving the best management and, where there may be one or more limiting factors, the biennial bearing habit of apples and reductions due to climatic conditions. Also marginal areas that experience winter injury, frost damage, or inadequate chilling to break the rest period have a low yielding potential.

Goals of Mineral Nutrients for Acceptable Crop Yields

Soil test values and soil tests are not reliable in determining the nutrient requirements for fruit trees. It is impossible to obtain a soil sample that represents

the total soil mass occupied by the roots. Under field
conditions there is a poor relationship between soil test
data and tree performance. Soil tests are useful for determining the pH (soil acidity or alkalinity) and hydrogen ions for liming. Soil pH per se has little effect on
tree growth; however, when soil pH is between 6 and 7 nitrogen is more readily available and there are fewer
deficiency problems.

Improved Varieties, or Hybrids

Due to the nature of the crop and vegetative propa-
gation, apple cultivars do not change drastically. Also,
with the exception of N, all other essential elements appear to be constant between the many cultivars. For most
elements there is a relatively wide range in concentration providing the nutrient levels are in balance. The
effect of nutrient imbalance is more pronounced at a low
level of nutrition than at a high level. Therefore, to
offset the effects of crop load, marked variation in soil
moisture, or any condition that would place a stress on
the trees, it is best to operate at a relatively high
level of nutrition.

Plant Tissue Nutrient Values Sought

The mineral element content of apple leaves has a
relatively good relationship to tree performance. The
most common leaf samples are those from well exposed mid-
shoot leaves of the current season's growth. The major
problem is obtaining a composite sample representative of
the block of trees. If the standard values are correlated to tree performance, then leaf samples can be used
to detect nutritional disorders, or as a basis for fertilizer recommendations. Although tissue analysis is a
recognized method of detecting nutritional disorders or
levels of nutrition, trees also should be evaluated for
visual symptoms and growth responses.

Nitrogen (N) is one element that shows a measurable
response with slight changes in leaf composition. When
the N level is below normal, yields can be reduced. Low
N can cause poor flower bud development, failure of weak
flowers to set, or a reduction in fruit size. Excess N
can also have detrimental effects. The fruit may have
poor color, be soft, and break down in storage. The N
level also has an effect on winter injury. Low N trees
do not develop as much hardiness to mid-winter temperatures as high N trees. However, high N trees are more
susceptible to early winter cold injury. In either case,
the trees are more susceptible following a heavy crop
load.

TABLE 28.2
Recommended leaf levels of essential elements for apples

Nutritional Element	Optimum Leaf Level (Dry Weight)	Comments
Nitrogen (N)		
Fresh Fruit		
Soft cultivars	1.8 – 2.1 %	Low levels – reduction of crop and fruit size
Hard cultivars	1.0 – 2.4 %	High levels – poor fruit color and storage quality
Processing cultivars	2.1 – 2.5 %	Low levels – reduction of yields and fruit size
Phosphorus (P)	0.9 – 0.4 %	High levels – interferes with zinc reactions
Potassium (K)	1.2 – 1.7 %	Low levels – reduced fruit color and photosynthesis
		High levels – adversely affects magnesium
Calcium (Ca)	1.0 – 2.0 %	Low levels – physiological disorders of fruit
Magnesium (Mg)	0.25 – 0.45 %	Low levels – premature drop of fruit and leaf scorch
		High levels – may adversely affect K and Ca
Manganese (Mn)	35 – 500 ppm[a]	Low – affects enzymatic reactions
		High – associated with internal bark necrosis of "Delicious"
Iron (Fe)	[b]>50 ppm	Not reliable, cannot separate active from inactive Fe. Rely upon leaf symptoms
Sulfur (S)	>150 ppm	Sulfate sulfur in terminal leaves
Copper (Cu)	5 – 50 ppm	Low – die-back of terminal shoots
Boron (B)	25 – 50 ppm	Low – internal cork and heavy fruit drop
Zinc (Zn)	>24 – ppm	High – may be due to spray contamination
Molybdenum (Mo)	>.5 – ppm	

[a]ppm = parts per million

[b] > = greater than

311

As noted in Table 28.2, a lower N level is recommended for the soft varieties and for "Golden Delicious." To maintain good fruit size, "Delicious" and "Empire" should have leaf N levels at the upper range for the hard cultivars.

Role of Nutrient Requirements for Apples

Animal manure. In most fruit producing areas the high cost of animal manure makes it impractical. Cover crops or sod are used to maintain the organic matter in the soil, and commercial fertilizers, per kg of active ingredients, are cheaper. Also, excessive application of animal manure increases fire blight disease (Erwinia amylovora).

General recommendations cannot be made for rates of application of fertilizers. Soil variability, tree age, crop load, etc. makes it necessary to adjust the fertilizer rates for each block of apples.

Nitrogen (N). The ammoniacal and nitrate forms are available in a number of different compounds. Under field conditions there is no information indicating a preference for one fertilizer over another.

Foliar (leaf) applied urea is useful to "fine tune" the N level. If the N is low because of choice or crop load, 10 to 13.5 kg/ha (9 to 12 lbs/acre) of feed grade urea can be applied at the time of the first and third cover sprays. Soil grade urea is high in biuret and is phytotoxic.

Phosphorus (P). Conclusive evidence has yet to be shown that an economical response in apples can be obtained from the application of P fertilizers. Phosphate fertilizers are applied but generally to stimulate the growth of cover crops.

Potassium (K). Potassium chloride (muriate of potash) is most widely used because of its low cost per unit of K. In some areas of the USA, K is readily fixed and rates of application must be high. Where the rate of application exceeds about 675 kg/ha (600 lbs/acre) of actual K on a coarse textured soil, a weaker salt of K (potassium sulfate or potassium phosphate) should be applied. Potassium nitrate could be applied, but at these rates the N application would be excessive.

Calcium (Ca). Calcium deficiency has not been shown on vegetative growth of apples under field conditions. However, low Ca in fruit increases the incident of "bitter pit", "York Spot", and internal breakdown of the flesh in storage. There is little if any relationship between soil, leaf and fruit Ca levels. The problem appears to be the movement of Ca through the fruit stem and into the flesh. Foliar applications of calcium chloride throughout the growing season will reduce the incidence of these problems, but generally does not result in complete control.

Magnesium (Mg). The most satisfactory control of Mg deficiency is the application of dolomitic limestone where there is a low soil pH. On slightly acidic soils, magnesium oxide is effective and on soils above pH 7.0 hydrated magnesium sulfate (Kieserite) is very effective. Where both K and Mg are low, sulfate or potash-magnesia (Sulpomag) is an excellent choice.

Sulfur (S). As would be expected, deficiency symptoms of S are similar to N. The S requirement for apples can generally be satisfied by sulfates in irrigation water or acid rain. However, any fertilizer containing S will correct the problem. In the state of Washington, the use of ammonium sulfate in alternate years has corrected the deficiency.

Manganese (Mn). Manganese deficiency is more common on alkaline soils, which favor the presence of manganic ions. Under acid soil conditions that favor the presence of manganous forms, which are readily available to plants, exchangeable Mn can be much lower before deficiency symptoms can be detected. In the milder stages, Mn deficiency-chlorosis of apples does not seem to be accompanied by loss of vigor or unfruitfulness. But under extreme deficiency, causing serious decreases in photosynthetic activity, there is tree stunting and loss of vigor.
In the humid apple producing areas, any of the Mn-based fungicide sprays will correct the deficiency. Under arid conditions a foliar application of zinc sulfate is an effective control.
Where the soil is high in exchangeable Mn, "Delicious" trees sometimes develop internal bark necrosis, or "measles". Liming is the obvious correction, but the reaction is slow. Where measles is a problem, N should be withheld and the trees severely pruned to stimulate vigorous vegetative growth.

Iron (Fe). Iron deficiency, or lime-induced chlorosis, is more common in semi-arid regions where irrigation water may be high in bicarbonates. Mineral soils have quantities of Fe far beyond the needs of the plant. Therefore, like Mn, problems are a function of the form of Fe. A tree may have an adequate level of Fe, but if it is in the ferric ion form, the tree will show typical interveinal chlorosis. Under acid conditions, most of the Fe is in the ferrous ion form, and there is seldom a problem with trees on acid soils.

Foliar applications of Fe have not been successful as the Fe remains immobilized at the point of contact. The most effective control has been by injecting slurries of chelated iron (EDTA) into the soil. Excessive irrigation aggravates the problem. Therefore, judicious use of irrigation water and good soil management practices will help alleviate the problem.

Boron (B). Boron deficiency can be prevalent in both semi-arid and humid regions. Unlike most nutritional problems that reduce yields, B deficiency generally causes a complete crop loss. The most common symptoms are pre-harvest drop of fruit and both external and internal fruit cork. Fruit symptoms appear before the level is low enough to cause stunting of shoots and terminal rosetting. When the level is very low there is dieback of the terminal shoot.

Boron can be added by either incorporating borax into other fertilizer materials or applying it directly to soil. The percent B in fertilizer grade borax varies between about 11 and 20%. The maximum application should not exceed about 2.25 kg/ha (2 lbs/acre) actual B every third year. Most growers prefer to apply B as a foliar application of 2.25 kg/ha of sodium pentaborate (Solubor or Boro Spray) in each of the "petal fall" and "first cover" sprays.

Copper (Cu). A few apple producing areas -- Australia, South Africa and the West Coast of the USA -- report Cu deficiency. Where Cu deficiency exists, fruit production is very low. The most striking symptom is dieback of shoots which had been making vigorous growth. In the following year growth resumes below the point of death, followed by dieback of the new terminals. Repetition of dieback causes the affected trees to have a bushy stunted appearance.

Copper deficiency can be corrected by a soil application of copper sulfate, or foliar applications of a bordeaux mixture of "fixed" coppers.

Zinc (Zn). Zinc deficiency is generally found in the same areas where Fe and Mn deficiencies occur. Zinc deficiency is synonymous with the descriptive names of "rosette" or "little leaf." When the deficiency is severe enough to cause rosettes of small narrow leaves early in the spring flush of growth, yields may be greatly reduced.

Zinc may be applied in several ways, soil application, injection, foliar applications and even zinc-coated glazing tacks. The most economical method for most soil conditions is spray application. Where there are visual symptoms, foliar applications of 1.3 kg of zinc chelate (EDTA) or 900 g zinc sulfate + 900 g lime per 450 liters of spray at "tight cluster stage," and again at the first or second "cover spray" should be sufficient. A post-harvest spray of 2.3 kg zinc sulfate + 2.3 kg lime per 450 liters of spray is also effective. If the Zn level is low, but not to the point of being deficient, one spray of either zinc chelate or zinc sulfate at the first cover spray will generally be sufficient.

Molybdenum (Mo). The only report of Mo deficiency is in New Zealand. A soil application of 2.4 kg/ha of sodium molybdate is sufficient to correct a deficiency.

FERTILIZATION PRACTICES

With a shift to higher density of apple trees per ha, methods of applying fertilizer materials have changed. The root systems of trees in a semi- to high-density planting overlap each other. It is no longer necessary to fertilize each tree. Fertilizer can be applied mechanically in bands or broadcast over the orchard floor.

For most soil applied fertilizers, time of application is not important. Except for N, applications can be made at the convenience of the grower. For N, the growth response and the effect on fruit storage quality limit the time of application. Nitrogen should be applied any time after the trees are completely dormant until about one month before "bud-break" in the spring. Due to the depth of rooting, leaching of nitrates has not been a major concern and many growers find the late fall or early winter period a convenient time to apply N. This is especially true for orchards which have a grass sod cover. Application at this time allows the N to be leached below the root system of most grasses and it is then available to the trees.

FRUIT MATURATION AND HARVEST

Nitrogen has more effects on fruit quality than any other element. Excessive N will increase total fruit tonnage, but it causes poor color, large soft fruit and poor storage and shelf life. Low K results in poorly colored fruit and a reduction in fruit size. However, above the level of deficiency, additional color cannot be obtained by adding excessive amounts of K. Premature ripening and excessive fruit drop are associated with low Mg and B. In addition to excessive fruit drop, low B also causes external and internal cork, which destroys the market value of the fruit. Low levels of Ca are known to affect fruit quality. Although the problem is known, a method of increasing the Ca content of the flesh has yet to be resolved.

The other essential elements affect the metabolic functions of the trees; however, they do not appear to have a direct effect on fruit quality, per se.

MEASUREMENT OF BENEFITS FROM ADEQUATE MINERAL NUTRITION

With apples it would be very difficult to place a dollar value on mineral nutrition because it is only one factor in maximizing production. Also the degree of pruning and the crop load both have an effect on the nutrient level of apple trees. Heavy pruning has the same effect as adding N, and an excessive crop load tends to lower the nutrient levels.

Balanced nutrition is the key to maximizing production, that is, maintaining the trees in the optimum range for each of the essential elements. Balanced nutrition is not easily accomplished. The trees are subject to climatic conditions which have a direct influence on the crop load. In many of the apple producing areas soil moisture cannot be controlled. When the soil is either too dry or too wet there is a direct influence on the availability and absorption of plant nutrients.

By the use of leaf analysis it is possible to obtain nutrient balance within a "least cost" fertilizer program. When it is known what the nutrient level is in the trees, then only those elements that are low enough to cause a stress on the trees or that would show an economic response need be applied.

29
PEACHES

Norman F. Childers

The peach (<u>Prunus</u> <u>persica</u>) originated in China near Sian at least as early as 2000 B.C. Three species are still found growing in the wild: <u>P</u>. <u>davidiana</u> grown in the north, which is used for a rootstock; <u>P</u>. <u>mira</u> on the Tibetan Plateau; and <u>P</u>. <u>ferganensis</u> in the Sinkiang province, both in West China. There are a few species growing in North China that will tolerate -10°C but the juice may be blood color and the quality is poor. Professor M. A. Blake in his research at the New Jersey Agricultural Experiment Station used these "blood" species in breeding for cold resistance.

Spreading from China, the peach now is grown around the world between latitudes 25° and 45°, above and below the Equator. These limits are extended in some areas by altitude, ocean currents, large bodies of water, use of low-chilling cultivars, and spray chemicals to break the rest period of the trees. The peach can be grown in most apple growing sections, but it tends to do better closer to the Equator or where temperatures are warmer in the summer and winter.

Table 29.1 shows the approximate peach production by country. Leaders are the USA, Italy, France, Japan, Spain, Greece and Argentina. Exact production figures are not available for China, but it has been low due to lower priority versus grain and soybeans since 1949. One cultivar, Tung-Tao, ripens in late season, but is of poor quality; it will hold up longer in storage than any known peach - over four months. Reliable figures for the USSR are not available, but it should be sizable in the southern regions.

The peach was brought to Florida, at St. Augustine, by Columbus and the Spaniards on the second and third visits where it spread from coast to coast in North America.

---

[1]Retired Blake Professor of Horticulture, Rutgers University; now Adjunct Professor of Fruit Crops, University of Florida, Gainesville, Florida, 32611.

TABLE 29.1
Approximate peach production for the world, 1975-80

Country	Thousands of Metric Tons	Country	Thousands of Metric Tons
USA	1322	Korea (Republic)	78
Italy	1283	Yugoslavia	70
Spain	355	Romania	60
France	346	Canada	50
Greece	332	Peru	35
Japan	274	Uruguay	25
Argentina	223	Korea (DPR)	25
Turkey	211	Israel	28
Mexico	197	West Germany	25
South Africa	168	New Zealand	20
Bulgaria	156	Morocco	15
Brazil	150	Czechoslovakia	14
Hungary	140	Lebanon	10
Chile	92	Pakistan	10
Australia	80	India	10

Other countries producing about 1000 mt or less include: Egypt, Libya, Algeria, Madagascar, Tunisia, Paraguay, Venezuela, China, Cyprus, Iraq, Syria, Republic of Austria, Belgium, and Germany (DR). Data are from FAO (USDA) Production Yearbook.

California is the largest peach producer in the USA, totaling (millions of kilograms) about 900 (675 clings, the rest freestones), followed by South Carolina, 114; Georgia, 68; Pennsylvania, 45; New Jersey, 41; Michigan 25; Arkansas, 18; Washington, 16; North Carolina, 14; Virginia, 9; and Colorado 4.5.

The peach, like the apple, is getting more competition from other fruits and vegetables such as melons and tomatoes in season. The per capita fresh consumption of peaches in the USA has dropped from 6 kg in 1947 to around 3 kg in the late 1970s. In addition, about 4 kg per capita are processed.

ECOLOGICAL FACTORS

Temperature is the key factor governing where peaches can be grown successfully. Man has tended to grow peaches farther north and south than they should be grown. Hence, there are problems with winter freezing of buds and wood, spring frosts and/or adequate chilling. In recent decades growers have abandoned unprofitable sites where only three out of five good crops were obtained. Today, a grower must get almost five out of five crops to make the business pay and the effort worthwhile. Breeders have been busy developing cold resistant rootstocks and cultivars. Breeders also are developing better quality cultivars that ripen over a long growing season.

Peach trees perform best in a sandy loam soil of moderate to good fertility. However, they can be grown in heavier soils provided drainage is satisfactory. The peach is very sensitive to excess soil moisture, particularly early in the season. Where the soil is shallow and the underdrainage slow, as in some sites in Australia and New Zealand, the soil is ridged 30 cm or so and the trees planted on top of the ridge with the field graded so water drains readily from the sodded row middles. Peaches can be grown in very light sandy soils low in organic matter, but special attention must be given to application of trace mineral nutrient elements and adequate water supply.

A rooting depth of at least 1 m is preferred but not always attainable. In back yards and limited size plantings, mulching with hay or leaves or other forms of organic matter to a depth of at least 15-25 cm under the spread of the branches can be done where irrigation is not readily available. If the mulch is brought from a fertile or well-fertilized field, it also may provide adequate fertility without addition of chemical fertilizers.

Most peach orchards today are provided with irrigation, which, if of the overhead type, is used also for temperature control. In regions where temperature extremes are not a major problem, the microjet irrigation system is used to save water and costs. Peaches tend to have better color and quality and are sweeter where the weather is sunny most of the time, but a preponderance of sunny weather may be accompanied by a drought, which necessitates irrigation. Some pests such as mites and mildew may become a problem with sunny dry weather. On the other hand, fungal and bacterial diseases are more a problem under humid conditions.

A soil pH of about 6.5 (near neutral in reaction) should be maintained in humid regions by liming soils and using high-magnesium lime (dolomite) where soil tests show low Mg. In desert regions where soil pH is 7.0 or above (neutral to alkaline), some difficulty is likely to occur with trace element deficiencies, notably Zn, Mn and Fe. The peach needs relatively little Cu and, in fact, is highly sensitive to Cu sprays, but there are reports of its need under field conditions. Peaches also are highly sensitive to excess B, but there are many cases of its reported deficiency. Where the soil pH is maintained above 6.0 (mildly acid), Mo is not likely to be in short supply. Sulphur deficiency, however, has been reported in Washington state where the soil is deficient, and no sulphur-containing fertilizers or sprays are used and where smoke from industry is non-existent.

The spray program must always be taken into consideration with respect to its containing particularly the trace elements. Zinc, S, Mn, Cu and Fe often are present in disease control spray programs. A Zn compound may be used in rodent control. Sometimes the spray program is the only or best method of getting trace elements into the peach tree. They may be "tied up" and unavailable when applied to the soil, particularly in high pH (alkaline) soils.

Soil tests are of limited value for tree fruits where roots may penetrate 1 to 2 m down. They do show pH and whether some nutrients or other elements may be in excess. Soil samples ordinarily are taken in the upper few centimeters of soil, 30 cm depth at most, and are not representative of the lower depths where the roots are feeding. Leaf and fruit analyses are more indicative of the nutrient situation to and in the tree.

Nematodes (root parasites) can be a definite factor in limiting growth and fruit production. Preplanting injection of a nematocide in the tree row is important where samples indicate a nematode problem. It can be drilled to a depth of 35-40 cm behind a subsoiler down the tree row. A single-prong subsoiler run down the row three times about 1 m apart or a three-pronged subsoiler will break any plow sole and provide better aeration and drainage for the roots. Subsoiling and the application of a nematocide are particularly important when peaches follow peaches in a cropping pattern.

The standard soil management system for peaches today is the use of herbicides (for weed control) down the tree row to the outside dripline of the branches, with the row middles occupied by an adapted sod crop. K-31 or Alta Fescue varieties of tall fescue are often used with a light fertilizer that is broadcast. This sod is thick and tough, helps to control weeds, and tolerates orchard machinery. Simazine is one of the better herbicides, but it should be used on a 3-year rotation with other herbicides to avoid soil buildup. Herbicide use usually reduces the need for N. Growth of trees seems to be better where one early cultivation just after bloom is used to avoid frost damage to bloom due to stirring the soil and blocking heat movement upward from the soil. Rotovator tillage to a depth of about 5 cm is best to avoid cutting roots. The herbicide is applied after cultivation.

Care must be taken to avoid damaging the tree bark with orchard machinery, and to control deer or any other pest that harms the wood, foliage or fruit. All the factors listed above will have an effect on fertilizer efficiency and the production of marketable fruit.

The peach can be grown successfully in the subtropics and tropics. The closer one gets to the Equator, the more altitude is needed to get the required chilling to break the rest period of the trees. The rest period is that period during dormancy when the tree cannot be made to grow even though temperature and other factors are favorable. There are cultivars bred in Georgia, Florida and Texas, for example, that have relatively low chilling requirements of 250+ hours below 5°C. Most cultivars (established varieties) need 800 or more hours of chilling during the dormant growth period. Factors or practices that help break the rest period are an extended drought during the dormant period or the use of K and urea or dinitro sprays shortly before bud break. Trees that have not had adequate numbers of chilling hours will leaf out, bloom and fruit irregularly and after a few springs die (See Childers, 1975, 1978). There is relatively little of the total world peach production in the tropics and subtropics due to the special handling needed and the uncertainties of quality production.

NUTRITIONAL REQUIREMENTS

The application of fertilizers is only one practice involved in the total nutrition and growth and fruiting response of a peach tree. The important interrelated factors are pruning; fruit thinning; spraying; drought; chilling requirement; such pests as insects, diseases and nematodes and viruses; mechanical damage to roots and trunks; and extremes in temperature and water supply. One must be conscious of all these factors when evaluating fertilizer need.

After a soil test has been made, and the site and soil discussed with extension personnel or a neighbor in the business, an estimate of the fertilizer to use on the young trees can be made. It may be only N if the soil has good native fertility. But even N may not be needed if the site and climate are similar to the Fresno area of California. There the river-deposited soil is rich in mineral nutrients and has not been leached appreciably in a low rainfall climate, so only Zn may be needed. If the site is in a humid region, and the soil is very light and sandy with low organic matter, then N, P and K may be needed, with possibly some or all trace elements in a more or less "shot gun" mix, formulas for which are available (Childers, 1966, 1978, or local fertilizer mixing plants). If uncertain of the amount to use based on tree size and age, it is well to start with a medium to low application and adjust later according to production and quality. Nitrogen is the key nutrient needed in most of the peach growing areas of the world, with adequate but not excessive available water throughout the season.

When the trees come into fruiting, the kind and amount of fertilizer to use will depend upon the current amount and quality of growth and fruiting. Weak trees are subject to cold damage either in wood or bud or both. Flower buds are the first to be damaged, leaf buds and wood next. Moderate growth should be maintained where the shoot growth on fruiting trees stops for the most part by about midseason. Trees that are dark green and growing vigorously in late season may have fruit with poor color, flavor, and low in sugar, and the trees may be subject to cold damage later. Excessive growth also demands extra pruning which can be costly. A keen grower will study his trees from season to season and adjust the kind and amount of fertilizer according to production and crop quality. It is unwise to make big shifts in the amount of fertilizer from season to season. Some spots in the orchard, usually due to soil variation, will be weaker or stronger, necessitating special attention.

In recent years there has been a trend toward summer shearing of the tops of bearing trees usually about mid-season, which in New Jersey would be about August 15. Shoot tips are removed and the carbohydrates manufactured by the remaining leaves go to increasing the size of the fruit with better color and quality. More light is ad-mitted to the fruit, which improves color. Regrowth is moderate if the shearing is properly timed. Light winter pruning is needed to thin-out growth. More fruiting twigs, which is the 1-year growth, are left if the flower buds in cross section are showing winter damage (the cen-ter of the flower or the female organ (the pistil) will die and turn dark first; it will otherwise develop into the fruit).

DEFICIENCY SYMPTOMS

Some nutrient deficiencies tend to develop first on the basal or lower leaves on a shoot; these include N, K, Mg and P. Others tend to appear first in the tip growth; these include Fe, Zn, B, Cu and S. Calcium, Mn, K and Mo may appear in both the tip and lower leaves. Some miner-al deficiencies may start to appear near the tip of a shoot when rains occur following a drought, and the shoot continues to grow, but the deficiency will not be evident until after another drought period. Water in the soil may help to "pick up" and transport more of the nutrient to the root, temporarily alleviating the situation.

Nitrogen (N). Leaves are small and uniformly light green. Leaf petioles are erect and of brighter color in the fall. Shoots are short and spindly. Bark is a lighter color, tending toward a yellowish orange. Root growth is abundant in proportion to top growth, the color is a healthy yellowish white. Fruit quality tends to be good to very good, with good red skin color and storage quality. Production is reduced, and this is the main problem. Trees low in N are more susceptible to winter damage. In fact, this is true for most nutrient defi-ciencies, but some more than others.

Potassium (K). Older leaves have marginal and tip scorch, upward curl, dark bluish green and speckling between the main veins. Tip leaves sometimes show yellowing that in some respects resembles Fe deficiency. The shoots and branches die back. Bloom is heavy in early stages of deficiency, but light in later stages. Fruit have poor color, acid taste, and are smaller. Roots are normal in color but small in volume.

Magnesium (Mg). Basal lower leaves may show marginal interveinal yellowing and dying with a tendency to drop in midseason or early fall, although there may be an early leaf drop or burning if the deficiency is acute. Fruiting is about normal, but the leaf scorch and leaf drop tend to be worse where the crop is heavy on the entire tree, or in areas of a tree where the crop is heavy. Fruit drops early. Growth is about normal except where deficiency is severe. When both Mg and N are in low supply, Mg shortage may be evident as a yellowing between the main veins and along the margins of the older leaves, and the leaves hang to the tree fairly well with less dying of the tissues compared with leaves with good N supply.

Boron (B). Young leaves are small, twisted, thick and leathery, malformed with large and somewhat yellowish and corky main veins. Leaves may "cup", with some mottled yellowing. The leaves form rosettes at the nodes with short or no internodes. Shoots die back and ooze sap, and winter-kill easily. The side shoots grow a short distance and die back as if attacked by the Oriental fruit moth larva. The conducting tissue of the foliage collapses under acute B deficiency, which causes wilting and dying of young lateral shoots. Bark is rough, lenticels large and raised. Multiple buds at the nodes may open and die shortly thereafter early in spring. In fact, the whole tree may die suddenly, even though it grew well the former season. Fruit shows gumminess, cork toward the center, blighted seeds and poor mealy tasteless quality.

Iron (Fe). Young leaves show a straw yellow chlorosis among a fine network of green veins, both large and small. Under severe conditions the tip leaves become a solid straw yellow with some marginal and tip burning in late stages of deficiency. Older leaves are a normal dark green. Shoot growth is only moderately affected under marginal conditions. Bloom is reduced. Fruit shows less green color when immature and the red may be faded at harvest. Fruit production is reduced under moderate to severe Fe deficiency. Shoots and twigs are more susceptible to winter injury. Deficiency is aggravated by soil pH around or above 7.0 (alkaline soils). Water sprouts during the growing season show Fe deficiency first and young leaves at the periphery of the tree may show interveinal mottling early in the season before the air temperature warms and soil temperature is still cold.

Zinc (Zn). Leaves appearing early in the spring are small, long, narrow, in rosettes, show wavy margins and may have a murky whitish green yellowing between the main veins, which themselves remain greener. As the season progresses these leaves become a darker green, but remain small and rosetted. Shoots show dieback with limited growth on the extremities of the tree but in some cases vigorous water sprouts develop. Amount of bloom is reduced, the fruit are of small lumpy size, normally colored, poor quality and the skin where present may be thick. Fruit production is reduced according to the severity. Both root and top growth are reduced.

Manganese (Mn). Young and recently matured leaves show interveinal lighter green areas as compared with the area around the main veins, many of which may remain a darker green, giving a herringbone appearance. Under severe deficiency conditions there may be interveinal spotting and dying, with early leaf drop. Shoots may die back and the tree as a whole has a thin foliar appearance early in the fall. Fruit color is faded and quality poor, with early drop and some skin cracking. Last year's wood during winter is a faded pink rather than a reddish cast. Under Mn excess as in very acid soils, bark tissue may show roughness similar to "measles".

Phosphorus (P). There had been only limited reports of P deficiency in deciduous orchards, including the peach, until researchers in Australia and New Zealand began reporting severe P deficiency under their soil conditions. They had previously reported the problem due to Zn shortages since the symptoms closely resembled Zn deficiency on mature trees.
    Leaves are small bluish-green with the lower main veins showing a purplish color. Defoliation of basal leaves may occur. Roots are brownish and of less volume. Shoots are thick, short, upright. Overall top is smaller. Fruits are soft, puffy and acid with small skin cracking. The ground cover under the trees is thin and of poor quality. Yields are down.

Copper (Cu). When the supply of Cu in the soil is low due to a low natural supply, a high soil pH, or excess heavy metals such as arsenic or P, Cu deficiency can be aggravated by heavy applications of N, which tend to disturb the nutrient balance between N and Cu. When Cu is in short supply, young leaves, as in the case of plum or apple, may be soft and whitish with little, if any, chlorophyll. In the case of the peach, the young leaves may be long and narrow, and mottled green and white with irregular margins. Some plants show a downward cupping of the leaves, with dying and tearing of the leaf margins giving a ragged appearance of the top of the plant and the outer growing areas. New stem growth is soft with the shoots assuming an "S" type growth. Blooming is reduced or stopped. Fruit are of poor quality, acid, and may show cracking of the skin.

Molybdenum (Mo). In the younger leaves on the apple, for example, there is a light greenish interveinal yellowing of the main veins and small veins remain green. The older leaves show a marginal burning due to the accumulation of nitrates under low Mo supply. But the non-burned areas remain a more or less normal green. This deficiency where present is usually associated with low soil pH. Raising the pH to 6.0-6.5 may correct the problem, using lime.

Sulphur (S). Sulfur deficiency has been reported in a few peach orchards in the northwest USA, including Washington state. It has not been reported in other areas possibly due to the fact that S is used frequently in pesticide chemical sprays and as an impurity in certain fertilizers. Tip leaves show a uniformly yellow color, somewhat resembling Fe deficiency, but with Fe shortage there may be a network of green veins on small leaves. There may be some "burning" of leaves where S is deficient.

Calcium (Ca). This mineral is more important as a nutrient in humid areas than formerly realized. Lime has been recommended for years in deciduous orchards in the eastern USA, mainly to correct soil acidity. Some of the favorable response in tree performance associated with liming may have been due to supplying Ca as a nutrient rather than a mere correction of soil pH, but this has been difficult to demonstrate. Greenhouse sand culture tests at Rutgers University have shown reduced yields and poor quality fruit due to low Ca in the nutrient solution. At harvest, soluble solids were lower and flesh firmer as ripening was retarded. Taste panel records showed poorer flavor and harder texture than trees with adequate Ca.

Young leaves are distorted with tips hooked downward and margins curled. Leaf margins may show thin chlorotic marginal bands, irregular ragged margins with some browning. Large dead spots may develop in the center of the spur or side shoot leaves, resulting in leaf drop. Growing points and shoot tips are stubby and weak with some dieback. Root growth is stubby, short and weak.

SUPPLEMENTAL READING

CHILDERS, Norman F. (ed) 1966. Fruit Nutrition -- Temperate to Tropical. Horticultural Publications, 3906 NW 31 Pl., Gainesville, FL 32601. 88 pp.

CHILDERS, Norman F. 1975. The Peach -- Varieties, Cultures, Pests, Marketing, Storage. Horticultural Publications, 3906 NW 31 Pl., Gainesville, FL 32601. 660 pp.

CHILDERS, Norman F. 1978. Modern Fruit Science. Horticultural Publications, 3906 NW 31 Pl., Gainesville, FL 32601. 975 pp., 8th Ed.

30
CITRUS

Robert C.J. Koo

Citrus is grown throughout the world in tropical and subtropical climates. The crops are grown primarily in a belt between 40° N and 40° S latitude, except at high elevations. Minimum temperature is the limiting factor, though the killing effect varies with variety, rootstock, absolute minimum and its duration, and dormancy of trees. Most citrus is produced in 20 countries grouped in four regions, North America, South America, the Mediterranean, and others. The USA is the largest single supplier of citrus fruit, notably oranges and grapefruit, while Brazil is the largest supplier of orange juice. The Mediterranean countries are important producers of oranges and they are the largest exporters of lemons. Japan produces about half of the mandarins in the world, with the Mediterranean countries and North and South America each supplying about 25% of the fruit.

World production of citrus is expected to increase in the 1980s as newer plantings in South America and the Mediterranean countries come into full production. Consumption of citrus will probably keep pace with the increase in supplies. The volume of processed citrus will probably continue to increase in the 1980s while the consumption of fresh citrus fruit should remain relatively constant.

MINERAL NUTRIENT REQUIREMENTS

Citrus appears to require the same complement of essential nutrient elements as other fruit trees. Perhaps the most distinctive thing about the nutrition of citrus is the large number of deficiencies of nutrient elements that have appeared under intensive cultivation. Foliage

---

[1]University of Florida, Institute of Food and Agricultural Sciences, Agricultural Research and Education Center, 700 Experiment Station Road, Lake Alfred, Florida 33850, USA.

symptoms of deficiencies of N, P, K, Ca, Mg, Mn, Zn, Cu, Fe, B and Mo have been recognized in the field as well as in artificial cultures. In order to obtain maximum yield of high quality fruit, it is essential to evaluate the nutritional requirements of citrus crops locally and provide a balanced fertilizer program.

Since mineral nutrition is a major factor in the production of citrus fruit, a brief account is given of the most important elements.

Macronutrients

Nitrogen (N). Nitrogen is the key element in citrus fertilization and has pronounced effects on the growth and appearance of the tree, fruit production and fruit quality. It is used more extensively than any other fertilizer ingredient in commercial production. Citrus will grow in many parts of the world on natural fertility alone, but consistent bearing and high fruit yields almost always require N fertilization. Nitrogen affects the absorption and distribution of practically all other elements.

Nitrogen application is needed where most commercial citrus is grown to compensate for root uptake, leaching and volatilization losses. No general recommendation can be given for N application. Nitrogen rate, time, form and methods of application are largely dependent on prevailing soil, tree and climatic conditions.

Deficiency symptoms: Dull-green, yellowish, small leaves; decreased growth; dieback, and gradual defoliation. Fruit production decreases. The color of the fruit rind tends to be pale and smooth and the juice tends to have lower soluble solids and acid contents.

Excess symptoms: Lush vegetative growth, delayed maturity, regreening of fruit, decrease in fruit quality (thicker rinds, less juice, more acidity). However, there is little effect over a fairly wide range of N rates and N sources.

Phosphorus (P). Citrus has a fairly low requirement for P. Phosphorus is relatively stable in soil and does not leach readily, except in soils with a very low pH. Fruit contains less P than other macronutrients. Phosphorus deficiency may occur in soils with a low total P supply and in calcareous or acid soils where P may be fixed in unavailable forms.

Deficiency symptoms: Phosphorus deficiency in citrus is very rare and has been reported only from a few places in the USA and in Africa. There is an excessive drop of older leaves during and after spring bloom. Older leaves have a bronze, lusterless coloration, the bloom is sparse, and spring flush leaves are weak. Phosphorus-deficient fruit has thick, coarse rinds with high juice acidity.

Excess symptoms: Phosphorus toxicity in citrus is unknown. Application of high levels of acidulated P reduces root growth and can produce minor element deficiencies, especially of iron and zinc.

Potassium (K). Potassium affects tree growth and fruit production and quality, but to a lesser degree than N. Potassium deficiency has been reported in Florida, California and South Africa. The most important cause of K deficiency is low exchangeable K or very low total K in the soil. Removal by fruit is another cause. Citrus fruit contains more K than any other element; it is estimated that 2 kg of K are removed in one ton of harvested fruit. Continuous citrus culture will deplete K in the soil unless more is added. Excess Ca on alkaline soils can also cause K deficiency.

Deficiency symptoms: Retardation of growth and leaf size, yellowing of leaves and leaf drop, necrotic areas and spotting on the leaves. Fruit is small and juice acidity is low.

Excess symptoms: Although K is readily absorbed, K toxicity symptoms--marginal yellowing and burning--have been found only experimentally. High levels of K may induce Mg deficiency.

Calcium (Ca). Calcareous soils are used to grow citrus in many parts of the world, so Ca requirements have received little study. On acid soils, liming is used extensively to reduce acidity and to supply Ca. Larger trees and more root growth may result from Ca fertilization on acid soils.

Deficiency symptoms: Loss of vigor, thinning of foliage, vein chlorosis, leaf yellowing, and decreased fruit production. Calcium deficiency is very rare because usually there is far more Ca in the soil than the trees need for nutrition. The amounts of K and Mg present influence Ca deficiency.

Excess symptoms: The main effect, because of associated pH changes, is usually minor element deficiencies. High Ca also depresses K and Mg levels in the tree.

Magnesium (Mg). According to some authorities, Mg acts as a carrier of phosphate. This belief arises from the fact that Mg is mobile within the plant and that the parts of the plant characterized by high phosphorus content, such as seeds and meristematic tissue, also have a higher Mg content.

Deficiency symptoms: Magnesium deficiency has been reported from many parts of the world. Symptoms appear first on mature leaves where yellowish areas develop on both sides of the midrib, leaving a well-defined, inverted V of green at the leaf base. Symptoms appear more commonly on branches bearing heavy fruit loads. Leaves may abscise and twigs may die back. Yields are reduced and alternate bearing is exaggerated, especially in seedy varieties. Magnesium deficiency can be induced by high levels of K and Ca.

Excess symptoms: Mg toxicity has only been shown in sand culture. The roots are stunted and the leaves show iron chlorosis.

Micronutrients

Leaf analysis and visual observation are the best guides for determining micronutrient requirements.

Iron (Fe). Iron deficiency occurs mostly on calcareous soils and poorly drained soils with poor aeration. Imbalance of nutrient elements, especially certain heavy metals, will induce Fe deficiency. There are differences in susceptibility of citrus varieties and rootstocks to Fe deficiency. Trifoliate rootstock is very susceptible to Fe chlorosis.

Deficiency symptoms: In early stages the only effect is decreased leaf size. The typical effect is chlorosis of the young leaves and a green network of veins on a lighter colored background. In severe cases the leaves become entirely yellow with only a tinge of green along the midrib, and brown discolorations on the most exposed parts of the leaves. Twig dieback will follow. Fruit set and yield decrease and the fruit tends to be small. Soluble solids and acidity of juice will decrease.

Excess symptoms: No iron excess symptoms due to soil conditions have been observed. Application of excess iron chelate can cause leaf burn and leaf drop.

Manganese (Mn). Widespread Mn deficiencies of citrus have been reported in many regions of the world, occurring in both acid and alkaline soils, and probably due to leaching losses in acid soils and very low solubility in alkaline soils.

Deficiency symptoms: Both young and mature leaves show light green areas between the leaf veins, with green bands of various widths on both sides of main veins and the midrib. Symptoms resemble zinc deficiency, but the contrast between light and dark green areas is not as great. Fruit may be smaller than normal. When acute, fruit may be soft and the rind may be pale in color.

Excess symptoms: Marginal yellowing of the leaf with the central area remaining green. Excess Mn is rare in citrus, usually resulting from excessive soil applications and sprays, and can be controlled by liming.

Zinc (Zn). Next to N, Zn deficiency is the most widespread of all nutritional disorders, and occurs under a wide range of soil conditions. Deficiency is especially prevalent in sandy soils with low Zn content. Soil alkalinity and high P content can also contribute. Disturbance of the root system will usually result in Zn chlorosis of the leaves.

Deficiency symptoms: Small, narrow leaves with "frenching", i.e., mottling of the terminal leaves with yellow or almost white blotches appearing between the veins. Weak new growth eventually dies back. If deficiency is severe, fruit yields are drastically reduced and fruit size is decreased. The fruit may be misshapen and its ascorbic acid content decreased. Symptoms often appear on trees suffering from a decline-type disorder known as 'blight' (YTD). Zinc sprays will correct the symptoms but will not alleviate 'blight'.

Excess symptoms: Zinc excess can induce iron chlorosis, leaf burn and defoliation.

Copper (Cu). Copper deficiency is rarely found in citrus because Cu is widely used as a fungicide. On sandy soils, excess Cu presents more of a problem than deficiency. Copper deficiency has been reported in high organic matter soils.

Deficiency symptoms: Large dark-green leaves (early stage) with twisted growth, dieback of terminal growth, multiple buds forming multiple shoots, gumming of young shoots, and finely netted venation or flecking of young leaves. Brownish or reddish gum pockets form on the rind of the fruit which degenerate with a scabby, cracked appearance.

Excess symptoms: Iron chlorosis, reduced vigor, dieback, and stubby roots are common on sandy soils with more than 50 kg Cu/ha. Application of iron chelate and lime will usually reduce toxicity symptoms.

Boron (B). Since citrus requires only small quantities of B with slight excess resulting in toxicity, caution must be exercised in B applications.

Toxicity has been reported in some arid regions where salinity is a problem or where B is supplied unintentionally in irrigation water.

Deficiency symptoms: Translucent, water-soaked flecks or spots are found in young, sometimes deformed leaves. Yellowing occurs along the midrib and lateral veins of mature or old leaves, with curling and deformation of older leaves. Dieback and multiple bud formation are symptomatic with gum formation and splitting of stems, branches and trunk. Enlargement, splitting, and curling of the veins on the upper leaf surface may also occur. Gum pockets and brown discoloration of the albedo occur on the fruit, which may be misshapen and hard with thickened rinds. Excessive drop of young fruit may also follow.

Excess symptoms: Apical leaf mottling with yellow spots on the upper surfaces, and brownish resinous gum spots on the undersides. Leaf drop and dieback may follow.

Molybdenum (Mo). Molybdenum deficiency is not widespread. Because Mo becomes less available as the soil becomes more acid, deficiency can be corrected by adjusting soil pH to 5.5-6.5.

Deficiency symptoms: "Yellow spot" is the most characteristic symptom found in the field in Florida. In the spring, water-soaked areas appear on the leaves, then in late summer develop into larger interveinal spots with gumming on the lower leaf surfaces. The gum spots may turn black. Defoliation occurs in severe cases. The fruit is affected only in severe cases when brown spots with a yellow halo develop on the rind. The discoloration does not affect the albedo.

Excess symptoms: Yellowing and dropping of older leaves with dieback and iron chlorosis are associated with excess Mo.

Chlorine (Cl) - Chlorine deficiency has not been observed in the field or induced in culture.

Excess symptoms: Chlorine toxicity is a severe problem in many dry areas where poor quality irrigation water must be used. Symptoms start with yellowing of the tip and bronzing of the leaves. As the toxicity becomes more severe, yellowing and necrosis start at the tip and proceed downward, especially along the margins. Heavier than usual N applications, frequent flushing of the soil, and resistant rootstocks are preventive measures against Cl toxicity.

RESPONSE OF CITRUS TO PLANT NUTRIENTS

Nitrogen and K are the two principal plant nutrients used in citrus, and more information exists on N and K than other macronutrients. Most of the investigations deal with fruit production and fruit qualities. Reports on P and Mg are not as plentiful because these elements are not commonly included in citrus fertilization programs. Where they are used, only small quantities are applied. Calcium in the form of calcitic or dolomitic limestone is usually applied on acid soils to maintain desirable pH ranges. The response of citrus to macronutrients is summarized in Table 30.1.

For most macronutrients, soil application is recommended because of the quantities involved. Foliar applications may be used where warranted, but not as routine practice. For example, foliar application of urea may be used when a quick buildup of N in the tree is needed. Similarly, polyphosphate, potassium and magnesium nitrates may be applied as foliar sprays where soil applications are ineffective.

Most micronutrients are applied to correct deficiencies of individual elements. Very little information in terms of fruit production and fruit quality is available on citrus response to micronutrients. Assumably, once deficiencies are corrected, fruit production and quality will improve. Information to substantiate these assumptions, however, is lacking for most micronutrients. Micronutrients should be applied on a need basis rather than as routine practice. Leaf analysis and visual observations are the best guides for micronutrient applications.

Except for Fe, foliar applications of micronutrients are usually more effective than soil applications. Foliar sprays of Fe salts are only partially successful. Soil applications of Fe chelates offer the most effective means of correcting Fe deficiency.

The suggested method, rates and timing of micronutrient applications are summarized in Table 30.2.

In order to obtain consistent results for leaf analysis, certain sampling and analytical procedures should be followed. Sampling of 4- to 6-month-old leaves from nonfruiting twigs is preferred. One hundred leaves sampled from 20 trees are considered adequate to represent 5 to 10 ha. Sample preparation and analytical methods are fairly well standardized. Ranges for mineral nutrition standards are listed in Table 30.3, and are applicable for citrus grown in most regions. Some adjustments may be necessary for application to local conditions.

TABLE 30.1
Response of citrus to macronutrients

Measurement	Element				
	N	P	K	Ca	Mg
A. Tree growth	+	+	+	+	+
B. Fruit production	+	0 to +	+	+	0 to +
C. Juice quality					
1. Juice content	+	+	-	0	0
2. Soluble solids (SS)	+	-	-	0	0 to +
3. Acid (A)	+	-	+	0	0 to +
4. SS/A	-	+	-	0	0
5. Juice color	+		0 to -		
a. Red	+	?	0 to -	?	?
b. Yellow	+	?	0 to +	?	?
D. External fruit quality					
1. Size	-	-	+	0	0
2. Weight	-	-	+	0	0
3. Color	+	-	-	+	+
4. Peel thickness	+	-	+	0	0
5. Peel blemishes					
a. Wind scar	-	0	0	?	?
b. Russett (Phyllocoptruta oleivora)	-	0	0	?	?
c. Creasing	+	0	-	?	?
d. Plugging	-	0	-	?	?
e. Scab (Elsinoe fawcetti)	+	0	0	?	?
E. Storage decay					
1. Stem-end rot (Phomopsis citri)	-	0	-	?	?
2. Green mold (Penicillium digitatum)	-	0	0	?	?
3. Sour rot (Geotrichum candidum)	0	0	0	?	?
F. Peel oil	+	?	-	?	?

Increase (+)   decrease (-)   no change (0)   no information (?)

TABLE 30.2
Guidelines for micronutrient application methods, rates and timing

Element	Method		Rate		Time	
	Foliar	Soil	Foliar	Soil	Foliar	Soil
			g/l^a	kg/ha		
Mn	yes	yes[b]	.9	10.0	When	Any
					trees	time
Zn	yes	no	1.2	--	have	
					the	
Cu	yes	yes	.9	5.4	most	
					fully	
Fe	no	yes	--	--[c]	expanded	
					new	
B	yes	yes	.06	1.0	leaves	
Mo	yes	no	34-36	--		

[a]Grams/liter of actual metallic content for all elements except Mo.; 34-36 g of Na molybdate per liter is recommended.

[b]Soil application of Mn is not recommended for calcareous soil.

[c]Fe application: Acid soil-FeEDTA 20 g Fe/tree, Alkaline soil FeEDTAOH 50 g Fe/tree.

TABLE 30.3
Leaf analysis standards for citrus based on 4- to 6-month-old, spring-cycle leaves from non-fruiting terminals

Element	Defi-cient*	Ranges			
		Low	Optimum	High	Excess
N (%)	<2.2	2.2-2.4	2.5-2.7	2.8-3.0	3.0
P (%)	<.09	.09-.11	.12-.16	.17-.29	.30
K (%)	<.07	.7-1.1	1.2-1.7	1.8-2.3	2.4
Ca (%)	<1.5	1.5-2.9	3-5	5-7	7.0
Mg (%)	<.20	.20-.29	.30-.49	.50-.70	.80
Cl (%)	?	?	.30-.40	.41-.60	.70
Mn (ppm)	<17	18-24	25-100	101-300	500
Zn (ppm)	<17	18-24	25-100	101-300	300
Cu (ppm)	<3.6	3.7-4.9	5-16	17-22	22
Fe (ppm)	<35	36.59	60-120	130-200	200
B (ppm)	<20	21-35	36-100	101-200	250
Mo (ppm)	<.05	.06-.09	.10-.29	.30-.40	.5

* < = less than.

SELECTED REFERENCES

CHAPMAN, H.C. 1968. The mineral nutrition of citrus. In: Reuther, W., L. D. Batchelor, H. J. Webber (eds.). The Citrus Industry, Revised ed., Vol. II:127-289. University of California Division of Agricultural Sciences, Berkeley, California.

EMBLETON, T.W., W.W. JONES, C.K. LABANAUSKAS, and W. REUTHER. 1973. Leaf analysis as a diagnostic tool and guide to fertilization. In: Reuther, W. (ed.). The Citrus Industry, Revised ed. Vol. III:183-210, Appendix I-447-495. University of California Division of Agricultural Sciences, Berkeley, California.

EMBLETON, T.W., H. J. REITZ, and W.W. JONES. 1973. Citrus fertilization. In: Reuther, W. (ed). The Citrus Industry, Revised ed., III:122-182. University of California Division of Agricultual Sciences, Berkeley, California.

KOO, R.C.J. 1979. Nuticao E adubacao Citros (Fertilization and nutrition of citrus). In: Yamada, T. (ed.). Nutricao Mineral E Aducacao Citros. Boletin Tecnico 5:99-122. Instituto da Potassa and Fosfato (EUA); Instituto International da Potassa (Suica). 13.400-Piracicaba-SP-Brasil.

REITZ, H.J., et al. 1979. Recommended fertilizer and nutritional sprays for citrus. Florida Agriculture Expt. Station Tech. Bul. 536C, 22 pp.

SMITH, P.F. 1966. Citrus nutrition. In: Childers, N. F. (ed.). Fruit Nutrition. p. 174-207. Somerset Press, Somerville, New Jersey.

31
BANANA

Robert L. Fox

Banana is the second most important fruit crop in world trade; the USA imported 2.46 million metric tons in 1981. But even that statement fails to convey the true significance of the crop. For countless people in Southeast Asia, the Pacific Islands, Central America, and Africa, locally produced bananas (including plantains) are staple foods. In the Cameroons, for example, the principal food crops, taro, cassava and plantain, are produced in approximately equal quantities, about 600,000 tons each per year.

Bananas became an appreciable commodity in world trade about 100 years ago. As demand for bananas grew, large companies entered into their production and distribution. Export production was centered in Tropical America where extensive areas of fertile soils developed from volcanic ash and alluvium were available for exploitation. Thus much of the commercial banana industry began on soils needing little or no fertilizer. In much of the industry, there was no urgent need to support research on banana nutrition. Native soil fertility was soon depleted in some soils, and in other areas the soils were infertile from the beginning. In either case, continued production demanded a large input of fertilizer since it was soon learned that the banana is a demanding crop for plant nutrients.

Although systematic research on banana nutrition and fertilization was underway in the 1930s, it is only since about 1960 that substantial detailed information has emerged. Before that time nutritional requirements and recommendations on fertilizer use were very general; usually some standard quantity of mixed fertilizer was specified per pseudostem or per unit land area. Such recommendations ignored two important considerations: (i) soils differ in their ability to provide nutrients, and (ii) individual banana plantings differ in their internal

---

[1]Professor of Soil Science, University of Hawaii, Honolulu, Hawaii 96822, USA.

nutritional requirements depending, among other things, on the yield potential (or yield sought), and on the efficient use of nutrients that are supplied by the soil or by fertilizers.

ECOLOGICAL ADAPTATION

Banana is grown in environments as distinctly different as the semi-arid, subtropical area along the eastern Mediterranean Sea and the humid equatorial tropics. This is not to say that either extreme is well suited for bananas. The humid equatorial tropics are favored by nearly ideal temperature regimes (temperatures below 24°C may be less than ideal) and uniform rainfall distribution (water received should approximately equal pan evaporation). However, less humid subtropical areas are plagued with fewer leaf diseases and are blessed with more sunlight than most areas in the true tropics. Centers of commercial production seem to have been dictated more by the convenience of the market place and availability of transport than by ecological considerations.

Nevertheless, there are limits beyond which banana production should not be pushed. Frost imposes an almost absolute limit on banana growing, and yet growers in several areas push that limit to the extreme. In contrast, Hawaii, which is within the Tropical Zone and is noted for its mild climate, grows commercial banana only in low elevations. Even then, if mineral nutrition is adequate, seasonal effects on yield are strong. Such effects tend to disappear at low fertility levels as the age of the planting increases, suggesting that factors other than climate become relatively more important in determining fruit production. Perhaps the most influential factor is not temperature, as is generally assumed, but rather it may be light, for which competition is keen in a dense, well-nourished stand of bananas, especially during the cloudy winter season. Evidence that such is the case is presented in Figure 31.1. In Hawaii, flowers that emerge during May and June mature fruit more quickly than flowers that emerge at any other time of the year. As levels of nitrogen fertilizer increase, the length of the maturation period also increases. Furthermore, the date of flowering associated with the minimum time required for maturation shifts from mid-June to early May. This has the effect of centering the maturation period on the longest days of the year, as indicated by the horizontal bars in Figure 31.2. We interpret this as added evidence that in the tropics, especially in a dense stand of heavily fertilized banana plants, low sunlight is a primary limiting factor in production. N. W. Simmonds (1966) has called attention to the fact that few bananas are produced in cool moist areas of the world. He assumed that this was because of a scarcity of such areas. If the previous discussion is valid, he should have added

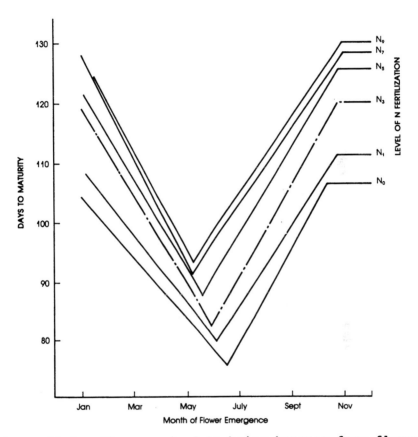

Figure 31.1. Time required to bring bananas from flower emergence to final harvest in relation to the date of flowering and level of nitrogren fertilization. The subscripts indicate the relative quantities of N applied. N_0 represents the no-nitrogen treatment; N_5 was designed to maintain 2.6% in leaf III.

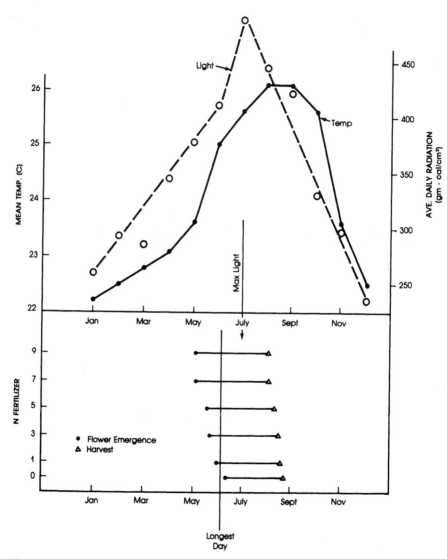

Figure 31.2. Date of banana flower emergence, date of harvest and duration of fruiting in relation to mean air temperatures and average daily radiation at Waimanalo, Hawaii. As the canopy becomes more dense (with increasing N fertilizer) the optimum time for flower emergence shifts so that the period of bunch development and maturation centers on maximum solar radiation and maximum day length.

light as a limiting factor because in the tropics, cool moist areas are usually under protracted cloud cover.

Banana production is directly related to bunch weight and stand density (number of plants per unit area). Production is inversely related to the length of time required for plants to produce fruit. Bunch size is related to the number of hands, the number of fingers per hand, and finger size. The number of hands and number of fingers per hand are determined at the time of flower differentiation which occurs 2 months or more before flower emergence. Thus, yield potential is set long before flowers emerge. High yield depends on maintaining plant vigor in every stage of development.

Much has been written about the importance of good nutrition during early stages of banana growth. This emphasis is justified, but there is some doubt about emphasizing the importance of potassium relative to nitrogen during early growth stages. Hawaiian experience has demonstrated that potassium deficiency symptoms intensify after flowering. As the developing bunch depletes K from the vegetative parts, leaves die abruptly, resulting in a green fruit bunch, sometimes with good specifications, atop a nearly dead plant.

Plant size, and ultimately bunch weight, are related to the number and size of functional leaves (leaf surface area). Many factors can influence leaf area -- disease, water availability, wind, light, temperature, and mineral nutrition. The problem is much less obvious when flowering is yet 1 or 2 months away and functional leaves are five to eight in number. Yet, this is a critical time. Healthy banana plants approaching flowering should have 10 to 12 functional leaves if the supply of nutrients, especially nitrogen, is adequate. In Hawaii, black streak disease reduces leaf surface area during winter months. Maximum bunch weights, however, have been attained during early winter, when days are short, skies tend to be cloudy, and disease has already begun to take its toll of leaves. This is because flower differentiation took place 5 to 6 months before harvest, perhaps in June or July when light, temperature and leaf area are most favorable. A detailed analysis of several thousand yield statistics and the corresponding weather data indicates that weather variables in the week nearest to the date of harvest have little influence on bunch weight, but at times more remote weather data relate more closely to bunch weight. Correlations are best between 13 to 25 weeks before harvest, which is the 2 to 3 month period centering on flower differentiation.

MINERAL NUTRIENT REQUIREMENTS OF BANANAS

The yield potential of bananas, when nutrient needs are fully met, vary greatly from area to area depending on weather conditions and planting (or stand) density.

In the Canary Islands, where mean winter temperatures are 16 to 17°C and mean temperatures of the hottest months are about 22 to 24°C, the time from appearance of the suckers to ripe fruit is 17 to 20 months. The time required from blossom to ripe fruit may be as short as 115 days during the coolest part of the year in that same area. For ideal conditions, no temperature would fall below 15°C or exceed 35°C, and for high yields temperatures above 24°C during most of the time are desirable.

Banana yields increase with increasing stand density. Chandler (1958) reckoned that about 740 bunches per hectare per year is as much as any but the best soil management combinations will average through a period of several years. If bunch weight is 32 kg, the per hectare yield would be 24 tons. Such yields or less are frequently reported. Experience in Hawaii has demonstrated that population densities considerably in excess of those usually maintained elsewhere can greatly increase yield potentials. Yields of approximately 100 tons (bunch weight) per hectare per year were realized over a period of 7 years by treatments which combined heavy N and K fertilization with high population densities on an irrigated soil that was naturally well supplied with Mg, Ca and P.

Yields of 30-35 tons per year are frequently reported for Tropical America and Taiwan, although 50 tons have been reported for a favorable site in Grenada. A 21-ton crop was assumed for purposes of calculation in Jamaica, but experimental yields of 32 tons have been reported there. About 50 tons is usually reported for the Coastal Plain of Israel. Experimental yields of 11 tons were reported for South Africa while in the Canary Islands a 70-ton crop was obtained on a plantation in which K and Mg were in favorable balance. A yield of approximately 80 tons was reported on Santiago Island, Cape Verde, for an experimental planting fertilized with Zn in addition to N and K.

It is difficult to know to what extent production could be increased by increasing the quantity of fertilizers supplied to bananas without at the same time removing other growth-limiting factors such as insufficient stand density. It is probable that substantial increases in production will require at least a proportionate increase in nutrient input. If such is the case, the quantity of nutrients in the crop (rather than concentration of nutrients in a specific tissue) becomes a matter of critical concern.

Nevertheless, critical concentrations are being used with some success for evaluating the need for fertilizers, if not for the quantities of nutrients required.

Tissue Analysis

Critical concentrations for the most frequently deficient nutrients in banana have been reasonably well defined over the past 25 years. Assuming that some allowance is made for relatively recent modifications, made necessary by greater yield expectations today than in former days, and considering the range of varieties and environmental conditions used for growing bananas, agreement is reasonably good on what the critical levels are. The following list should serve as a guideline for the midsection of the third most recently unfurled leaf at flowering:

N	- 2.8	to	3.0%
P	- 0.18	to	0.2%
K	- 3.2	to	3.5%
Mg	- 0.3	to	0.6%
Ca	- 0.6	to	1.0%
S	- 0.22	to	0.25%

The International Group on Mineral Nutrition of Bananas has proposed a standard sampling technique which, if used, would facilitate comparisons of results among banana-producing areas. This recommended procedure for sampling was omitted by Lahav and Turner (1983). The sample is the center one-half of 10-cm sections from both sides of the midrib taken from the midpoint of leaf III after one to three male flowers are exposed.

Although there is generally good agreement on appropriate critical concentrations of mineral nutrients in specific tissues, correlations between the concentrations of nutrients and response, as measured by yield increases associated with appropriate fertilizer treatments, have not been especially encouraging. This has led to suggestions from several sources that the internal nutrient requirement should be specified as <u>quantity</u> of nutrient in the plant or in the entire crop rather than concentration (<u>intensity</u>) of nutrient in a specific tissue.

The adequacy of some nutrients may best be presented as ratios. Such ratios have been calculated for K/Mg and K/Ca + Mg. In the Canary Islands a 70-ton yield was associated with a K/Mg ratio of 0.77, symptoms that could be corrected with Mg fertilizers were associated with a K/Mg ratio of 2.0 (calculated on an equivalence basis). If these ratios have general application, a plant containing 0.5% Mg and 3.2% K would be Mg deficient, and to arrive at a ratio of 0.77 -- as in the case of the 70-ton yield -- would require about 1.3% Mg. Such levels are rare indeed, which casts some doubt on the validity of K:Mg ratios, and it also suggests that more attention should be given to Mg deficiency -- or at least to K/Mg imbalance. This discussion may have some bearing on the unusually high sustained yields at the Waimanalo Experimental Farm, Hawaii (approximately 100 tons/ha/yr). The soil there is 50% saturated with Mg++ and cation exchange capacity is 50 meq/100 g. A K/Mg ratio of about 0.7 (all leaves) has been recorded there at harvest for plants with bunch weights in excess of 36 kg.

Excellent yields have been obtained in Israel from plants with a K/Ca +Mg ratio of about 1.0. The K:Mg ratio seems to be a sensible ratio to monitor because its use will help guard against serious depressions of Mg and Ca uptake that are associated with luxury uptake of K.

The uneven distribution of N and S in a profile of banana leaves leads to some anomalies in the N:S ratio. For leaf II, a N:S ratio of 18 to 1 was associated with maximum growth but for leaves zero, I, IV and V, the ratio was 10 to 6 and for leaf III the value was 13 to 6. Conditions which provide borderline conditions of N nutrition will probably give a N:S ratio in leaves II and III of approximately 10 to 6 which seems to be a more consistent value for well-nourished plants.

Luxury uptake of potassium. Luxury uptake of K requires special attention because K dominates the cation balance in most plants and especially so for bananas. Also, luxury uptake during the latter part of the vegetative phase can produce unrealistically high estimates of critical levels for K. In some cases, 5% K in leaves, or more, has been recommended as being required for maximum yields.

Critical concentrations are usually based on relative yields as a function of concentration in a specific leaf tissue at, or shortly after, flowering. The best treatment, it is assumed, will generate 100% yield. A least squares curve is fitted to the data. An example of one such curve based on Taiwan data is presented in Figure 31.3. Such curves probably are valid if the K supply is constant as a function of time; but if K supply is built up by repeated applications of K, or if some other growth factor(s) becomes more limiting than K, then what was once a deficiency situation may become one of luxury uptake, and low relative yields will be obtained with high levels of tissue K. The crop may continue to respond positively to the fertilizer because an unfavorable K supply was corrected during early growth.

An alternate procedure for interpreting such data is to enclose the data points in a response envelope. An example of this procedure is also presented (by heavy dashed lines) in Figure 31.3. The indicated critical concentration for maximum yields was shifted from >5% to 4%, a more reasonable number although still high by most standards.

Internal nutrient contents vs. internal nutrient requirements. The uncertainty that exists about the critical percentage of N, P or K in banana leaves emphasizes a weakness in the use of foliar analysis alone to assess

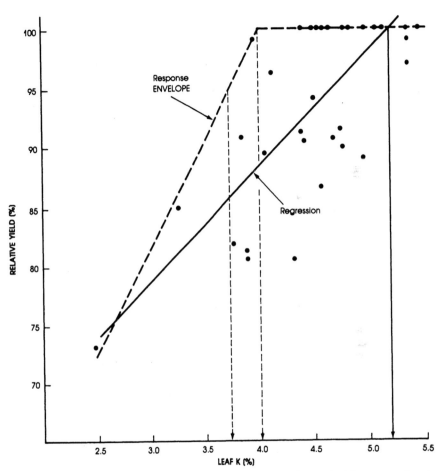

Figure 31.3. Two methods of interpreting foliar analysis data to obtain a critical level of K in banana leaf III. Linear regression frequently leads to over-estimated internal requirements for nutrients. Based on data from Taiwan by C.T. Ho.

fertilizer requirements. For example, plants that are growing slowly for some reason, even because of K-deficiency, may contain acceptable concentrations of K, although the external supply of K may be low. But small plants do not produce large bunches of bananas, even if leaf K percentages are above a critical level. Consequently, leaf composition often correlates poorly with fruit yields.

We now believe that a closer approximation of the true nutritional status could be had by determining the current quantity of nutrients in the plant or crop, (the Internal Nutrient Content), and comparing this quantity with the quantity of nutrient in a plant (or crop) capable of producing some desired yield. The quantity of nutrient in the plant or crop associated with near maximum yield has been named the Internal Nutrient Requirement. It can be expressed on the basis of a single stem, a single mat, or expanded to a hectare if the number of plants per unit area is known.

One method of expressing this is the nutrient content of the standing crop per ton of fruit produced per year. The internal nutrient requirement expressed in this manner will vary with the variety being produced and the general level of production being attained (the adequacy of all factors of production). The internal nutrient requirement is a way of evaluating the internal efficiency of nutrients. High internal nutrient requirements suggest internal inefficiency of one or more nutrients, but other factors may be involved also.

Such a system of analysis has an advantage over the critical concentration approach because it facilitates predicting quantities of fertilizers needed. If the internal nutrient requirement is known from experience or research, and the internal nutrient content can be estimated from appropriate samples, a deficit (or excess) can be calculated and expressed quantitatively. Appropriate corrections are made by multiplying deficits by fertilizer efficiency factors.

Internal nutrient requirements based on internal nutrient contents of high-yielding stems have been estimated for bananas. Certain assumptions have been made. The most important are: (i) yields are high (100 tons bunch wt/ha/yr), which is made possible by high stand density (2800 bunches/yr) and moderately heavy bunches (36 kg); (ii) growing conditions are reasonably good throughout the year; (iii) nutrient contents of the standing crop are critical for production; (iv) nutrient accumulation increases linearly from the end of the juvenile stage until harvest (this seemed to be a valid assumption for high-yielding banana in Grenada; (v) about 50% of the nutrient is removed in the bunch; (vi) recycling of nutrients from standing residue is complete in 3 months; and (vii) nutrients in the corm can be ignored.

An evaluation of the internal nutrient content of an experimental 5-year old planting at Waimanalo, Hawaii, is presented in Figure 31.4. The important implications of the evaluation are: (i) the internal nutrient content of banana is no greater than in a good crop of maize, therefore heavy, infrequent fertilization will not be an efficient use of fertilizer; (ii) nutrient uptake is a relatively slow and continuous process, therefore fertilizers should be applied frequently; (iii) gross nutrient removal, considered on an annual basis, is relatively great; fertilizers must be applied on a sustained basis; and (iv) the internal nutrient content (current contents) is about 53% of the total annual content of all harvested plants.

The annual content (kg/ha) of harvested plants (100 tons bunch-weight) at Waimanalo, Hawaii, is approximately as follows:

N	P	K	S	Mg	Ca
265	36	760	16	189	109

Such data can be made the basis of a start-up fertilizer program. Because only about 2% of the nutrients is in the suckers, these values are reasonable estimates of the nutrients that must move into the crop during the 9 months beginning with the vegetative stage. However, because nutrient uptake is somewhat sluggish, fertilizer must be given some lead time -- we suggest 1 to 3 months depending on the mobility of nutrients in the soil. Thus nutrients required during the small vegetative stage should be applied by the onset of that stage, and so on. (Such generalizations apply only to nutrients that are mobile in the soil, of course.) The percentages of the total nutrients absorbed during four stages of development are presented in the tabular material of Figure 31.4.

We assume that the nutrients required for full production potential should be in place by the end of the first year. By that time, the crop will have accumulated the nutrients required for full production. Production during the first year will probably not equal the second or third, but given a reasonable planting density, the nutrient requirement will be as great as any year thereafter, because nutrients applied in excess of requirements for the plant crop will go into mat development with relatively little recycling.

The External Nutrient Requirements--Sufficiency of the Soil Nutrient Supply

Two general concepts can be used to develop nutrition from the soil: (i) a quantity approach, best exemplified by extractable nutrients (often incorrectly referred to as available nutrients; and (ii) an intensity

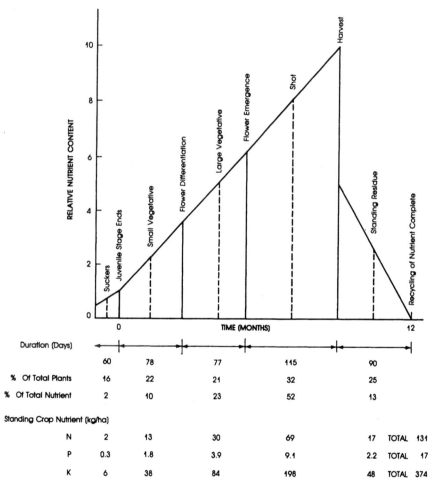

Figure 31.4. A schematic presentation of the various components of a mature planting of bananas giving the approximate interval of time in each stage of development, also the proportion and absolute quantities of nutrients residing in the various components. Favorable year-round growing conditions are assumed.

approach, of which nutrient concentration in the soil solution is one example and ratio of nutrients in the soil is another.

In Hawaii, during the establishment and early production phase, K fertilizer was needed to maintain 3.2% K in the mid-section of leaf III of plants which grew in soils containing 0.3 meq K/100 grams soil, whereas K was only slightly deficient in banana plants growing in soil that contained 0.8 meq K/100 g. The deficient soil supported a K concentration of about 0.03 meq/l (in a supporting electrolyte of 0.00125 M $CaCl_2$) and the energy of exchange (1364 log Ca+Mg) was - 3800 calories per mole. The nearly adequate soil supported a K concentration of 0.1 meq/l. Reports from Tropical America suggest that exchangeable K should be about 0.5 meq/100 g and that Mg should be about 0.8 meq/100 g.

Because K added as fertilizer usually accumulates near the surface of the soil (where it may, or may not, be available to the crop) it is difficult to evaluate results of soil K analyses from established stands. In Hawaii, low N plants (1.8 to 2.1% N in leaf III) scarcely responded to exchangeable K in the range 0.4 to 2.3 meq K/100 g soil but in the same experiment, plants that were intermediate in N content (1.8 to 2.1%) responded positively to increased exchangeable K over the range 0.4 meq to about 1.4 meq/100 g. Plants in the highest N category responded to increasing exchangeable K over the entire range 0.4 to 2.3 meq/100 g.

There is reason to believe that subsoil K will correlate with K nutrition better than K in the surface soil. Although banana has a reputation of being a "surface feeder", roots are relatively scarce in the surface 10 cm of soil.

There is ample evidence that good banana production can be obtained over the pH range 5.0 to 8.2. However, low pH almost always indicates problems with cation balance, phosphorus solubility, inadequate nitrification and excesses of Mn, Fe and Al. There is evidence that yields are substantially better when banana is grown on soils that are in the pH range 5.5 to 6.0 as compared with 4.5 to 5.0, or even 5.0 to 5.5. Nevertheless, most of the liming experiments with banana (including plantain experiments in Puerto Rico) indicate that lime was ineffective for increasing yields.

Banana apparently secures adequate phosphate from soils that would be deficient for many species. Simmonds (1966) regrets that mild phosphorus defining symptoms are not known. We suppose that an effective mycorrhizal association is responsible for this. However, it should be remembered that banana accumulates the P that it requires over an extended period of time, that a relatively small quantity of P is exported with the fruit, and that P moves readily from the mother plant to the suckers. All of this suggests that P deficiency, if it is observed at

all, will be seen early in the development of the plant
crop. In the Windward Islands, soils that test 6 ppm P
using Truog extractant are deemed to be P deficient, and
soils in the 6 to 10 ppm range are suspect. In Hawaii,
high yields have been attained for 10 years without phosphate fertilizer. The level of P in leaf tissues has
declined slightly over the years, however. When the experiment was initiated, the indicated soil solution concentration was 0.05 ppm P. Such concentrations are generally more than adequate for agronomic crops but would
be deficient for many vegetable crops.

EFFICIENCY OF FERTILIZER USE

Except in the most general terms, very little has
been written on the subject of fertilizer efficiency in
banana production. From work done in French Guinea and
Guadeloupe, apparently 100% of the applied K and at most
40% of the applied N are utilized. At Waimanalo, Hawaii
(clay soil), estimates of N recovery in harvested bananas
in relation to N applied as fertilizer (urea) indicate
approximately 65% efficiency. It is supposed that much
of the N was lost as volatized NH_3 that resulted from
spreading the fertilizer on bare ground, weeds and trash.
A reasonable estimate of mean fertilizer N efficiency in the tropics is 50%. As a first approximation,
estimates of fertilizer N requirements based on the internal nutrient requirement can be increased by a factor
of two.
Potassium and phosphorus pose different problems.
Although leaching can be a problem, even of P on quartz
sands and some organic soils, the usual problem is one of
restricted mobility. Nutrients accumulate near the soil
surface where they are inaccessible to many roots. On
this account, the efficiency of P and K may be no better
than N and sometimes much worse.

TIME, FREQUENCY, AND METHOD OF FERTILIZATION

Although some early recommendations were for three
fertilizer applications per year, it is now generally
agreed that nitrogen should be applied frequently for bananas, even as frequently as every month for irrigated
bananas. The early stages of growth are critically important because fruiting depends upon the development of
meristematic tissue during the early months of the
plant's life, development that is influenced by mineral
nutrition. Although particular importance is sometimes
given to K in the early growth stage, work in Hawaii
seems to justify stressing a special importance of K on
the maturation process instead. Sometimes levels of
plant K, although they are low by any standard, are adequate to produce a 6 to 8 hand bunch, but do not provide
enough K to bring the bunch to maturity. The potassium

demand of the bunch is so great that many or all of the
leaves die, leaving an immature bunch of bananas with no
supporting leaves. Thus there is good reason to empha-
size the desirability of a continuous supply of nutrients. There are other reasons for emphasizing a steady
supply of nutrients. The mother plant, and younger
plants that grow along side to produce later ratoons,
share a common pool of nutrients in the soil and in the
rhizomes of the mat. To fertilize specifically for one
stage of growth is difficult if not impossible.

The banana root system is normally very extensive.
This author has observed abundant roots in his compost
heap from a neighbor's banana plant at least 6 meters
away. A 2-month old plant may have roots extending out
nearly 2 m. Thus, unless some factor interferes with
root development, placement of fertilizer in close prox-
imity to the mat probably is an inefficient use of fertilizer anytime after the first 2 months.

Although banana root systems have been described as
being superficial, and roots can be observed under trash
on the surface of moist soil, there are usually few primary roots in the surface 5 cm. This does not necessarily mean that roots are not extracting nutrients from near
the soil surfaces; radioactive P studies demonstrate that
they do. But mycorrhizal-forming fungi can greatly enhance the effectiveness of roots in extracting P, and
these usually go undetected. In Hawaii, although K fertilizer has built very large reserves of exchangeable K
in the surface 5 cm, plants frequently contain less than
the desired contents of K.

FERTILIZER CONSUMPTION IN RELATION TO NUTRIENT
REQUIREMENT

The annual nutrient content (yield) of all harvested
banana plants from a productive planting is reasonably
constant. The following values are for Waimanalo,
Hawaii:

Nutrient	Complete Plant	Bunch
	- - - - - - - kg/ha/yr - - - - -	
N	265	132
P	36	18
K	760	357
S	16	4
Ca	109	12
Mg	189	28

It is instructive to examine the nutrient yield of
banana as given above, with levels of fertilizer that
have been required for near maximum (95%) production
(2.6% N, 3.2% K in leaf III). As a mean of 7 years, the

values were: N, 390 kg/ha/yr; K, 810 kg/ha/yr. The apparent efficiency of fertilizer, assuming complete recycling of nutrients from the residue, was 34% for N and 44% for K. Much of the unused K accumulated in the soil as exchangeable K.

Banana fertilizer application rates were compiled for 19 countries by C.C. Weir in 1974. The approximate averages as kg/ha/yr were:

$$N - 211$$
$$P - 35$$
$$K - 323$$

The data suggest that for near maximum yields, bananas are being underfertilized by a factor of approximately two.

CONCLUDING REMARKS

Banana is an important food crop of the world. It is a staple food in several tropical nations; and yet, fundamental knowledge on banana nutrition is deficient in many regards. Root distribution, root activity and mycorrhizal relationships, and associated nutrient uptake are poorly understood.

Banana requires large quantities of potassium and nitrogen relative to other nutrients -- a fact that has dominated research and practice to such an extent that little is known about requirements for other macronutrients, and much less so about the micronutrients. Deficiencies of phosphorus, sulfur and magnesium are reported infrequently, and acute zinc deficiency is rumored, but we have little solid information on which to judge the likelihood that deficiencies of these will be encountered. Clearly banana nutrition is a subject that should be more intensively studied and the results more widely published.

SUPPLEMENTAL READING

CHANDLER, W.H. 1958. Evergreen Orchards. 2nd Ed. p. 535. Lea and Febiger, Philadelphia.

FOX, R.L., R.A. LOWER, and R.M. WARNER. 1980. Premature leaf senescence of banana - a symptom of potassium exhaustion. Illustrated Concepts in Tropical Agriculture No. 18 Dept. of Agronomy and Soil Science University of Hawaii, Honolulu.

FOX, R.L., R.A. LOWER, and R.M. WARNER. 1978. Fertilizer requirements for near-term and long-term banana production in Hawaii. Proceedings, Hawaii Banana Industry Association 10th Annual Conference. pp 3-11.

353

FOX, R.L., R.A. LOWER, and R.M. WARNER. 1980. Nitrogen
and potassium interact to shape a yield response
surface. Illustrated Concepts in Tropical Agriculture No. 19. Department of Agronomy and Soil
Science, University of Hawaii, Honolulu.

GARCIA, V., E. FERNANDEZ CALDES, C.E. ALVAREZ, and J.
ROBLES, 1978. Potassium-magnesium unbalances in
banana plantations of Tenerife. Fruits 33:7-13.

HO, C.T. 1969. Study on correlation of banana fruit
yield and leaf potassium content. Fertilite
33:19-29.

LAHAV, E. and D.W. TURNER. 1983. Banana Nutrition.
International Potash Institute, Bulletin 7. pp 62.
Berne, Switzerland.

SIMMONDS, N.W. 1966. Bananas. p. 512. Longmans,
London.

TWYFORD, I.T. and D. WALMSLEY. 1973-1974. The mineral
composition of the Robusta banana plant. 1. Methods
and plant growth studies. 2. The concentration of
mineral constituents. 3. Uptake and distribution of
mineral constituents. 4. The application of fertilizers for high yields. Plant and Soil 39:227-243;
41:459-470; 471-491; 493-508.

WARNER, R.M. and R.L. FOX. 1977. Nitrogen and potassium
nutrition of Giant Cavendish banana in Hawaii.
Journal American Society Horticultural Sciences
102:739-743.

WEIR, C.C. 1974. Application of fertilizers to bananas
in Jamaica. Annual Report, Research and Development
Department. Banana Board Jamaica 1974. pp. 111-114.

32
WALNUTS

K. Uriu and D. E. Ramos

Walnut (genus <u>Juglans</u>) is known for the nuts and timber it produces. There are many species but it is the Persian walnut (<u>Juglans regia</u>), which is best known and grown most extensively in the world as a food crop. It is also commonly known as the "English" walnut but the name Persian is preferred because it indicates the area in which the species probably originated. It is widely cultivated with commercial production in France, Italy, Germany, Turkey, China, India, Chile, Iran, and the USA, the leading producer. Practically all of the USA production is in California, which now has over 80,000 bearing hectares with an average yield of about 2200 kg/ha (in shell basis).

The kernel of the walnut is high in fat and is valued for its high calorie content. The edible part being a seed, it is also high in proteins, minerals such as K and P, and vitamins such as thiamine.

ECOLOGICAL ADAPTATION

Walnuts are sensitive to both low and high temperatures. High summer temperatures can result in sunburning of the hull and dark shriveled kernels. Some damage to exposed nuts occurs at 38°C (100°F) and severe damage at 40-43°C(105-110°F) or higher, particularly when the trees are under moisture stress. In the spring, once growth begins, walnut leaves, shoots, blossoms, and nuts are easily killed if temperatures drop below freezing. Young vigorous Persian walnut trees, which continue to grow into late fall, are subject to killing of the new shoots by frost. In the winter when the trees are dormant, cultivars grown in California can tolerate -11 to -19°C (12-16°F) without serious injury. Some cultivars grown in colder parts of the world (carpathian walnuts) can toler-

[1]Professors, Department of Pomology, University of California at Davis, Davis, California 95616.

ate much colder temperatures when the trees are dormant. The amount of winter chilling needed to break the rest or dormant period is an important climatic factor that determines where walnuts may be grown satisfactorily. Most walnut cultivars require 800 hours or more of temperatures below 5°C during the winter months. If this condition is not met, the results can be seriously delayed bud opening, poor crops, and dieback of branches.

SOIL REQUIREMENTS

Walnuts do best on soils where roots can develop uniformly to a depth of 3 to 4 m or more. This requires soil of medium texture free of hardpan or layers of high clay content. Soils of this type are typically associated with alluvial fans of rivers and streams.

WATER

A mature walnut orchard requires about one million liters of water per ha (4 to 4.5 acre-feet of water per acre) per year if the trees are to produce the maximum number of high quality nuts possible. The period of irrigation is from late spring to late summer prior to harvest. However, late fall or winter irrigation may be necessary in areas of low average rainfall or in very dry years. Good quality water is also essential. Water that contains excess quantities of boron, chloride, carbonates, bicarbonates, or other salts is not suitable for walnut production.

CULTIVARS

There are many cultivars now available to growers. Varietal characteristics such as leafing out date, fruitfulness, tree vigor, blooming habit, shell seal, insect and disease resistance, sensitivity of the nut to heat, and kernel percentage help determine whether a cultivar is well adapted to a particular location. Cultivars that leaf out early may be damaged by late spring frosts. Also, they are exposed to more spring rains and therefore are most susceptible to walnut blight. Thus, cultivars are chosen to fit the conditions of a particular location.

ROOTSTOCKS

Persian walnut cultivars planted from seed do not produce trees and nuts that are true to horticultural type. For this reason, Persian walnut cultivars are propagated on rootstocks to maintain their distinctive characteristics and to maintain desirable traits. Rootstocks used include Northern California black (Juglans hindsii), paradox hybrid (Juglans hindsii x Juglans regia), and in rare cases English walnut (Juglans regia). Walnut rootstocks vary in their ability to resist diseases and nematodes and in the vigor they impart to the scion. Black walnut seedlings are moderately vigorous and are highly resistant but not immune to oak root fungus. On the other hand, they are susceptible to crown and root rot, root lesion nematode, and blackline, a graft union disorder caused by a virus which invades the J. regia scion following infection through diseased pollen. Paradox hybrid seedlings usually grow more vigorously especially on less fertile soils than do black walnuts. Also, it is better able to withstand waterlogged conditions and is generally more resistant to crown and root rot than is black walnut. In addition, paradox hybrid seedlings show moderate to high resistance to root lesion nematode. However, they are more sensitive to soil salinity and more susceptible to crown gall than is black walnut rootstock.

In California, walnuts are grown under intensive farming conditions. They are usually grown in large blocks. However, some farmers also grow almonds since the two nut crops require the same type of harvesting equipment. Sometimes a walnut grower will also grow other fruits such as prunes/and or peaches or row crops. The average walnut yield depends largely on the cultivar. The older cultivars such as Franquette are rather low yielding, producing at best about 2200 kg/ha (in-shell). The newer, more precocious varieties can produce 4400 kg/ha or more and up to 6600 kg/ha under ideal management conditions. The above figures are for California, which has probably the highest producing orchards in the world.

NUTRIENT REQUIREMENTS

For maximum yield, mineral nutrient levels must be such that no deficiency of any kind is present. Trees which exhibit adequate growth and large, dark green leaves generally are adequately supplied with all necessary nutritional elements. Lack of a particular element is usually expressed by characteristic deficiency symptoms. Thus, if one is familiar with deficiency symptoms, a suspected nutritional disorder can be diagnosed by visual observation. If this cannot be done, one can resort to soil or plant analysis.

SOIL ANALYSIS

Soil analysis has not been very useful since there is poor correlation between soil analysis values and the presence of leaf symptoms. Also, since walnut roots are so extensive and widespread, taking a single soil sample representing all the soil from which the tree derives its nutrients is nearly impossible. It is also often difficult to diagnose a deficiency from soil analysis results alone because reliable soil-standard values for walnuts are lacking. Of course, if a soil analysis shows an extremely low value for an element, it can be assumed that the tree will be lacking in that element. However, because deficiencies are poorly correlated with soil analysis values, this method is of limited usefulness.

Soil analysis is, however, useful in diagnosing toxicity caused by excess amounts of some elements such as sodium, chlorine, and boron. It is also useful in locating where these excesses are present in the soil. Knowing the location of a toxic material will help in deciding what corrective treatments to use and how to apply them.

Sampling procedure. In general, soil samples can be taken at any time. Although no major changes occur for most soil elements during the season, some precaution should be taken for a few of them. Nitrogen, for example, is subject to loss from denitrification under prolonged wet soil conditions. In many soils, nitrate, sodium, chlorine, and--to some extent boron--can be leached by winter rains and irrigation. Interpretation of soil analyses for these elements needs to take into account the time and quantity of prior irrigation or rain.

Because walnut tree roots spread through a large volume of soil and soil variability can be quite high, 3 to 10 spots around a tree need to be sampled. Most soils differ in composition with depth; surface soil can be quite different from the soil 30 cm down and the soil further down could be different again. Thus, samples of the surface 15 to 30 cm as well as each subsequent 30 cm downward should all be kept separate for analysis. However, samples taken from different spots around the tree can be composited if the soil at a given depth on one side of the tree is not too different from that at the same depth on the other side.

Each sampling situation will differ. Some orchards need to be sampled down only 1.3 m; others may require sampling at more than 3.5 m depth. Even deeper samplings may be necessary if one is trying to determine the depth at which a high concentration of a toxic element might be located.

Interpretative guides. As mentioned previously, no good soil analysis standards are presently available for many elements. However, for toxicities, critical levels may be approximated.

In the case of excess salts in the soil, the approximate conductivity of the saturation extract (ECE) at which yields have been reduced range from about 2.5 to 4 mmbos/cm (at 25°C) for many fruit crops. (The symbol, mmbos/cm is a measure of the concentration of soluble substances in soil extracts or in irrigation water).

Soil chloride tolerance of many fruits ranges from 7 to 25 meg/liter in saturation extracts. No specific value is available for walnuts, but nut crops are generally regarded to be chloride-sensitive.

Excess sodium causes structural deterioration of soils when the exchangeable sodium levels are greater than about 15% (10% for fine textured soils, 20% for coarse textured soils).

Excess boron is found in many areas in California. Walnut is sensitive to B and 1 ppm B in the soil can lead to visible leaf injury.

PLANT ANALYSIS

Plant analysis is very useful in the diagnosis of mineral deficiencies and toxicities. The plant tissue generally sampled and chemically analyzed is the leaf. Leaf mineral composition is dependent on many factors such as its stage of development, the climatic condition, the availability of mineral elements in the soil, root distribution and activity, irrigation, and so on. The level of the elements in the leaves is the result of all the factors that influence the absorption of mineral elements from the soil by the trees. In other words, the tree integrates all the factors and the composition of the leaf reflects this integration. Desirable concentrations of different elements, the critical levels below which deficiency occurs, and the levels above which toxicity can develop have been established (Table 32.1). The analytical results then can be compared with these values to determine the deficiency or toxicity. Plant analysis also can help one decide when measures should be taken to prevent deficiencies and toxicities from developing.

Seasonal use and level of mineral elements. In interpreting leaf analysis, a knowledge of the seasonal use pattern of mineral nutrients is helpful. At bud break in the spring, when root activity is still minimal, many of the elements stored in the stem and root tissues become available to the rapidly developing buds. Much of the N, P, Zn and perhaps K that is initially required comes from the stored supply and is redistributed to the growing points. Subsequently as root activity increases, these as well as all other mineral elements are increasingly taken up from the soil.

The results of leaf analysis do not necessarily show these use patterns of nutrients. They merely indicate the concentration levels of the elements since analysis results are always expressed as percentages (or parts per million) of each element in a given weight of dried leaves.

TABLE 32.1
Critical nutrient levels (dry-weight basis) in walnut
leaves sampled in July

	Concentration in dried leaves
Nitrogen (N)	
Deficient below	2.1%
Adequate	2.2 - 3.2%
Phosphorus (P)	
Adequate	0.1 - 0.3%
Potassium (K)	
Deficient below	0.9%
Adequate over	1.2%
Calcium (Ca)	
Adequate over	1.0%
Magnesium (Mg)	
Adequate over	0.3%
Sodium (Na)	
Excess over	0.1%
Chlorine (Cl)	
Excess over	0.3%
Boron (B)	
Deficient below	20 ppm*
Adequate	36 - 200 ppm
Excess over	300 ppm
Copper (Cu)	
Adequate over	4 ppm
Manganese (Mn)	
Adequate over	20 ppm
Zinc (Zn)	
Deficient below	15 ppm

ppm - parts per million

Mineral nutrient concentrations in leaves change as
the leaves first emerge, expand to full size, and then
finally reach senescence in the fall. The least change
in concentration occurs in mid-summer for many of the
elements. Leaf samples should be taken during this period because the critical levels established through experimentation and observation are based on samples taken
at that time. However, if one wants to take comparative

samples, that is, compare a good tree against a poor one, leaf sampling can be done at any time, provided one is aware of how the concentration of each element changed during the season. Generalized concentration curves for the season are shown in Figure 32.1.

Sampling procedure. The usual leaf sampling method for walnuts is to take the terminal leaflet of the compound leaf from spurs or the middle of moderately growing shoots. Since walnut leaflets are large, 20 to 30 taken at random around the tree are sufficient for a sample. Sometimes when toxicities are suspected or tip or marginal necrosis is found, the necrotic tissue can be removed from the rest of the leaflet and analyzed separately.

Interpretative guides. The correlation between leaf analysis levels and expression of a deficiency symptom ranges from poor to excellent for the different elements. It is excellent for deficiencies of Mg, Mn, P, and B. It is also excellent for excesses of chlorine, sodium and B. It is good for K and N, fair for Cu, Zn and Ca, and poor for Fe. Leaf analysis is very useful for all the elements when interpreted together with visual symptoms.

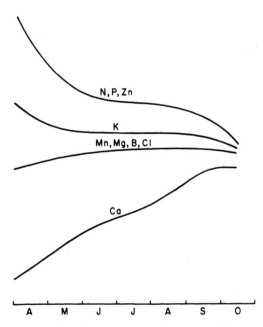

Figure 32.1. Generalized shapes of concentration curves of mineral nutrients in the walnut leaf during the growing season. Curves show trends, not actual levels (April through October).

FOLIAR DEFICIENCY SYMPTOMS

The third approach in diagnosing nutritional problems is to look for abnormal symptoms in foliage or growth. However, by the time symptoms appear, the tree would have suffered somewhat from the deficiency or toxicity in question. Also, experience is required to learn properly the symptoms for all the deficiencies and toxicities. Borderline deficiencies as well as deficiencies of two or more elements occurring in the same tree are difficult to diagnose. However, recognition of symptoms obviously can be very useful when accompanied by soil or leaf analyses. Sometimes all three approaches need to be followed to diagnose successfully a particularly difficult problem.

To be successful, diagnosis by observing symptoms must be done periodically during the growing season. One deficiency may show symptoms early in the growing season whereas another deficiency or a toxicity may show symptoms late in the season.

Nitrogen (N). Almost all trees can become N deficient if this element is withheld. Nitrogen deficiency is very difficult to diagnose by symptoms alone. Early in the season, leaves may be pale in color and small sized with reduced shoot growth. In the fall, the leaves senesce early, becoming yellow and defoliating earlier than normal. However, other conditions such as insufficient soil moisture may also produce similar symptoms.

Zinc (Zn). Zinc deficiency symptoms appear early in the season, especially if the deficiency is severe. The first evidence of Zn deficiency is delayed opening of the buds along the distil ends of shoots. Sometimes the delay can be as long as a month. When they open, the leaves are little, chlorotic, and appear in clusters, giving rise to the term "little-leaf." Leaves that are slightly Zn deficient are only slightly reduced in size but show scattered chlorotic areas between the veins in blades giving them a mottled appearance. Often portions of the leaflet margin fold upward to form a distinct wavy leaf margin. As the season progresses, the subsequent shoot growth tends to mask the Zn deficiency symptoms that were present earlier, making the evaluation of deficiency more difficult when done late in the season. Nuts along the Zn deficient parts of the shoots are markedly smaller in size than normal.

Boron (B). Boron deficient trees are stunted with weak shoot growth characterized by short internodes, which gives the tree a bushy appearance. Leaves often are chlorotic and misshapen, sometimes reduced to small, bractlike forms. In severe cases, terminal dieback occurs.

Boron <u>toxicity</u> symptoms begin appearing in mid- to late-summer after excess B has accumulated in the leaves. First the tip and then the margins of the leaflet become necrotic. In severe cases the necrotic tissue progresses into the interveinal areas and the margins of the leaflets curl upward. In very severe cases, most of the leaves on the tree can be so severely scorched that the tree appears as if it had been injured by a fire.

<u>Chlorine (Cl)</u>. Excess Cl results in terminal and marginal necrosis of leaflets. Symptoms appear in mid-summer and progressively become worse with time. In the field, Cl injury is difficult to distinguish from B excess since the latter produces a similar scorch pattern. Also, many trees observed with high Cl content may also contain high B. Leaf analysis is essential in this situation. Also, analysis of irrigation water for B and Cl will show whether excess amounts of those elements are coming from the water.

<u>Potassium (K)</u>. Potassium deficient leaves begin showing symptoms in early to mid-summer. Leaves become pale (resembling N deficiency) and later the edges of the leaves fold upward and curl in. The exposed undersides of the curled part of the leaflet develop a grayish cast (bronzing). Generally, most of the affected leaves are along the middle of the shoots. Leaf size is reduced as well as shoot growth. Invariably, overall nut size also is reduced.

<u>Manganese (Mn)</u>. Manganese deficiency symptoms begin to show in early to mid-summer. The characteristic symptom on the leaflet is chlorosis that develops between the main lateral veins and extends from the midrib to the margin. The chlorotic pattern produces a "herring-bone" effect. The chlorosis gives the leaves an overall pale or slightly yellow appearance. Generally, leaf size is unaffected unless the Mn deficiency is very severe, in which case leaf size could be smaller and the yield reduced.

<u>Iron (Fe)</u>. Iron deficiency symptoms appear early in the season. The entire leaf is uniformly yellow with the terminal leaves affected more than the basal. Sometimes the leaves are so chlorotic that they appear white. Often the basal leaves on shoots are green and only the leaves towards the terminal are chlorotic. In severe cases the chlorotic leaves may develop scorch and abscise. Iron chlorosis symptoms frequently are associated with calcareous soils (lime-induced chlorosis) or heavy, poorly drained soils.

Magnesium (Mg). Magnesium deficiency usually is associated with low pH (strongly acid) soils. Symptoms appear in mid- to late-summer after considerable shoot growth. The basal leaves, especially those on vigorous shoots, begin to show a distinct chlorotic margin at the apical and lateral edges of the leaflets, eventually leaving an inverted V-shaped green area along the basal part of the midrib. As the season progresses, the chlorotic areas along the margins become necrotic and dark brown in color.

Copper (Cu). Copper deficiency symptoms appear in mid-summer. Scorch develops on leaves near tips of shoots and defoliation follows. Slight shrivelling of the shoots occurs and some small dark brown spots appear on the shoot near the tip. Terminal dieback subsequently occurs in late-summer. Kernels are often badly shrivelled.

Phosphorus (P). Phosphorus deficiency is rare. Trees usually show very weak shoots and have sparse foliage with smaller than normal leaves. These leaves are yellow and develop necrosis in irregular areas followed by early defoliation.

TREATMENTS FOR SOME OF THE MORE COMMON DEFICIENCIES AND EXCESSES

Nitrogen (N). In mature walnut orchards a yearly application of 170 to 225 kg N/ha has successfully prevented N deficiency. Nitrogen must be applied yearly since the tree will remove the N applied, and N--being mobile in the soil--may be lost. The usual time for N application has been in the fall before the winter rains.

Zinc (Zn). Zinc deficiency can be corrected by several methods. (i) Zinc grazier's point (an old treatment--now seldom used) can be inserted into trunks or limbs up to 5 cm in diameter. For larger trees, galvanized iron pieces 2 to 2.5 cm wide and 5 cm long should be driven well into trunk or branches. Pieces should be spaced in rings with individual pieces about 5 cm apart in an irregular pattern. Treatment generally lasts for several years and can be made at any time of the year. (ii) Soil treatments have given variable responses. In sandy soil, trenching or injecting 2 to 9 kg zinc sulfate per tree, depending on size, has given good response. Broadcasting 1 kg of zinc chelate (zinc EDTA) before an irrigation also has corrected deficiency. Soil treatment becomes less effective in fine textured soils. (iii) Foliar application of zinc sulfate (36% zinc) at 0.45 kg per 400 liters of water (1 lb/100 gal) or zinc chelate (zinc EDTA) at 0.9 kg per 400 liters of water (2 lb/100 gal) have corrected deficiencies. The first spray should be applied soon after full bloom when the "feathers" on pistillate flowers have turned brown. Depending on the severity of the deficiency a second and third application should be made at 2 to 3 week intervals. This probably is the most practical correction for Zn deficiency in walnuts.

Boron (B). Boron deficiency can usually be corrected very easily by the application of 1 lb (0.45 kg) of borax per tree broadcast onto the soil surface. This treatment should last from 3 to 5 years before reapplication is necessary.

Potassium (K). A massive application of a K fertilizer is usually needed in the correction of K deficiency. Potassium sulfate can be drilled into the soil along the side of the tree at a depth of about 15 to 20 cm (6 to 8 inches). This treatment may last for several years before another application is needed.

Manganese (Mn). Manganese deficiency can be corrected easily by the application of a foliar spray of manganese sulfate, 3 to 5 pounds (1.35 to 4.7 kg) per 100 gallons (400 liters) of water, or manganese chelate (manganese EDTA) at 2 to 3 pounds (0.9 to 1.35 kg) per 100 gallons (400 liters) of water. This spray should be applied early in the season soon after leafing out.

Magnesium (Mg). Magnesium deficiency is usually corrected with large amounts of dolomite applied to the soil surface. This treatment may take some time before an effect is obtained. In the interim a spray of magnesium nitrate (magnesium sulfate plus calcium nitrate) can be applied in the spring.

Copper (Cu). Copper deficiency can usually be corrected with the application of a Bordeaux spray.

Boron, chlorine and sodium excesses. If any of these elements are present in high amounts in the irrigation water and are causing toxicity, the water source needs to be changed. If toxicity is from an excess in the soil, irrigation water in excess of the needs of the tree can leach some of the toxic elements and thus reduce injury to the tree.

BENEFITS FROM ADEQUATE NUTRITION

Actual information on the benefit to be derived from the correction of various mineral deficiencies is difficult to find. With some elements it is obvious that the correction of deficiencies would result in considerable economic benefit. Zinc deficiency results in stunted growth and in the production of small nuts. Potassium deficiency also results in small-sized nuts. It is obvious that the correction of these deficiencies would result in increased yields and therefore, in increased economic returns.

It appears that the walnut is not sensitive to high N application. In many other fruit crops high N results in delayed maturity, but with walnuts there appears to be no influence on maturity by N. Besides with walnuts there is no incentive to be late or early in maturity. The quality of the nut is based primarily on color and the presence of mold or worms. The adverse effect on color is by high temperatures and not by nutrition. The increase in the amount of mold is caused by wet conditions rather than nutritional disorders. Worms increase more by delaying the harvest than by poor nutrition.

In the case of toxicity, for example excess B and salinity, the crop is somewhat reduced. The correction of these toxicities can increase the crop. However, to correct toxicities, these elements must be leached with excess water. Labor and water costs go up and it becomes very difficult to assess the ultimate economic benefit.

UTILIZATION OF INCREASED PRODUCTION

Walnut acreage and total production have been increasing rapidly in recent years, especially in California. The USA supply exceeds the domestic demand and exports now account for about 40 to 50% of USA production. In order to utilize increased production of walnuts, new uses must be found and/or domestic or export consumption increased.

33
MANGO
Donald L. Plucknett

Mango (<u>Mangifera indica</u>), a relative of cashew and pistachio nuts, is perhaps the most popular fruit in the sub-humid tropics and subtropics. A native of the Indo-Burma region, mango has been grown for thousands of years in India and neighboring countries. It has spread world-wide in the tropics and subtropics and is esteemed for its delicious colorful fruit.

Mango is grown both as an orchard crop and as a tree in the home garden. The fruits are mostly consumed near the place of production, but a small amount does enter international commerce. The present world production of mango is estimated by FAO at about 14 million mt. India is the biggest producer with more than one million hectares of the crop producing over 9 mt of fruit. Almost 80% of the world mango crop is produced in Asia.

Mangoes are eaten mostly as fresh ripe fruit, although some people prefer them green. The most common use is as a dessert fruit, but pies, chutney, pickles, jams and other products are also made. Also, the fruit is sometimes processed and sold as canned juice, canned purees, or other products.

THE MANGO TREE

Mango is an evergreen tree that may become very large with age. Seedling trees can be enormous, but grafted or budded trees are usually smaller. The trees have a distinctive dense growth habit and are usually broad and round-topped or oval in shape. The leaves are lance-shaped, commonly 25-30 cm long, dark green and glossy.

[1]Scientific Adviser, Consultative Group on International Agricultural Research (CGIAR), the World Bank, Washington, D.C. 20433, USA.

Growth in mangoes is not continuous, but rather occurs in "flushes" of growth followed by periods of inactivity. Parts of a tree may be in different stages of growth at the same time, for example one side may be actively growing while the other side may appear to be dormant. New leaves are a reddish-copper color, and usually hang limply from the shoot. As the leaves mature they become dark green and quite rigid.

The flowers are borne in large panicles at the end of terminal branches that were produced during the previous season of growth. This feature is a factor in the cyclical bearing behavior seen in most mango varieties. Flowering occurs over several months, usually in late winter and early spring. Trees may bloom two or three times during a season. Several thousand small flowers may be produced on a single panicle, but only a few will produce fruit.

The fruits are highly variable in size, shape and quality, depending on the variety. Fruits may be oval, round, heart-shaped, kidney-shaped, or long and slender. Individual fruits may weigh from a hundred grams or so to more than 2 kg. Skin color when mature ranges from green to red or orange or purplish, with some fruits showing multiple-colored patterns. The flesh is usually yellow or orange in color and can range from quite fibrous and "stringy" to smooth and melting. The single seed is large and flattened, with a fibrous woody husk enclosing a white kernel.

There are hundreds of varieties of mango, each originating from a single seedling and then cloned usually by grafting or budding. Improved varieties are best to plant since seedling orchards are too variable to produce uniform, high quality fruit.

CROP GROWTH REQUIREMENTS

Mango grows best in areas where rainy periods coincide with vegetative growth and dry conditions prevail during flowering and fruiting. The tree will grow vigorously and become quite large in the humid tropics, but fruit production will usually be very infrequent and unsatisfactory because of the wet conditions during the reproductive period. Hot dry sunny conditions during fruit formation and ripening usually result in large harvests of delectable sweet fruit.

Mango responds well to irrigation in arid areas such as in Egypt where the crop is grown entirely under irrigation. Generally, irrigation should be withheld or reduced at or about the time of flower-bud differentiation.

CYCLICAL FRUIT PRODUCTION

The most important problem in mango production is its cyclical nature of bearing. Mango tends to have good and bad years, usually referred to by growers as "on" and "off" years. During an "on" year a heavy crop will be produced, often creating surpluses. During an "off" year, production may be too small to meet expenses, despite the usually higher prices.

Understanding the cause for on- and off-year bearing of mango is important; however, little scientific information is available concerning the problem. Some generalizations can be made: (i) some varieties are not so variable in fruiting habit as others; (ii) a definite pattern of alternate-year bearing does not occur, rather a heavy crop is interspersed with one or several years of light harvests; (iii) the chief factor affecting flowering is the amount and stage of maturity of the terminal shoots; (iv) usually shoots that are at least 8 months old and which have not grown actively for 4 months or so will produce flowers; (v) good vegetative growth the previous year stimulates flowering; (vi) a stress or shock to the tree may help to produce a good fruit crop the next year; and (vii) withholding nitrogen and water during the autumn, or before or during the period when flower-bud differentiation takes place, can be beneficial.

Good management can help to reduce the on-year, off-year bearing pattern of mango, and good nutrition of the tree can play a role in this. The main aim is to keep from over-stimulating the trees during off-years and to prevent the tree from producing mostly flowers and fruit during the next on-year, and stimulating enough vegetative growth during an on-year to ensure sufficient flower and fruit production the next bearing season. To do this, trees are fertilized less during off-years, and receive more nitrogen fertilizer in early summer in order to stimulate vegetative growth on parts of the tree that have experienced early spring flowering.

NUTRITION OF MANGO

Mango grows rapidly and responds to fertilization, but it is not a highly exacting crop as regards nutrition. Its root system is deep and extensive, and it has the ability to recover and grow well following periods of neglect. Therefore, if soil and tissue analyses are to be used effectively, they should be practiced over a period of years rather than at a single given time.

Mango has not received as much research attention as have other tree crops. Most research on mango nutrition has been done in Florida, India, Israel and South Africa.

VISUAL DEFICIENCY SYMPTOMS

Dr. Ramon Valmayor of the Philippine Council for Agricultural and Resources Research and Development (PCARRD) has summarized the visual deficiency symptoms of mango, and these are presented below.

Nitrogen (N). Yellow undersized leaves, severe retardation of growth. Mature trees have higher than normal levels of P, K, and Ca. Twigs sometimes also become yellow in color.

Phosphorus (P). Retarded growth; premature dropping of older leaves, following partial die-back from the tip. Under severe deficiency, only small green younger leaves are borne at the tips of the branches. Some branches may also die back.

Potassium (K). Older leaves develop small yellow spots (freckles) distributed irregularly over both the under and upper leaf surfaces. With time the spots enlarge, coalesce, and produce necrotic areas along the leaf margins. Affected leaves stay on the tree and do not fall until completely dead.

Magnesium (Mg). Considerable reduction in growth, and premature defoliation. A few older leaves that persist show a yellowish brown chlorosis, featured by a green wedge down the central part of the leaf and a bronzing starting from the edge of the leaf, with a rounded margin between each pair of lateral veins. If severe, the chlorosis extends to the midrib, and the margins both die.

Manganese (Mn). Reduced growth, but leaf symptoms appear very late, starting on fully extended but tender younger leaves about 1 month old. Leaves show a yellowish green background with a fine network of green veins, this pattern being most marked on the upper surface and disappearing after a few weeks. Mature leaves are thicker and more blunted.

Sulfur (S). Leaves show a similar appearance to phosphorus deficiency. Tree growth is retarded and defoliation begins. Mature leaves develop very deep green color and marginal necrosis, and later fall from the tree. Sulfur deficiency scorch develops along the sides of the leaf while in phosphorus the scorch develops to the leaf tip.

Zinc (Zn). Symptoms start when leaves are very young and still pink in color. Leaf blade thickens and does not reach its normal size. As the leaf matures, leaf shape is distorted due to bending of the leaf margin up or down, also causing the tip to bend accordingly. In some cases the veins may become prominently yellow, and the tip may curve back toward the branch on which it is borne. Interveinal areas may be normal green or mottled. Leaves are usually smaller than normal, both in length and width, sometimes by as much as half normal size. The thickened leaf blade may be quite brittle. Twigs with affected closely spaced leaves show a rosette appearance, the well-known "little leaf" symptom of Zn deficiency. If not severe, only a portion of the tree may show the symptoms. If severe, nearly all the new growth is affected, and new flushes fail to appear. Some twigs die back. In nursery plants and young transplants, all new growing points are affected and remain stunted. Flower panicles of trees showing "little leaf" symptoms are usually small, irregular in shape, and often contain many deformed and drooping spikes.

Copper (Cu). Deficiency symptoms may accompany Zn deficiency. Shoots produced on long drooping, S-shaped branches of previous growth are weak, lose foliage, and die back.

Other elements. Deficiencies of Ca, B, and Fe do not cause distinctive leaf symptoms. Trees low in Ca are small and slightly paler green in color than normal trees.

PLANT TISSUE ANALYSES

Some tissue analysis work has been carried out in Florida: Table 33.1 presents a summary of the results of that work.

TABLE 33.1
Desirable ranges of five mineral elements in mango leaves in Florida (Young and Sauls, 1981)

Element	Desirable Range
Nitrogen	1.0 to 1.5%
Phosphorus	0.08 to 0.175%
Potassium	0.3 to 0.8%
Calcium	2.0 to 3.5% (acid soil)
	3.0 to 5.0+% (alkaline soil)
Magnesium	0.15 to 0.4%

The Florida workers point out that to be effective, leaf analysis should be done on standardized samples taken over a period of several years from orchards that are managed under a known fertilizer program.

SOIL ANALYSIS

Mango can be grown on many types of soils, but it cannot withstand prolonged waterlogging. It has a deep, extensive root system that appears to adapt to a wide range of soil conditions. Soil analyses do not appear to be as useful for mango as plant tissue analyses, and no standards for soil analysis have been established. For best results, soil analysis should be used in conjunction with plant analysis.

NUTRITION AND CROP MANAGEMENT

Mango has three major growth stages; (i) nursery, (ii) non-bearing stage, and (iii) bearing stage. Nutritional needs are quite different in each of these and must be considered separately.

Nursery Stage

Both seedlings and grafted or budded trees are grown in the nursery. Young trees may be grown either in pots or in special beds in the field. The main objectives of the nursery stage are rapid growth of seedlings to be used as rootstocks and good establishment of grafts or buds.

Either basal applications of farmyard manure or frequent light applications of inorganic fertilizers are used to promote rapid, healthy growth. Nitrogen is very important in the nursery, and most of the frequent fertilizer applications that are made will be nitrogen fertilizers.

Non-Bearing Stage

After the trees are transplanted in the field, a new phase of growth begins. Mango trees are usually spaced widely, from 8 to 11 m or so, to allow for their eventual large size. Thus large areas of the field may remain essentially unused for several years while the trees are growing and maturing. Intercropping is often practiced during this time to occupy the open space and to provide additional income.

The objective during the non-bearing stage is to obtain rapid healthy growth of the young tree. This stage may last from 4 to 8 years, depending on variety and growing conditions. Most trees will produce fruit by the fourth or fifth year.

Nitrogen, calcium and phosphorus especially are required during this period of rapid growth. Liming is beneficial for acid soils. In Florida dolomitic limestone is used to provide both calcium and magnesium.

Bearing Stage

This stage begins when fruit production begins, usually by the fifth year. From this point fertilizer applications should increase each year until the trees reach 12 years of age. After that, fertilizer applications tend to level off. In Florida, when bearing first starts, individual trees producing about 25 kg of fruit are fertilized with about 5 kg of an 8-0-8-2 ($N-P_2O_5-K_2O-Mg$) fertilizer. As trees increase both in size and fruit production, fertilizers are increased until about 12-15 kg per tree of the mixture is being applied to trees each producing 250-300 kg of fruit.

SOFT-NOSE DISEASE

A nutritional disease, "soft-nose", has been found in Florida. It involves a physiological breakdown of the tissues of the smaller end of the fruit away from the stem. It may be the same as the "tip pulp" disease known in India.

Soft-nose has been associated with high nitrogen fertilization on acid soils in Florida. Affected trees usually have low calcium in the leaves. High calcium in the leaves, even when nitrogen fertilization is rather high, tends to reduce the incidence of soft-nose.

SUPPLEMENTAL READING

Indian Council of Agricultural Research 1967. The Mango, A Handbook. ICAR, New Delhi.

SINGH, L.B. 1968. The Mango. Leonard Hill, London.

SINGH, R.N. 1978. Mango. Indian Council of Agricultural Research, New Delhi.

VALMAYOR, Ramon V. (no date). The Mango: Its Botany and Production. College of Agriculture, University of the Philippines, College, Laguna, Philippines.

YOUNG, T.W. and J.W. SAULS. 1981. The Mango Industry in Florida. Bulletin 189, Florida Cooperative Extension Service, Institute of Food and Agricultural Sciences, University of Florida, Gainesville, FL 32611. 70 pp.

34
COTTON

Ulysses S. Jones

Cotton (there are four major species, but <u>Gossypium</u> <u>hirsutum</u> is most important) has been grown for at least 5,000 years. World production has increased from 50.5 million bales (one bale = 480 lbs = 218 kg) in 1965 to a mean of 65.3 million bales for 1979-81. During 1979-81, the USSR, the USA, and China produced about 13.9, 12.2, and 11.4 million bales per year, respectively. India, growing 6.1 million bales per year, ranks number four in world cotton production, and Pakistan, site of the ancient (3000 B.C.) Indus River Valley cotton industry, ranks number five with 3.1 million bales per year. Turkey, Egypt, and Brazil round out the top eight producers, each producing more than 2 million bales of cotton lint per year. Mexico produces about 1.5 million bales annually. By 1990, it is estimated the world will need 75 to 80 million bales of cotton lint.

Egypt, the USSR, and Mexico show mean yields of 941, 930, and 912 kg/ha, respectively, for the years 1979-81. Turkey is fourth with 740 kg/ha and the USA and China are fifth and sixth with 515 and 496 kg/ha, respectively. Pakistan, Brazil, and India round out the top nine with 317, 290, and 166 kg/ha, respectively.

Good quality irrigation water, fertilizer, and pesticides, given in order of importance, are perhaps the greatest constraints to higher yields per hectare.

ECOLOGICAL ADAPTATION

Cotton is a tropical plant, many types of which are adapted to temperate climates. It does best with high temperatures, abundant soil moisture, and high soil fertility.

Except for cotton produced in southern Texas, the USA crop is grown roughly between 30° and 37° N latitudes where most of the solar energy is received between March

---

[1]Professor, Department of Agronomy and Soils, Clemson University, Clemson, South Carolina 29631, USA.

21 and September 21. In Turkey, however, cotton is produced as far north as Bursa, above 40°N latitude, and in the Philippines and Central America it is grown roughly between 10° and 15°N. In the Southern Hemisphere it is grown near 25°S latitude in Brazil, Paraguay, Argentina, and Australia.

The soil characteristics that determine adaptability for cotton culture are water, temperature, aeration, texture, nutrient status, and toxic substances. Perhaps the best way to describe adaptable soils is to list soils that are <u>not</u> adaptable and therefore constitute hazards. These are:

o   Soils on which water and wind erosion is likely.

o   Soils on which excess water is caused by a high water table or temporary flooding from overflow.

o   Soils on which root development is limited by salinity, sodium, acidity, stoniness, shallowness, low fertility, or low moisture-holding capacity.

o   Soils on which extremes of either precipitation or temperature limit use.

Getting good plant stands is highly correlated with soil conditions. Soil temperature at 10 cm depth should be about 20°C for a period of three or four consecutive mornings before planting cotton seed. Lime, fertilizer, and herbicide (to control weeds) must be properly applied to promote early growth. Very acid soils deactivate some herbicides and are more conducive to growth of soil-borne fungi causing seedling disease. Incorporating preplant herbicides too deeply or at high rates on soils with less than 2.0% organic matter may cause plant stand losses.

When stand losses have occurred on cool damp soils in previous years, it is good insurance to use a soil fungicide to control seedling diseases plus a systematic insecticide in the seed furrow to control insects that damage seedlings. Delayed seedling emergence because of deep planting, wet soil, soil crusting, or low temperatures increases the chances of fungal infection and stand losses.

Land is prepared for cotton by first forming a raised bed, about 15 cm high, with appropriate tillage instruments. The raised bed is allowed to "settle". Just before planting, the raised bed is cut down to a slightly raised bed with a power-driven rotary hoe or a smoothing tool pulled by a work animal. The smoothing tool can be something as simple as a log of 10-12 cm diameter and long enough to drag over as many of the 95-

to 105-cm rows as desirable. The seeds are then planted at the rate of 3 to 4 seed per 30 cm of row in the moist, freshly prepared seedbed.

Because the packer wheel on a planter presses the soil down firmly over the seedbed, a mulch is formed on silty and sandy soils above the seed. A firmed soil mulch dries more slowly than loose soil. Under these conditions cotton planted 2 to 4 cm deep will germinate to a good stand without precipitation or irrigation. Obviously clods from clayey soils will not form a mulch under the packer wheel, making it necessary to irrigate or have rainfall on these soils before the cotton comes up. Thus, soil texture influences the adaptation of a soil to cotton culture.

To get a good stand, and subsequently a higher yield, most cotton grown in unirrigated humid regions, such as the Mississippi Delta, has migrated to the fields having sandy loam, fine sandy loam, and silt loam textures. Unfortunately such coarse textured soils permit percolation and mass flow of water through the profile, and essential nutrient elements are carried away. Water percolating freely through the soil profile influences the kind, amount, timing, and placement of fertilizer nutrients for cotton.

PLANT NUTRIENT UTILIZATION

By projecting regional yields per hectare from previous years to future realities, perhaps a 1500 kg/ha lint yield (4500 kg/ha of seed and lint) should be considered in describing mineral nutrient utilization by cotton.

In detecting nutrient deficiencies that affect cotton, consideration should be given to the amounts of nutrients contained in the crop (Table 34.1). The amounts of the major (N, P, K) and secondary (Ca, Mg, S) elements, together often referred to as macronutrients, are more significant in making fertilizer recommendations than the amounts of micronutrients. Micronutrients (Mn, Fe, B, Cu, Zn, Mo) are removed by the crop in amounts less than one kg/ha, and in some cases less than one-tenth kg/ha. Such extremely small amounts are hard to quantify in terms of deficient and excessive levels by plant analyses; guidelines for sufficiency levels have been developed and are given in Table 34.2.

The goal for the grower should be to provide a balance of mineral nutrients that the plant needs to produce highest lint yields.

Visual Deficiency Symptoms

Nitrogen (N). Deficiency symptoms are general yellowing of the older leaves, stunted growth, and few vegetative and fruiting branches. Stored in the organic

TABLE 34.1
Approximate amounts of the various soil-derived nutrient elements contained in the above ground parts of cotton plants producing at the level of 1500 kg/ha of lint (about 7.5 bales per ha)

Element	In a 4500 kg yield of seed and lint	In a 6000 kg yield of stalks, leaves, and burs	Total contained
	(kg)	(kg)	(kg)
Nitrogen (N)	120	105	225
Phosphorous (P)	18	13	31
Potassium (K)	38	87	125
Calcium (Ca)	6	84	90
Magnesium (Mg)	12	24	36
Sulfur (S)	7	15	22
Manganese (Mn)	0.11	0.15	0.26
Iron (Fe)	0.18	0.24	0.42
Boron (B)	0.06	0.12	0.18
Copper (Cu)	0.02	0.03	0.05
Zinc (Zn)	0.07	0.09	0.16
Molybdenum (Mo)	0.005	0.01	0.01

fraction of the soil, N is usually the most limiting nutrient element in cotton production. It can be added in fertilizers in amounts needed (Table 34.3). Small amounts, 15-30 kg N/ha, may be added in rainfall or irrigation water and by free-living soil microorganisms that can convert N_2 gas in the air to a form usable by plants.

Phosphorus (P). Especially important in early growth and in flower production and seed development. A deficiency results in dark green stunted plants, small leaves, and later maturity. Like N deficiency, the symptoms first appear on the lower or older leaves and progress upward on the stalk.

Potassium (K). In K deficiency the older leaves are chlorotic, droopy, and have yellow spots between the veins. The edges turn yellow, then brown, curl downward, and die. Farmers sometimes refer to it as "rust", and it has been confused with pathological disorders.

Calcium (Ca). The relationship of Ca to soil pH and the importance of both for healthy root development cannot be overemphasized. Cotton roots fail to grow into subsoils low in Ca. The difficulty in getting a plant stand in acid soils during cool wet seasons has already been discussed. A low Ca supply results in rather large plants and few fruiting forms. No well-defined field symptoms of deficiency are known. A disorder known as "crinkle leaf" associated with acid soils is due to ex-

TABLE 34.2
Guide to deficient and excessive concentrations of soil-derived elements in top mature leaves of cotton at first bloom, and low soil test values

		Plant Content	Soil Content	
	Deficient (D) or Excessive (E)	Critical Level	Low Soil Test Value	Extracting Solution
Major Nutrients		(% Dry Wt.)	(ppm)	Total in soil
Nitrogen (N)	D	3.30	1000	Total in soil
Phosphorus (P)	D	0.25	10+	0.05N HCl in .025N H_2SO_4 [a]
Potassium (K)	D	1.50	35	"
Calcium (Ca)	D	2.00	200	"
Magnesium (Mg)	D	0.30	20	"
Sulfur (S)	D	0.20	14	0.5N $NH_4C_2H_3O_2$ in 0.25N CH_3
Micronutrients		(ppm)	(ppm)	
Manganese (Mn)	D	25	1.0	DPTA+$CaCl_2$(pH 7.3)
	E	400		
Iron (Fe)	D	40	2.0 [b]	Water
Boron (B)	D	20	0.35	Hot Water
	E	150		
Copper (Cu)	D	4	0.20	DTPA+$CaCl_2$(pH 7.3)
Zinc (Zn)	D	15	2.0	0.1N HCl [c]
	E	850		
Molybdenum (Mo)	D	0.5	-	-
	E	898	-	-

[a] In calcareous soils, extraction with 0.5M $NaHCO_3$ buffered at pH 8.5 predicts response at 8 ppm P.

[b] Fe (II) in calcareous soil solution.

[c] In California 0.5 ppm of dithizone extractable soil zinc is dividing line between deficiency and sufficiency.

TABLE 34.3
Fertilizer nutrients recommended for cotton as related to grouping of soils by identifiable characteristics and soil test levels determined by extracting with 0.05N HCl in 0.025N H_2SO_4.

	RECOMMENDED NUTRIENTS IN KG PER HECTARE											
Cotton Crop	Organic matter (%) present in soil sample				P (ppm) present in soil sample				K (ppm) present in soil sample			
Land Class	0-2	2.1-3.5	3.6-4.5	>4.5	0-6	7-10	11-15	16-20	<35	36-64	65-78	>78
	Fertilizer-N (kg/ha)				Fertilizer-P (kg/ha)				Fertilizer-K (kg/ha)			
I[a]	80	60	40	20	60	40	20	0	60	40	20	0
II[b]	60	50	30	10	50	30	20	0	50	30	20	0
III[c]	60	40	20	0	40	20	20	0	40	20	10	0

[a]Class I Land: Little or no land slope; good internal drainage, depth, water relations, etc.

[b]Class II Land: Generally steeper land slopes; some less desirable characteristics.

[c]Class III Land: Deficient soil texture, drainage, topography, or some other known condition constitutes a production hazard.

> = greater than
< = less than

cessive Mn in the plant and can be corrected by raising the soil pH by applying finely ground limestone.

Magnesium (Mg). On a soil low in Mg and very high in Ca (such as Houston clay), cotton fertilized with Mg blooms earlier; this indicates a relation of Mg to reproductive processes and maturity similar to the P function in plants. Magnesium deficiency symptoms show up as purplish-red leaves with green veins developing later into bronze-colored leaves on the lower branches.

Sulfur (S). Deficiency may be identified by pale green to yellow leaves near the top of cotton plants. The color is similar to N deficiency, but N deficiency symptoms begin near the bottom and not at the top. The plants are characteristically small and spindly with short, slender stalks. Until young cotton develops a large enough root system to probe the S-rich subsoil, the entire plant may appear yellow all over. Often, upon entering the subsoil (15-20 cm or so), roots take up S rapidly and S deficiency symptoms disappear quickly.

Micronutrient Deficiency Symptoms. Micronutrient deficiency symptoms have been estimated for cotton. Soil conditions most likely to be subject to micronutrient deficiencies are described in Table 34.4. Micronutrients are required in concentrations as small as one-fifth ppm (Table 34.2) and as much as 2 ppm of some micronutrients in solution may prove damaging, if not lethal, to cotton. Therefore limited amounts need to be added to prevent damage to the crop.

The safe maximum amount of Mo to be added is considered to be about 0.2 kg/ha; safe maximum amounts of Zn, Mn, Fe, and Cu are about 10 kg/ha, and for B 1 kg/ha. If micronutrients must be added to cotton, Mo is suggested at a rate of 0.1 kg/ha, B at the rate of 0.5 kg/ha, Cu and Mn at the rate of 3 kg/ha, and Fe and Zn at a rate of 5 kg/ha.

Deficiencies of B, Zn, Mn, and Fe have been found and described in the field (Table 34.4). No field plant deficiency symptoms of Cu and Mo have been described for cotton, but these elements are essential for growth.

Improved Cultivars

Agricultural cultivars of Gossypium hirsutum, commonly called upland cotton, are characterized as annual subshrubs with few vegetative branches and short- to medium-fine fibers borne on seeds within rather large, rounded, usually 4 to 5 loculed (locks) capsules (bolls).

The older agricultural cultivars of G. barbadense, referred to in the USA as American-Egyptian or Pima cotton, are inherently perennial shrubs, but they behave as annuals under American Cotton Belt conditions. They bear

TABLE 34.4
Soil conditions most conducive to micronutrient deficiencies, and key to micronutrient deficiency symptoms in cotton

Micronutrient	General Soil Type and Conditions	Key to Deficiency Symptoms
Boron (B)	Acid leached soils, coarse-textured, sandy soils, peats and mucks, droughty conditions, overlimed acid soils	Short leaf petioles with dark green rings, excessive shedding of buds and young bolls, ruptured nectaries, small bolls, bushy vegetation, delayed maturity
Copper (Cu)	Sandy soils, peats and mucks, overlimed acid soils	None described in field even on organic soils
Iron (Fe)	Alkaline soils, particularly when cold and wet; excess phosphorus	Yellowing of cotton leaves at top of plant following irrigation
Manganese (Mn)	Sands, mucks, and peats, alkaline, particularly calcareous soils	Leaf cupping and interveinal chlorosis, veins remain green; starts in young leaves
Molybdenum (Mo)	Highly weathered acidic leached soils, acid soils	None described in field even on acidic leached soils
Zinc (Zn)	Calcareous soils after leaching and erosion, acid leached soils, after heavy phosphorus, fertilization, coarse sands, subsoil exposed by land leveling	Pronounced interveinal chlorosis differs from manganese in that leaves are more misshapen; tips of leaves elongated and parallel

long, fine fibers on seeds within small to medium, tapering, usually 3 to 4 loculed capsules. Modern Pima cultivars of G. barbadense are known to have been introgressed
considerably with G. hirsutum. In these cultivars lint
yields have been increased but fiber length has decreased. The USA cotton crop is approximately 99% G.
hirsutum and 1% G. barbadense.

Many factors govern the choice of a cotton cultivar. Although yield is usually the first factor considered, other important characteristics are certain quality
measurements and market acceptability. Resistance of
cultivars to fusarium wilt disease and root-knot and
other types of nematodes is a primary consideration if
maximum production is to be obtained, especially on certain lighter textured soils.

Other plant characteristics to evaluate in selecting
a cultivar are storm resistance (loss of seed cotton from
bolls to storms), plant type, relative maturity, habit of
growth, and resistance to insects and diseases. Staple
(lint) length, gin turnout (lint separated by spinning),
fiber quality, and spinning properties are also important, since they affect prices, harvesting costs, and
market acceptability.

Plant Density

Cotton plant populations from 20,000 to 50,000 per
acre (48,000 to 120,000/ha) produce the same amount of
cotton in rainfed regions. Above 68,000 plants per acre
(163,000 plants/ha), cotton yield and picking efficiency
decrease in dry years.

Some growers plant two rows and skip one to produce
a 2 X 1 pattern on 40-inch (100 cm) rows. When combined
with dry years in rainfed regions, or limited irrigation
in arid regions, yields are significantly greater than
with solid plantings. Under full irrigation, solid-
planted cotton can produce higher yields than any other
row pattern. Thus where water is scarce, the skip-row
pattern conserves moisture and boosts yields.

Whether solid planting or skip-row pattern are used
and moisture is not limiting, farmers try to add enough
fertilizer for top yields obtainable under these
conditions.

Effective Use of Fertilizers and Soil Amendments

Nutrient needs for cotton worldwide are rather
consistent because only four species of the many original
wild types were domesticated. Nutrient requirements for
good lint quality are the same as those for high yield
(Table 34.3).

Nutrition influences quality of cotton fiber to the
extent that it influences earliness of maturity, which in
turn affects late season storm losses and late season in-

sect pest buildup. Some growers permit a slight N deficiency to develop in late season to encourage the top of
the plant to mature (cut out) early. This has recently
been practiced in the USA because DDT (insecticide),
which was formerly used to control late season buildup of
bollworm (Heliothis) and tobacco budworm (H. virescens)
has been banned by the U.S. Environmental Protection
Agency. Pyrethroids have successfully replaced DDT in
some areas.

Defoliation of the plant by chemical sprays before
harvest assists the grower to produce good quality clean
seed cotton, but defoliation is made more difficult if
large amounts of vegetative growth have been encouraged
by excessive levels of fertilizer N. Phosphorus sufficiency promotes early maturity, and K sufficiency promotes resistance to fusarium wilt.

Fertilizers

Table 34.3 gives a typical set of fertilizer recommendations for cotton, based on soil analysis and other
identifiable soil characteristics. For example, if the
soil was Class I and the soil sample contained 0.2% organic matter, 7-10 ppm P and 36-64 ppm K, the fertilizer
recommendation for cotton would be 80 kg N/ha, 40 kg
P_2O_5/ha, and 40 kg K_2O/ha.

These nutrient amounts can be obtained in the equivalent of 400 kg/ha of a 20-10-10 fertilizer or by applying 400 kg/ha of ammonium sulfate (80 kg/ha N), 90 kg/ha
of triple superphosphate (41 kg/ha P_2O_5), and 70 kg/ha of
muriate of potash (42 kg/ha K_2O) either in three separate
applications or by prior mixing of the three ingredients
and applying them together in one operation. Urea (46%
N) or ammonium nitrate (33.5% N) could also supply the 80
kg N/ha requirement. Ordinary superphosphate (20% P_2O_5)
at a rate of 200 kg/ha could also supply the 40 kg/ha
P_2O_5 requirement.

The choice of fertilizer materials to provide a giv-
en nutrient should be based on the cost per unit of actual nutrient.

Soil Amendments

Soil amendments used on cotton include ground limestone to neutralize acid soils, gypsum to supply S for
acid soils and to reclaim sodic-alkali soils, elemental S
to neutralize alkaline soils, and human or animal manures
to supply needed nutrients based on soil test. Dolomitic
limestones supply Ca and Mg.

Methods of Fertilizer Application

Manures, soil amendments, and fertilizers are efficiently used as preplant broadcast applications. Nitro-

gen fertilizer is most efficiently used when added nearer to the time it is needed by plants, and P fertilizer is most efficiently used by applying it in a band 5 cm to the side and 5 cm below the seed at planting time.

Most growers in humid rainfed cotton regions apply a multi-nutrient fertilizer like 14-14-14 in the furrow, or broadcast it and then bed over it and plant seed about 7-8 cm above the fertilizer. Later in the season, at about the 8-10 plant leaf stage, the additional N fertilizer needed is sidedressed.

Virgin soils in humid regions are naturally acid, and must be limed to about pH 6.0 for good cotton growth. Nitrogen fertilizer (Table 34.3) so essential for profitable cotton, will create more soil acidity. Therefore, in humid regions, liming acid soils to correct acidity is the backbone of soil fertility.

Micronutrients

Micronutrients may be added to the soil, or sprayed on the foliage. Because the amount to be added is so small and must be spread over a large area, a diluent is almost always needed. The multi-nutrient fertilizer or other soil amendments added in large amounts are often used as diluents, but careful mixing is necessary for good results. Water is the diluent for foliar application.

Foliar application of B is more efficient on cotton grown in sandy soils, permitting rapid loss of soluble sodium borates to the drainage water. Boron is often mixed with liquid insecticides and applied on foliage before bloom stage of plant growth.

Iron is occasionally deficient in calcareous (high lime) soils. However, adding Fe to the soil does not eliminate the deficiency because the soluble Fe compounds convert to non-available insoluble soil Fe complexes. Thus, Fe must be applied to the leaves in a water solution for good results.

Zinc and Mn may be added to the soil or the foliage (leaves), whichever is more convenient, as long as the rate does not exceed about 10 kg/ha. Molybdenum and Cu deficiencies for cotton are unknown in field soils.

Costs and Returns

Since most farmers have to borrow money to produce a crop, they must use land, labor, and capital in their most productive combinations. They would be wise to present to their lender a well-organized plan of operation showing how much they will need and how it will be used. Estimated production costs (in percent) of rainfed, mechanized cotton in Southeastern USA during the 1980s was 1.2% for seed, 11.8% for fertilizer and lime, 23.5% for chemicals (weed and insect control and defoliation), 4.3%

for motor fuel, 7.9% for machinery repairs, 6.7% for labor, 3.7% for interest, 8.6% for land (at $35/A or $84/ha), 20.4% for overhead, and 12.0% for ginning. In contrast, fertilizer costs account for 30.1% of costs to produce maize; and 17.6% of costs to produce soybeans, considerably above the 11.8% for cotton.

For the same years and the same rainfed location, it is estimated that cotton will return to land, labor, and management about $188/A ($464/ha); maize will return $19 and soybeans $82. Obviously cotton should not be neglected when the fertilizer money is being spent.

FUTURE PROSPECTS

Prospects are bright that cottonseed may be even more useful to humans by breeding glandless (gossypol-free) cultivars. Cottonseed meal might then potentially become a useful food source for humans. Although cotton, like most industries, is beset by problems, it is likely that this fiber of antiquity will continue to be used for a long time to come.

35
RUBBER

Camillus Silva

The current world requirement of elastomers has been estimated at between 14 and 17 million metric tons. If the present 43% share of natural rubber is to be sustained, world production of natural rubber would be not less than 6 million mt. At present production rates there is a shortfall of one million mt. Analyzing the situation more closely, we find that this estimate of demand for natural rubber will have to be met largely by Southeast Asian countries that today provide about 87% of the current world requirement.

ECOLOGICAL ADAPTATION

For optimum growth and yield, rubber (Hevea brasiliensis) plants require an evenly distributed annual rainfall of 2000-3000 mm. In areas where rainfall is lower or unevenly distributed, trees become stunted with crooked trunks and lower branches. In general, growth of rubber is retarded in regions with a pronounced dry season. Excessive rainfall is also not desirable as it favors soil erosion, loss of nutrients through leaching, and makes tapping of mature trees difficult in rainy weather. Rubber is not recommended in elevations over 750 m because of disease problems.

Tropical temperatures of about 21°C to 35°C without wide variations are suitable for rubber. The mean temperature recorded in the major rubber growing areas is 27°C with a maximum of 38°C and a minimum of 20°C.

The natural vegetation of the rubber-growing areas of the humid tropics is jungle, where rich and evergreen vegetation gives an impression of high soil fertility. This is deceptive, for most soils of the area are highly leached and contain only low levels of plant nutrients.

---

[1]Soil Scientist, Rubber Research Institute, Sri Lanka. Additional Director, National Fertilizer Secretariat, Implementation, Colombo, Sri Lanka.

The most common rubber soils are Entisols, Inceptisols, Ultisols or Oxisols.

The desirable soil properties required for optimum growth of rubber are: (i) soil depth up to 100 cm free of hardpan or rock outcrops, with homogenous properties; (ii) well drained; (iii) good aeration; (iv) good structure; (v) friable to firm consistency; (vi) good water holding capacity; (vii) no peats or acid peats; (viii) texture with minimum 35% clay and about 30% sand; (ix) gently sloping or rolling terrain (2-9° slopes with an an upper limit of 16°); (x) water table deeper than 100 cm; (xi) pH 4.5 to 5.5; and (xii) cation exchange capacity more than 10.

MINERAL NUTRIENT REQUIREMENTS

Potential yields when nutrients needed are fully met. Yield is a result of the combination of soil factors, suitability of clones, and proper nutrition. The rubber tree is tapped in four stages: on the virgin bark during the first six years of tapping (Panel A), on the virgin bark from the seventh to twelfth years of tapping (Panel B), on the renewed bark from the thirteenth to eighteenth years of tapping (Panel C), and on the renewed bark from the nineteenth to twenty-fourth years of tapping (Panel D). The yield potentials of these four different stages vary. Yields of 1,450 kg/ha/yr on Panel A and 1,650 kg/ha/yr on Panel B have been recorded in Malaysia with nutrients supplied as required.

Average yields now realized in producing countries. Yields in the major rubber producing countries vary widely (Table 35.1), and plant nutrition can be an important factor in explaining the differences (Table 35.2).

TABLE 35.1
Yields of rubber in several producing countries

Major Producers	Average Yield	Minor Producers	Average Yield
Malaysia	864 kg/ha	Philippines	1,207 kg/ha
Sri Lanka	654 kg/ha	Ivory Coast	1,000 kg/ha
India	607 kg/ha	Liberia	692 kg/ha
Indonesia	350 kg/ha	Nigeria	188 kg/ha
Thailand	205 kg/ha		

Source: Natural Rubber - Organization and Research in Producing Countries by J. Keith Templeton.

Present levels of mineral nutrition. Recommended levels of nutrients in the three major rubber growing countries, Malaysia, India and Sri Lanka, are given in Table 35.2.

TABLE 35.2
Recommended levels of plant nutrients in three major
rubber-producing countries

Country	Composition (%) of Mixture				Quantity of Mixture Per Tree Per Year
	N	P_2O_5	K_2O	Mg	
India	10	10	10		1 kg (mature trees)
	15	10	6		
	12	12	12		
Malaysia	8.8	16.2	3	2.1	1 kg (mature trees)
	13.0	8.6	3.6	2.1	
	8.4	14.4	7.2	2.1	
	10.7	10.4	7.2	2.1	
Sri Lanka	11	11	5	+	750 g (immature trees)
	9	11	11	+	
	15	15	7	+	1 kg (mature trees)
	12	14	14	+	

+ Magnesium is recommended as a separate application.

DIAGNOSIS OF NUTRITIONAL DISORDERS

Nutritional disorders of hevea rubber can be diagnosed either by the observation of visual symptoms, or by
soil and plant tissue testing. The visual method of
diagnosing a deficiency is entirely qualitative. However, growth may be severely retarded before the actual
visual symptoms are visible and it may often be too late
for remedial action. Under most circumstances rubber is
adapted to low levels of available nutrients. In view of
the fact that most cultivars exhibit symptoms of nutrient
deficiencies at a very late stage of deficiency, diagnosis of nutritional disorders in mature rubber trees is
identified on a routine basis by soil and plant tissue
analyses.

Soil Testing

Critical levels of soil parameters used in India and
other Southeast Asian countries are given in Table 35.3.
Soil chemical analysis is used to determine "available
nutrients", exchangeable cations, and total nutrient content of the soil. Interpretation of soil analytical data
is often difficult for diagnosing nutrient disorders.

TABLE 35.3
Soil test standards and critical leaf nutrient levels for
rubber

Nutrient	Low*	Medium	High**
SOIL			
Organic carbon % used as a measure of availability of N	<0.75	0.75-1.50	>1.50
Available phosphorus (P) (mg/100 g soil)	<1.00	1.00-2.50	>2.50
Available potassium (K) (mg/100 g soil)	<5.00	5.00-2.50	>2.50
Available magnesium (Mg) (mg/100 g soil)	<1.00	1.00-2.50	>2.50
LEAF			
Nitrogen (N) %	<3.00	3.00-3.50	>3.50
Phosphorus (P) %	<0.20	0.20-0.25	>0.25
Potassium (K) %	<1.00	1.00-1.50	>1.50
Magnesium (Mg.) %	<0.20	0.20-0.25	>0.25

Source: Handbook of Natural Rubber Production in India.
Rubber Research Institute of India. Rubber Board,
Kottayam 686009 (1980).

* < = less than   ** > = greater than

Plant Tissue Testing

Chemical analysis of plant tissues, especially
leaves and to a lesser extent the analysis of bark and
latex, is employed widely in determining the nutrient requirements of rubber. Critical nutrient concentrations
in leaves have been determined for most nutrients and are
summarized in Table 35.3. Time of sampling, type of leaf
to be sampled and correction for age of leaf, using Ca as
the index, have been worked out in Malaysia.

Visual symptoms - Visual symptoms as indicators of deficiencies or toxicities in rubber have been determined in
sand and nutrient solution cultures as well as in the
field. Detailed descriptions of deficiency symptoms with
color photographs have been published by Shorrocks (1964)
and can be summarized as follows:

Nitrogen (N). Reduced growth of the entire tree,
reduced leaf size, leaf number and trunk girth. Leaves
pale yellow-green, but no pattern of chlorosis develops,
the whole leaf becomes first yellow-green and later yellow in color.

Phosphorus (P). Bronzing of part of the under sur-face of the leaf, leaf tip frequently dies back.

Potassium (K). Development of a marginal and tip chlorosis, followed by a marginal necrosis.

Magnesium (Mg). Development of chlorotic intervein-al areas on the leaves, spreading inward from the leaf margins, giving a "herring-bone" pattern.

Calcium (Ca). The first symptom is the development of a tip and marginal scorch, usually white to light brown in color, often over a large portion of the leaf.

Sulfur (S). A gradual uniform yellowing of the en-tire leaf which is much reduced in size; later a tip scorch develops which results initially in a cupping of the leaf, and later affects the whole of the distal end of the leaf.

Manganese (Mn). An overall paling and yellowing of the leaf with bands of green tissue outlining the midrib and main veins.

Iron (Fe). A general leaf chlorosis; with increase in severity the entire leaf assumes a pale lemon-yellow to white color.

Boron (B). Leaves are distorted, reduced in size and somewhat brittle; there is not loss of color and oc-casionally the veins may appear to be wider than normal.

Molybdenum (Mo). A very pale brown scorch develops around the leaf margins, particularly in the region of the leaf tip.

Zinc (Zn). Leaves become very much reduced in width relative to length; leaves may become twisted and the margins appear wavy or undulating; there is also a gen-eral chlorosis of the leaf, but the midrib and main veins remain dark green.

Copper (Cu). Scorching of the leaf margin at the tip, with subsequent upward cupping of the leaf tip; the pale brown marginal scorch often spreads down the leaf from the tip, resulting in leaf shedding.

IMPROVED VARIETIES OR HYBRIDS IN RELATION TO NUTRIENT REQUIREMENT

The breeding of new rubber cultivars in most coun-tries has focused on disease resistance and high yields. The nutrient requirements of the high yielding varieties have been determined by field experiments and more so by using soil and foliar analyses. Adjusting the nutrient requirements of wind-resistance clones, reducing N to reduce foliage, and increasing K in the mixture to resist wind damage are a few recent innovations.

ROLE OF NUTRIENT REQUIREMENTS IN THE CULTURE OF RUBBER

Increasingly, farmyard manure and compost are being recommended to satisfy partially the nutrients required for growth and yield. In India the recommendation is for the incorporation of 2000 kg/ha of compost or well-decomposed cattle manure in seedling nurseries. In particular, similar practices are recommended for small-holder growers in Southeast Asia.

Routine maintenance applications of fertilizer to immature rubber usually follow the schedule dictated by the date of planting. For mature rubber it is preferable to apply fertilizer during the season when the leaves are falling from the tree, in order that they may be absorbed at the time of refoliation when the tree has a relatively large requirement for nutrients.

Broadcast applications of fertilizer are recommended for mature rubber. For immature trees, the method of application during the first few years is very important as the plants have a limited root system. The principle to follow is that fertilizer is added to the area of soil where the concentration of active feeding roots is high. The recommendation in India is typical of what is suggested in other countries (Table 35.4).

TABLE 35.4
Recommended fertilizer placement practices in India

Year of Planting	Age in Months at Manuring	Distance from the Tree (cm) to be Left Unmanured	Band Width of Fertilizer (in cm)
1st	3 - 5	7	30
2nd	9 - 15	15 - 30	45
3rd	21 - 27	45 - 60	60
4th	33 - 39	75 - 90	60

Source: Handbook of Natural Rubber Production in India.

The fertilizer should be forked into the top 5-8 cm of soil. Where mulching is used, the mulch is placed over the fertilizer and not forked in.

After the first 2 years of growth, fertilizer is applied twice a year and timed according to the monsoonal rains or the moisture status of the soil. Rates of application of fertilizer are related to plant population and are calculated in amounts per tree.

TABLE 35.5
Nutrients removed through stimulation, and the effect of stimulation on nutrient content in leaves

Treatment	Yield kg/ha/year	Nutrients Removed kg/ha/year				Leaf Nutrient Content %			
		N	P	K	Mg	N	P	K	Mg
Unstimulated	1,390	9.4	2.3	8.3	1.7	3.55	0.32	1.44	0.22
2,4,5-T	1,660	11.9	3.1	11.1	2.1	3.54	0.31	1.41	0.23
Ethrel (10%)	2,570	23.9	7.2	22.3	4.1	3.40	0.32	1.26	0.21

Source: Proceedings of the RRIM Planters Conference, 1971.

Both seedling and budded trees can be regarded as tappable when they have attained a girth of 50 cm at a height of 90 cm measured from the ground (seedling trees) and from the highest point of the graft union (budded trees). Extraction of latex by tapping retards growth, therefore the initiation of tapping must be done at the correct time.

The period between planting and first tapping is generally 5 to 7 years. Although considerable research efforts have been made to shorten the length of the unproductive period, conventional practice has centered mainly on improving nursery techniques, breeding of rapid-growing clones, manuring, and other field management practices. In many of the better rubber-producing countries, plantations are tapped in 5 years and this is achieved by the use of correct types and quantities of fertilizer.

Higher yields of rubber have been obtained by the stimulation of latex flow by applying yield stimulants such as 2,4-D, 2,4,5-T and Ethrel. Such additional yields are expected to increase the nutrient losses from the soil because of the greater volume of latex being removed from the trees. Table 35.5 gives an estimate of nutrients removed through stimulation and its effect on leaf nutrient content.

If stimulation is practiced, crop nutrient requirements are increased. It is estimated that for every 1,000 kg of increase in latex yield through stimulation, 12 kg of N, 4 kg of P, 12 kg of K and 2 kg of Mg must be replaced for the trees.

MEASUREMENT OF BENEFITS FROM ADEQUATE MINERAL NUTRITION

The yield range among major rubber producers is very wide (Table 35.1). Yields attained in Malaysia are largely a result of a vigorous replanting program. Malaysia has also benefited from a large progressive estate sector. Sri Lanka which has a strong estate sector and India which has pursued a vigorous replanting program also obtain good average yields. Indonesia and Thailand on the other hand have low yields mainly because most of their rubber is grown on smallholdings with low management levels.

GENERAL REFERENCES

RUBBER RESEARCH INSTITUTE OF INDIA. 1980. Handbook of Natural Rubber Production in India. Rubber Research Institute of India, Rubber Board, Kottayam 686009.

SHORROCKS, V.M. 1964. Mineral Deficiencies in Hevea and Associated Cover Plants. Kynock Press, Birmingham, England.

36
Oil Palm

R. L. Richardson

The term "oil palm" generally refers to <u>Elaeis guineensis</u>, the "African oil palm," but should also include <u>Elaeis oleifera</u>, the "American oil palm." These palms are among the 273 genera and approximately 4000 species of mostly tropical "trees" which make up the Palm family.

Although most palms store oil in their seeds, the oil palm also produces oil in its fleshy-fibrous mesocarp which is about 48% oil on a fresh weight basis in the African species. Palm mesocarp oil is used principally as shortening and sometimes is fractionated to produce a liquid fraction used in cooking oils and margarines, and a solid fraction used as soap stock. Palm kernel oil is similar to coconut oil in its physical properties and has similar uses, e.g. fine soaps, margarine, cosmetics, lubricants, and confections.

Palm kernel meal is used in livestock feeds, and fiber and shell by-products are utilized as fuels in palm oil extraction plants to generate steam and electricity for the oil extraction process. Natives of regions where oil palm occurs naturally also use the palm for thatching, fencing and palm wine.

Total world production of palm oils is nearing 6 million tons, of which over 60% is exported. In spite of these impressive volumes, palm oil represents only about 12% of world trade in vegetable oils, a figure estimated to increase to 15% by 1985.

Malaysia is the world leader in production of palm oil, followed by Nigeria and Indonesia.

Although natural groves of oil palms are still utilized for oil production, far more important is the considerable plantation industry built around this crop.

---

[1]United Brands Company, Compania Barranera de Costa Rica, San Jose, Costa Rica.

World production of palm oil is almost exclusively
from the African oil palm, but the American-African in-
terspecific hybrid is planted in some restricted areas
because of its disease resistance. In addition, this hybrid is being studied extensively due to its characteristics of slower trunk growth and better oil quality.

ECOLOGICAL ADAPTATION

The African oil palm is indigenous to the west coast
of tropical Africa from Guinea (10°N) to Angola (7°S)
with considerable inland penetration in the Republic of
Zaire. The American oil palm is indigenous in Central
America as far north as Honduras (15°N) to the Amazon
basin in Brazil (5°S).

The oil palm is a completely tropical plant requiring mean temperatures of 25 to 27°C. More important,
monthly mean minimum temperatures below 19°C are deleterious to oil palm growth and yield. In controlled environment trials with oil palm seedlings it has been shown
that growth stops at 15°C and that growth at 20°C is
roughly 3 times that of growth at 17.5°C.

Oil palm growth and yield are highly dependent on
the water supply, requiring 125-150 mm per month for optimum growth. Periods of prolonged drought are highly
detrimental to yield if irrigation is not practiced.
Since in most of the oil palm growing areas in the tropics water deficits are experienced during part of the
year, annual yield is often well correlated with the magnitude of the accumulated annual water deficit.

High levels of solar radiation are thought to be
necessary for optimum yields in oil palm. The better oil
palm growing regions usually receive more than 1800 sunshine hours annually. Solar radiation measured by pyra-
nometer in gram calories per cm^2 per day shows better
growing conditions during periods with over 370 Langley's
per day than during periods below this level. Periods of
inadequate climatic conditions cause stress on the palms
which through bunch failure, inflorescence abortion, and
effects on sex-determination of developing inflorescences
affects future yields. Consistent climatic cycles at a
locality induce annual yield cycles.

The yield difference between high and low producing
periods is a measure of the poor growing periods. When
over 12% of the total annual yield is concentrated in one
month, seasonally poor growing conditions are indicated.

Oil palm is grown on a very wide range of soils from
sands to heavy clay in texture, neutral to acid in pH,
and flat to rolling in topography. Almost any soil,
which with prevailing climatic conditions can maintain an
adequate supply of water and oxygen for the roots, can
grow oil palm. Clearly, best yields at lowest cost are

obtained on deep, well-drained loams, with flat topography and good natural nutrient levels.   These soils
reduce the necessity for terracing, intensive drainage
systems and expensive fertilization programs.
     The oil palm is a perennial crop whose economic life
is determined by the height of the palm and the accompanying difficulty in harvesting.   Usually the palms are
felled and replanted when they reach heights of 10-12
meters.   The age at which palms reach this height varies
from 20 to 25 years depending upon the growing conditions
of the locality and the variety planted.

MINERAL NUTRIENT REQUIREMENTS

     Many examples of yield increases through fertil-
ization are reported for oil palm (Table 36.1).   In many
cases these yield increases have an economic benefit
above the cost of fertilization.   In general, however,
the magnitude of these yield increases is less than yield
differences due to variations in the annual water deficit
among growing areas.   Put in another way, much oil palm
is grown under sub-optimal growing conditions.
     Under near ideal growing conditions, yield of fresh
fruit bunches (FFB) can reach 28 to 30 mt/ha, yielding
about 6 to 6.8 mt of mesocarp oil and 1.2 to 1.3 mt of
palm kernels per ha (195-210 kg FFB/palm).
     Average yields of 18 to 24 mt of FFB per ha, yield-
ing 4 to 5.2 mt of oil and 0.8 to 1.1 mt of palm kernels
are more common in good growing areas (126-168 kg
FFB/palm).
     Lower yields of 10 to 16 mt of FFB per ha are common
in areas where growing conditions pose serious
limitations to yield (70-112 kg FFB/palm).
     Fertilizer rates in general usually tend to be 2 to
3 times higher in Asia with good growing conditions and
higher yields than in Africa with often mediocre growing
conditions and yields (Table 36.2).   While this
difference in usage is probably due to the increased
response to high rates of fertilizers under good growing
conditions, the case of N is special.
     Except for results from Zaire, N additions to adult
palms in the rest of Africa rarely give yield responses.
It has been assumed that this result is due to some
natural microbiological nitrogen-fixing system in these
areas where oil palm is endemic.

TABLE 36.1
Some representative yield increments of oil palm due to
fertilizer application (after Corley et al., 1976)

Country	Nutrient	Rate Applied (per palm)	Yield of FFB[1] (kg/palm)	
			No fertilizer added	Fertilizer added
Indonesia	N	4 kg ammonium sulphate	106	169
Brazil	N	1 kg urea	85	93
Nigeria	N	3 kg ammonium sulphate	67	72
Indonesia	P	3 kg triple superphosphate	111	156
Brazil	P	1 kg triple superphosphate	58	121
Nigeria	P	3 kg ground rock phosphate	66	72
Indonesia	K	2 kg potassium chloride	134	129
Brazil	K	1.5 kg potassium chloride	85	93
Nigeria	K	2 kg potassium chloride	52	72
Cameroon	Mg	1 kg magnesium sulphate	67	91
Brazil	Mg	0.5 kg magnesium sulphate	88	90

[1] FFB = Fresh Fruit Bunches.

Table 36.2
Commonly used fertilizer rates for adult oil palms
(kg/ha/year)

Element	Africa	Asia
N	0	80 - 100
P	0	20 - 30
K	70 - 165	140 - 240

Oil palm is a heavy user of N and K. Total nutrient
uptake under Malaysian conditions is as follows:

Element	kg/palm/year	kg/ha/year
N	1.29	184
P	0.18	26
K	1.79	256
Mg	0.31	44

Soil Analysis

Soil nutrient status is a valuable general guide to
oil palm nutrition. The following "ratings" of soil
nutrient levels are for Malaysian soils (0-40 cm depth).

Approximate Ratings for Oil Palm Soils

Parameter	Very Low	Low	Medium
N (%)	<0.05	0.05-0.08	0.11-0.25
P (ppm)-Bray#2	11	11-26	30-50
Exch. K (meq)	0.15	0.15-0.26	0.30-0.50
Exch. Mg (meq)	0.15	0.15-0.26	0.30-0.50
C.E.C.	<3.0	3.0-7.9	8.0-14.9

Foliar Analysis

Foliar analysis is widely used in oil palm to detect
nutrient deficiencies and as a guide to fertilization, in
spite of some recognized short-comings. For precision,
strict adherence to recommended leaf sampling techniques,
with respect to sample size, leaf sampled, time of year,
and sample preparation techniques is necessary. Desired
levels for leaf No. 17 on adult palms are as follows:

Element	Malaysia	Africa
	per cent of dry matter	
N	2.75	2.50
P	0.18	0.15
K	1.25	1.00
Mg	0.25	0.24
Ca	0.60	0.60
S	-	0.20

	parts per million (ppm)
B	18
Zn	15
Cu	5
Mn	200
Fe	80
Mo	1

Visual Symptoms

Deficiency symptoms on nursery and field palms have been well described. These symptoms usually indicate rather severe nutrient deficiencies and under good estate management would have been detected by foliar analysis and corrected before reaching the stage of visual symptoms.

Deficiency symptoms can be summarized as follows:

Nitrogen (N). Uniform pale green color of all leaves; leaflet mid-ribs and rachis deep yellow to orange.

Phosphorus (P). Typical symptoms have not been described for P.

Potassium (K). Deficiency symptoms vary; it is thought that corresponding levels of other cations may be involved in producing these symptom differences: (i) confluent orange spotting - chlorotic spots changing to orange develop, enlarge and coalesce, and finally become necrotic on the older leaves of the palm; (ii) mid-crown yellowing - a distinct and uniform ochre color, except for a normal green band along the leaf midrib, appears on the leaves around position No. 10, hence "mid-crown"; (iii) orange blotch - large yellow or orange patches appear on the leaflets of older fronds, the patches contain a mass of minute orange spots.

Magnesium (Mg). The older fronds turn chlorotic yellow, then orange; typical of Mg deficiency is a "shade effect" of the chlorosis in which areas of leaflets shaded from the sun retain their normal green color.

Sulphur (S). Interveinal chlorotic streaking on the youngest expanded leaves; the leaflets finally become a pale yellow, broken by green veins; older leaves have only slight interveinal chlorosis.

Boron (B). A number of leaf deformities have been attributed to B deficiencies of different levels of severity; these have been described in order of lesser to greater severity of deficiency as: (i) rounded frond tip (short distal leaflets); (ii) leaflet shatter (weak midveins); (iii) hook leaf (leaflet hooking); (iv) bristle tip (veins of distal leaflets only); (v) fish-bone leaf (poor lamina development on whole frond); (vi) little leaf (severely restricted development of rachis and leaflets); (vii) bud rot (new severely reduced fronds distorted into a twisted mass at the center of the crown -- may be lethal).

FERTILIZATION

As a result of processing FFB for oil extraction, considerable quantities (22% of FFB) of empty bunches (peduncle, spikelets) are produced. Generally these are incinerated and the resulting ash is used as K fertilizer. Chemical analyses of bunch ash indicate 6 to 7% P_2O_5, 34 to 38% K_2O and 4 to 5% MgO. The application of empty bunches directly to the field appears to be a better solution than incineration and use of the ash, but has resulted in serious insect problems (flies) on the decomposing organic matter.

Placement of commercial fertilizers should be in areas with the greatest concentrations of feeder roots. Clearly, in young palms fertilizers should be applied thinly and evenly in a circle from the base of the trunk to the "drip circle" formed by the tips of the palm fronds. In adult palms, however, experiments have not shown consistent advantages of any particular placement method, hence the fertilizers are usually broadcast.

Young palms are usually fertilized every 6 months, but the period is generally extended to annual applications in mature palms. More frequent applications on mature palms may be justified on coarse soils under conditions of high rainfall, especially with respect to N.

Fertilizer applications should be avoided during seasons of heavy rains, and N fertilizers should not be applied in the dry season. Ideal application time is during seasons of light rains preceding the annual cropping peak, if the two coincide.

Response to fertilizers -- in terms of leaf nutrient levels and the correction of deficiency symptoms -- is usually relatively rapid with young palms (2 months). Nutrient deficiencies in old palms with massive amounts of trunk tissue are corrected much more slowly, and yield response is usually not seen until 12 to 18 months after application.

Ammonium sulphate and ammonium nitrate are the most common N sources used for oil palm. Several Malaysian experiments have shown better yield response with these fertilizers than with urea applied at equal rates of N. The considerable loss of ammonia through volatilization is the suspected cause for the inferior performance of urea.

Potassium chloride is used almost exclusively as the K source in oil palm cultivation. The use of potassium sulphate may be warranted on some high pH soils.

Rock phosphate is the least expensive and most widely used P source but in high Ca soils superphosphate or triple superphosphate are preferred.

Magnesium is usually supplied to palms in the form of kieserite.

Except for the cost of harvesting fruit, fertilizer costs are the most important cultivation expense and may represent as much as 20% of total farm direct costs.

Optimum fertilizer rates depend principally on the yield response, but also on palm oil and fertilizer prices.

Fertilization necessary to maintain optimum palm growth during the coming-to-bearing phase produces higher early yields. Where edaphic and climatic conditions permit a yield response to fertilizers, the response curve generally shows large yield increments with increased fertilizer at low rates of application, then lower and finally no yield increments at higher rates of application. The point of maximum economic benefit -- which is usually realized at slightly sub-optimal yields -- must be calculated for each specific plantation situation.

Prices on the world market for vegetable oils are controlled more by the supply and demand of soybean oil, which makes up about half the volume, than that of palm oil. Since oil palm is a perennial crop with slow yield response to fertilizer application, little can be done to manipulate productivity to correspond to periods of high palm oil prices, even in countries with closed markets in vegetable oil.

SELECTED REFERENCES

CORLEY, R.H.V., J. J. HARDON, and B.J. WOOD (eds.). 1976. Oil Palm Research. Elsevier Scientific Publishing Company, Amsterdam.

HARTLEY, C.W.S. 1977. The Oil Palm. Longman, London.

TURNER, P.D. and R.A. GILLBANKS. 1974. Oil Palm Cultivation and Management. Incorporated Society of Planters, Kuala Lumpur, Malaysia.

37
COCONUT

Donald L. Plucknett

Coconut (<u>Cocos</u> <u>nucifera</u>) may be the most extensively grown tree crop in the tropics. There are probably more than 6 million hectares of coconut in the world; about 90% of this crop area is in Asia and Oceania. Major producers of copra and coconut oil are the Philippines, Indonesia, Sri Lanka, Mexico, Malaysia and the islands and territories of Oceania.

Truly a wonderful plant, coconut supplies man with many products including food, fiber, and raw materials for crafts, fuel and construction.

Although coconut is often considered a plantation crop, most of the world production comes from small farms. It is grown mainly on islands, peninsulas, and along coasts. Coconut is often a neglected crop, and its management is commonly poor. This is especially serious when one considers that the trees are long-lived, from 60 to 80 years or more, and that it often occupies some of the best lands of coastal areas. Better management of the crop and more intensive land use in coconut areas are necessary. Improved and effective plant nutrition is a key factor in improved management systems.

THE COCONUT TREE

Coconuts usually take 6 or 7 years to reach production from transplanting, although some of the newer dwarf varieties may begin producing in 4 years. The trunks usually do not form until 4 or 5 years after planting. Older palms may attain great heights, 20-25 m or more, bearing whorls of large fronds and fruit bunches at the top of the trunks. Coconuts have an economic life of about 60 years, although they may continue to produce for 80 years or more.

---

[1]Scientific Adviser, Consultative Group on International Agricultural Research (CGIAR), The World Bank, Washington, D.C. 20433.

The palms are monoecious, bearing both male and female flowers in the same inflorescence. Most of the fertilized young fruits (buttons) fall long before maturity. It takes about 12 months for the remaining fruits to mature. Normal annual nut production per palm is about 20 nuts at 8 years of age, 40 at 12 years, 60 at 16 years and 100 at 20 years.

Coconut roots are coarse and fibrous with few branches. Their depth and lateral spread vary greatly with soil type. In sandy soils the diameter of spread may be as much as 10 m. Roots will not develop in waterlogged soils. Most roots are found in the top 1 m of soil.

An individual coconut at harvest is the outcome of events over a 3.5 year period. The primordium of an inflorescence forms about 32 months before its opening and the primordium of the spathe (flower sheath) begins about 15 months before opening. After the spathe opens, it takes about 12 months for female flowers to develop into ripe fruit. The components of yield in coconut are: the number of female flowers formed about 2 years before the crop matures, fruit set - determined by the number of female flowers pollinated about one month after the spathe opens, the amount of shedding of young fruits (button shedding), number of filled fruit at maturity and kernel (copra) weight per nut. From this discussion it can be seen that growing conditions over a 42 month period, from first initiation of the inflorescence to ripe nuts, can greatly influence yields.

CLIMATIC AND SOIL REQUIREMENTS

Coconut grows best in areas receiving 1,250 to 2,500 mm of well-distributed rainfall. Below 1,250 mm, unless irrigation is practiced or seepage of ground water occurs, periods of water stress can result.

Coconuts grow on a wide range of soils, from coastal sands to very heavy clays. Probably the best soils are well-drained alluvial deposits along rivers or estuaries. Sandy soils, oxisols, and ultisols are well drained but usually infertile, and will require fertilization. The palms can tolerate a wide range of soil acidity, from pH 5 or so to over pH 8 for coral sands. Most soil nutrient deficiency problems can be overcome by use of inorganic or organic fertilizers.

NUTRITION OF COCONUT

Coconut is frequently a neglected crop and its plant nutrition is often inadequate. Also, the crop usually does not respond immediately to fertilization, and an improved nutritional program, including soil and plant analyses, must be carried out over several years to obtain best results. With balanced nutrition, leaves will green up and grow larger and increase in number, more flowers and fruits will be produced, and nut and copra yields will rise.

There can be significant differences in the response of coconut to fertilization depending on local growing conditions, climate and crop variety. Field trials should be conducted in major growing regions to determine the soil and plant levels at which the crop does best.

Coconut has not received as much research attention as it deserves. Research on coconut nutrition has been carried out in India, Ivory Coast, Jamaica, Mozambique, the Philippines, Sri Lanka, and in the islands of the South Pacific.

NUTRITIONAL REQUIREMENTS

Coconut nutrient requirements differ with age. Young palms respond readily to N, P and K, and especially to N and P. Nitrogen deficiency in young palms results in yellowing of plants, particularly the lower leaves, and K deficiency results in necrotic tips. Phosphorus increases leaf collar girth and leaf number, while K increases collar girth.

Young palms must grow rapidly, and NPK fertilizers improve growth, promote early bearing, and result in higher yields. Palms that are not fertilized when young do not reach the yield potential of fertilized trees, even if fertilizers are applied later.

Mature palms in full production require higher rates of fertilizer, particularly K. Indeed K is considered to be the dominant nutrient for coconut, and palms not receiving it may soon show deficiency symptoms, thinness, and a tendency to become diseased. Nutrient removal by palms producing 40 nuts per year can be in the order of 95 kg N, 20 kg P, 110 kg K, 86 kg calcium and 34 kg magnesium. Nitrogen has a significant effect on both nut and copra yields, while K can often have a limiting effect on yields.

DETECTION OF NUTRIENT DEFICIENCIES

Detecting and determining nutrient deficiencies in coconut is not easy because the crop is so long-lived and responses to fertilizers may not be detected for a year or more. Visual symptoms are useful as indicators of problems, but used alone are of less advantage. Soil analyses can help to determine which nutrients are likely to be limiting or in excess and can help to provide guidelines for laying out field trials to determine the major needs. Soil analysis for coconut is not an overly powerful tool, however, because a palm may take a year or more to reflect changes in nutrition, while for some nutrients soil analysis may reflect seasonal differences in nutrient supply.

The major tool in coconut nutrition is foliar analysis, coupled with detection of visual symptoms and soil analysis. Foliar analysis has the advantage that it measures the amount of nutrients in the leaves of coconut, thereby giving a more direct picture of how the plant is doing.

The reference leaf for older palms is leaf number 14, counting down from the youngest leaf in the apical whorl. In palms less than 4 years old, leaf 4 is commonly used.

In some cases, analysis of the "water" of the nut is used to help determine nutrient status. Composition of the water correlates quite well with foliar analysis for some elements. In Papua New Guinea, sulfur-deficient palms had 8 ppm S or less in the water of the nuts.

VISUAL SYMPTOMS

Nitrogen (N). Symptoms of deficiency are typical; the plants are stunted and generally yellow, with lower leaves being most affected. Older leaves are golden yellow in color.

Phosphorus (P). Specific symptoms are not easy to detect; however, growth, leaf size and leaf number are reduced. Phosphorus deficiency is not common and is usually associated with soils in which severe P deficiency or fixation is present.

Potassium (K). If the deficiency is severe, the tree appears generally yellow and sickly, the trunk is slender and carries only a few short leaves, which in turn bear short leaflets. The canopy is quite open and much sunlight reaches the ground. Three gradients of deficiency symptoms can be distinguished: (i) the vertical gradient, in which lower leaves show more pronounced yellowing than upper leaves; (ii) a longitudinal gradient, in which leaflets show more marginal yellowing and necrosis near the tips than in the middle of the leaflets or near the rachis, and (iii) a transverse gradient in which yellowing is more pronounced along the edges of leaflets than along the midrib. In most cases yellowing is not uniform but is accompanied by irregular brown blotches.

Sulfur (S). Deficiency causes uniformly yellow or orange leaves, with necrosis following, resulting in death of leaflets and leaf tips. Arching of leaves can be common. Vegetative growth is reduced. In older palms the leaves tend to bend above their normal abscission point, causing dead leaves to hang downward in a skirt-like pattern around the trunk. In severe cases few live leaves are present, and these are usually small and stunted. Few nuts are produced, and these yield normal appearing but rubbery copra upon drying.

Zinc (Zn). The major symptoms are dwarfed or deformed young leaves, and the classic "little leaf" condition.

Boron (B). Boron deficiency can affect the young leaves and has been implicated in a "crown rot" in India.

TISSUE ANALYSES

As was previously stated, most tissue testing of coconut depends on values determined from leaf number 14, and is expressed as percentage of dry matter or in parts per million (ppm). The critical values are as follows:
N - 1.8 to 2.0%; P - 0.12%; K - 0.8 to 1.0%; Ca - 0.3 to 0.5%; Mg - 0.2 to 0.3% Fe - 50 ppm; S - 0 15%, Mn - 60 ppm; Cl - 0.50-0.55%; and Na - not yet determined.
Sulfur deficiency can also be determined by analyzing the "water" of the nuts. In trials in Papua New Guinea, deficient palms yielded nuts in which S content of the water was below 8 ppm, while nuts of S-treated palms had S levels between 10 and 60 ppm in the water.

FERTILIZATION OF COCONUT PALMS

It is essential to fertilize young palms so they will mature faster and bear more heavily. Poorly fed young palms will never catch up with those that receive good nutrition.
In most cases it will not be necessary to apply phosphorus throughout the life of the crop. Applying P to young palms for the first 5 years should be sufficient, after that only occasional applications to replace the P being removed should be adequate.
Young palms should receive increasing amounts of fertilizer as they grow. A suggested amount for year one is 60g N, 20g P, 120g K per palm; this amount can be doubled in the second year and doubled again in year three. Rates for year four are 360g N, 100g P, 750g K. By year five a young palm should be receiving about 500g N, 140g P and 1000g K. Of course the exact amounts and ratios of fertilizers to be used will depend on local conditions, but such levels should ensure an adequate plane of nutrition for the palms.
The NPK requirements of adult palms are difficult to state with certainty, for they vary so much depending on soil, spacing, climate and other factors. However, some general guidelines can be given. Mature palms should each receive NPK fertilizers annually in the following amounts; 500-700g N, 120-140g P (P may not be needed every year for some soils) and 700-1000g K.
Fertilizers are usually applied in a circle about 3 m in diameter around the palm. It is considered best to incorporate the fertilizer by hand hoeing, discing or burial.

SELECTED READING

CHILD, R. 1964. Coconuts. Longman, London.

FREMOND, Y., R. ZILLER, and M. DE NUCE DE LAMOTHE. 1966. The Coconut Palm. International Potash Institute, Berne, Switzerland.

PLUCKNETT, D.L. 1979. Managing Pastures and Cattle Under Coconuts. Westview Press, Boulder, Colorado.

THAMPAN, P.K. 1981. Handbook on Coconut Palm. Oxford and IBH Publishing Co., New Delhi.

38
CACAO

Percy Cabala Rosand, Marcia B.M. Santana,
and Charles J.L. de Santana

Cacao has higher mineral nutrient requirements than other tropical perennial crops that are able to grow in poor acid soils. When considering cacao nutritional requirements, it is also necessary to take into account genetic characters and the environmental conditions in which the tree is growing, especially regarding the degree of shade. Generally, Amazon varieties are more nutrient demanding than Amelonado varieties. Cacao without shade also demands higher amounts of nutrients than shaded trees. In soils with low fertility, the shade works as a buffer, and at the same time that it reduces metabolic activity it also reduces both nutrient uptake and soil degradation. These aspects are relevant because they point out the possibility of growing cacao in acid and poor soils by using lime and fertilizers. In this respect, recently some Oxisols and Ultisols have been used with relative success for cacao. Possibly, if the use of fertilizers becomes limited for economic reasons, increasing the degree of shade may become advisable to reduce not only production costs but also to increase crop longevity.

ECOLOGICAL ADAPTATION

Soil Adaptation

Cacao is a very demanding crop. In West Africa or Brazil where the more important cacao regions are located, plantations are mainly in soils of high to medium fertility and without any physical limitations. Also, attempts to establish cacao in poor soils without use of fertilizers were abandoned, or less demanding crops were substituted. In southern Bahia, Brazil, a nutritional

[1]Principal Research Scientist and Senior Research Scientists, Soil Science Department, Centro de Pesquisas do Cacau, CEPLAC, Itabuna, Bahia, Brazil.

assessment of soils carried out in an area of 117,000 ha showed that soil fertility was higher in productive plantations than in decadent plantations or those planted on non-traditional soils (Table 38.1).

Several methods have been suggested to classify soils into categories of suitability for cacao. For Brazil four categories were considered, taking into account some physical and chemical properties. In this classification a suitable soil for cacao should have, besides good physical properties, a pH about 6.2, total bases about 12 meq/100 g, base saturation about 70 percent and organic matter never below 3.5%.

Today, with modern management practices and fertilizers, cacao is expanding on low fertility soils. High production can be obtained on Oxisols with use of fertilizers and other improved practices. In some Oxisols in southern Bahia (Tabuleiro soils), response to fertilizers is greater in the first years, with a drop in production in later years as a probable result of micronutrient deficiencies.

In the Ivory Coast, natural soil fertility is not considered to be a limiting factor in establishing cacao, and in fact different kinds of soils -- even chemically poor soils -- are being used to expand the crop. Attention must be paid to such soils, because production will be limited if the plantation is not protected by adequate shade or if other environmental conditions are not suitable.

Nowadays, with continually increasing fertilizer prices, choice of areas for expansion must fall on lands of high natural fertility, mainly in areas without roads and far from commercial areas. Research attention has been given not only to try to reduce fertilizer use but also to increase efficient nutrient use by cacao.

Soil Acidity

The ideal soil for cacao is around pH 6.5 and whenever possible the choice soils are those in which pH varies between 6.0-7.5. In soils above pH 7.5, micronutrient deficiencies, especially iron and zinc, can occur.

In southern Bahia the most productive plantations are in soils with pH 6.0-6.5. In decadent plantations or in areas where cacao has not yet been established, soil pH values are mostly below 5.0 and are always associated with aluminum in the complex and low amounts of exchangeable calcium and magnesium.

Cacao within limits is tolerant to aluminum. In solution culture, 15 ug ml^{-1} of Al in the solution is the upper limit for cacao; higher levels affect plant growth and drastically reduce P and Ca uptake. In Trinidad and Guyana, in acid soils where the level of Al induced poor root formation in maize, cacao is known to grow well. Also, in the Ivory Coast quite high yields are obtained in sedimentary soils with pH 4.5-5.0.

TABLE 38.1
Chemical analysis of soil samples taken from cacao plantations, replanted areas and planted areas in marginal
soils in southern Bahia, Brazil.   (After Cabala et. al.
1975)

Zone Sampled area (ha)	% area planted	pH	Al	Ca	Mg	K	P ppm	%
			---meq/100g----					
North								
34,609[a]	30.6	6.2	0.1	6.7		0.16	2.5	1.5
513[b]	0.4	5.9	0.2	5.7		0.16	1.9	3.4
3,405[c]	3.0	5.8	0.3	5.2		0.20	2.4	2.4
Center								
52,719[a]	34.6	5.9	0.2	7.5		0.15	2.8	2.6
1,298[b]	0.8	5.7	0.4	6.1		0.16	2.4	6.1
2,891[c]	1.7	5.6	0.5	5.8		0.19	3.4	7.9
South								
19,092[a]	18.0	5.7	0.6	5.9		0.15	2.7	9.2
541[b]	0.5	5.4	0.8	5.8		0.14	1.9	12.1
1,676[b]	1.6	5.4	1.1	4.3		0.19	3.2	20.4

[a]Cacao plantations.
[b]Replanted areas.
[c]Areas in marginal soils.

It has been suggested that cacao, coffee and tea are
more tolerant of acid soils than other tropical crops,
although very acid soils should be avoided. Furthermore,
cacao is included among species that are able to increase
the soil pH in the soil-root interface and therefore at-
tenuate Al toxicity.

There is not much evidence to support extensive use
of lime in cacao plantations. However, cacao seedlings
(5 months) grow better when lime is applied in soils with
Al saturation over 50%. Below this level, P alone results in greater growth than when lime is added. It may
be that Al toxicity is important in early development of
cacao but is not a problem in adult plantations.

Organic matter

Soil organic matter is one of the most important
parameters in the selection of suitable soils for cacao.
In kaolinitic soils, organic matter can increase the
cation exchange capacity, improving conditions for cacao
growth and high yields. The first 15 cm of topsoil
should contain at least 3.5% of organic matter and have
C/N ratios around 9. Higher C/N ratios may indicate a
shortage of N or low mineralization of soil organic matter. Those values, however, do not include forest soils

with high levels of organic matter, because when these soils are exposed to sunlight a rapid oxidation of organic matter takes place and the natural fertility drops.

Levels of nutrients

A nutritional assessment was carried out in southern Bahia, and available levels of P and K, extracted by Mehlich solution, as well as exchangable Ca and Mg were established (Cabala et al. 1974). Table 38.2 shows those values for soils of high, medium and low fertility.
Concerning relationships of cations, it is suggested that the Ca/Mg ratio in the topsoil (15 cm) should be around 4 and the Ca+Mg/K ratio should not be below 25. High levels of K in relation to Mg may also cause difficulties.

TABLE 38.2
Parameters and levels considered in order to group soils for cacao according to the fertility of the topsoil (0-20 cm layer) (After Cabala et. al. 1974)

Soil Parameters	Units	Degree of fertility		
		high	medium	low
pH (water 1:2.5)	-	7.5-6.0	6.0-5.0	< 5.0
Organic matter	%	> 3.5	3.5-2.5	< 2.5
100-%Al (Ca+Mg saturation %)	%	100-90	90-75	< 75
P (Mehlich)	ug g^{-1}	> 15	5-15	< 5
K (Mehlich)	meq 100g^{-1}	> 0.30	0.30-0.1	0 <0.10
Ca+Mg	meq 100g^{-1}	12-6	6-3	< 3

> = greater than   < = less than

NUTRIENT CONTENT

All major elements and micronutrients are essential for cacao, but very little is known about absorption, transport and metabolic processes. It is difficult to understand these processes due to the influence of other factors and also the characteristics of cacao as a perennial crop which grows under a semi-forest ecosystem.
Only in a few cases have symptoms of Fe deficiency been recorded in West Africa and Trinidad. Zinc deficiency has appeared in West Africa and Dominican Republic, and in very rare cases B and Mn deficiencies were found. However, none of these deficiencies was important enough to justify the use of micronutrient fertilizers over extensive areas.

In southern Bahia, soil chemical analysis showed
that suitable soils for cacao contained higher levels of
micronutrients than non-traditional soils. However, in a
few cases Zn and sometimes Fe deficiencies were recorded,
especially in Tabuleiro soils (Haplorthox).

Some environmental and plant factors can affect the
uptake of nutrients, and therefore mineral composition of
cacao. Generally, shaded cacao contains higher levels of
N and K than unshaded, which in turn has higher levels of
P, Ca and Mg.

There are differences among cacao varieties in relation to nutrient absorption, and more productive cultivars may be more demanding for nutrients. Nitrogen
source can influence chemical composition of cacao;
plants fed nitrate-N contain higher levels of cations
than those fed with ammonium-N.

The nutrient content of seeds, husks and the whole
fruit has been determined, and in spite of the limitations of this procedure -- mainly because of methods of
chemical analysis and clonal differences -- this approach
gives an idea of the degree of nutrient extraction related to cacao yields (Table 38.3).

Table 38.3
Content of nutrients in dry beans and husks of cacao

	N	P_2O_5	K_2O	CaO	MgO
			kg		
Beans (1,000 kg dry beans^{-1})	21.8	9.7	14.8	1.6	4.5
Husks (1,000 kg dry beans^{-1})	13.2	2.6	35.7	4.6	3.9

In the nursery and seedlings the leaf is the most
important plant part concerning nutrient uptake. In production the leaves, branches and stems take up equal
amounts of nutrients, with the exception of K and Zn
which are found mainly in the branches.

In Malaysia, one ha of cacao (4-6 years old) required the following total amounts of nutrients for maintenance and to produce 1,000 kg of dry beans per year: (i)
N-469 kg/ha; (ii) P_2O_5 - 121/kg/ha; (iii) K_2O - 824 kg/
ha; (iv) CaO - 529 kg/ha; and (v) MgO - 211 kg/ha. It
was also estimated that 94% of the nutrients taken up by
cacao are related to growth and maintenance of structural
parts, while only 6% are involved with fruit production.
Adequate nutrition is necessary, however, to maintain vigorous growth and high yields. Successive harvests without addition of fertilizers will exhaust the soil even-

tually and reduce nutrients below adequate levels. On the other hand there is a strong limitation in using fruit analysis as an indicator for nutrient requirements. However, analysis of nutrient content of seeds can help to determine nutrient removal.

FERTILIZER EXPERIMENTS

Fertilizer experiments have been carried out for more than 50 years, and initially were done in shaded plantations considering only the physical response to fertilizers -- and in some cases to the addition of lime. Generally, the response to fertilizers in shaded cacao was small and uneconomic, and for this reason fertilizer use was not recommended to farmers.

A new perspective appeared later when shade was taken into account as a factor in fertilizer experiments, resulting in drastic changes in levels of productivity in relation to those initially considered for cacao (Table 38.4). Even with these new results the use of fertilizers in commercial cacao was not expanded much due to limitations presented by foliar analysis and a lack of research on correlation and calibration of soil tests.

In the past ten years, however, with improved research, fertilizers can be recommended on the basis of soil tests, and, in some cases, used together with leaf analysis.

Correlation and calibration of soil test methods

In Brazil, soil test methods are used. For this purpose 0-20 cm composite samples are collected. The Mehlich method, used widely in Brazil to extract P and K, has limitations in soils fertilized with rock phosphate; in these cases P availability is overestimated. This limitation is greater when poor rock phosphates are used (Table 38.5).

The Mehlich method is promising for measuring K availability in cacao seedlings, but isn't always reliable to predict cacao response to K fertilizers, even in soils with medium or low levels of K.

Response to lime

The effect of soil acidity on cacao and the degree of tolerance to Al has been evaluated in laboratory, greenhouse, and field experiments (Table 38.6). In soils where levels of both exchangeable Al and Al % saturation are high, there are no remarkable detrimental effects on growth, especially when P was added at the same time. Nevertheless, attention must be paid when Al saturation is very high; in this case growth decreases considerably.

Table 38.4
Yields of cacao dry beans obtained in the first three years of a shade and fertilizer experiment in southern Bahia, Brazil

	Shaded				Not Shaded			
	No fertilizer		With fertilizer		No fertilizer		With fertilizer	
	kg/ha	%	kg/ha	%	kg/ha	%	kg/ha	%
1964	781.92	100	774.68	99.07	912.24	116.67	876.04	112.04
1965	1245.28	100	1295.96	104.07	1491.44	119.77	1773.80	142.44
1966	1049.80	100	1187.36	113.10	1491.44	142.07	1889.64	180.00
Mean	1025.67	100	1086.00	105.41	1298.37	126.17	1513.16	144.83

Table 38.5
Yields of cacao dry beans (kg/12 trees) and phosphorus extracted by Mehlich solution from soils after additions of different phosphate fertilizers

Treatments	Means		Adjusted	Extracted Phosphorus ug/g				
	Blank period (1969–70)	After treatments (1971–75)	Means	0–5 cm	5–10 cm	10–15 cm	15–20 cm	20–40 cm
Control	22.73	23.44	25.50	5	2	2	1	1
NK (Addition)	26.69	28.28	28.08	3	2	1	1	1
Triple superphosphate + K	24.23	32.73	33.93	68	31	13	11	1
Diammonium phosphate + NK	27.08	33.99	33.57	24	26	17	9	2
Single superphosphate + NK	29.80	41.00	39.03	98	30	6	3	1
Olinda phosphate rock + NK	35.56	39.23	33.97	111	17	6	3	1
Thermophosphate + NK	26.15	32.54	32.64	74	20	12	2	1
Bonemeal + NK	22.76	30.09	32.13	79	5	2	2	4
Araxa phosphate rock + NK	22.63	28.61	30.73	111	21	7	3	1
Itambe apatite + NK	25.76	27.48	27.81	101	10	3	2	1

LSD 5% (Means after treatments) = 7.34
LSD 5% (Adjusted means) = 6.13

Table 38.6
Ranges of phosphorus, potassium, calcium+magnesium, and aluminum; and criteria to recommend lime and fertilizer in cacao plantations in southern Bahia (CEPLAC, 1978)

Element Ranges	Criteria				Fertilizer mixtures	Dolomitic limestone
	P	K	Ca+Mg	Al		
Phosphorus (ug/g)						
low < 5	low	low	–	–	A(11-30-17)	–
med 6-15	low	med	–	–	B(13-35-10)	–
high > 15						
Potassium (meq/100 g)						
low < 0,12	med	low	low	–	C(12-19-24)	–
med 0,13-0,30	med	med	med	–	D(16-24-15)	–
Calcium-magnesium (meq/100g)						
low < 3,0	–	–	low	low	–	nil
med/high > 3,0	–	–	low	med/high	–	1,5 x Al
Aluminum (meq/100g)						
low < 0,5	–	–	med/high	low	–	nil
med/high	–	–	med/high	med/high	–	[Al-0,3(Al+Ca+Mg)]x1

In soils with low Ca and Mg levels, small additions of dolomitic limestone could be beneficial because of the addition of Ca and Mg as nutrients. In these soils, however, levels of exchangeable Al from 0.6 to 1.0 meq/100g represent high saturation as a result of low levels of Ca and Mg and low cation exchange capacity. In this context it is difficult to determine the limit between direct effects of Al on cacao growth and that resulting from low Ca and Mg levels.

FERTILIZER RECOMMENDATIONS

To advise on the use of fertilizers in cacao, soil analysis -- and soil analysis plus tissue analysis -- can be used. The following factors should be taken into consideration before fertilizers or lime are used in cacao: (i) prior to fertilizer application, shade should be partially removed; (ii) the nutrients removed by yields up to 1,200 kg/ha must be replaced; (iii) not more than 10-20% of total profit expected should go to fertilizer and lime; and (iv) careful consideration should be given to fields to be fertilized and to other practices that should be carried out at the time of fertilizer applications.

Soil Analysis

Soil analysis is used only in Brazil and Ivory Coast. At least eight single samples per ha should be collected to obtain a composite sample. Sampling should be done every three years in old plantations and after five years in new plantations.

Leaf Analysis

Although leaf analysis is suitable to predict nutritional deficiencies and to advise on fertilizer use in tropical crops like sugarcane, coconuts and oil palm, its use in cacao does not appear to be promising. There is a great difficulty to establish suitable correlations between leaf composition and degree of soil fertility.

Cacao leaf composition is influenced by factors of variety, age, position, flushing, flowering and fruiting. External factors such as light intensity and seasonal variation also can influence leaf composition.

Nitrogen, P and K increase with time until the 9th or 10th week, then there may be a reduction related to formation of new tissues. Also, levels of N, P and K drop gradually during the fruiting period during heavy harvests. Nutrient movement from leaves to fruits may also occur.

The source of nutrient also has a strong influence on leaf composition; the same happens with light intensities. Shaded cacao may contain higher concentrations of N and K than unshaded cacao in which the levels of P and Ca may be higher.

For Nigeria the best period of leaf sampling is between April and May, while for Ghana this period would be between March and April, just before the rainy season.

Soil analysis, leaf composition and cacao response to fertilizers were investigated by Wessel (1971); significant responses to N were recorded when N in the leaf was below 1.8%, but there was no response when the levels were below 1.8 and 2.0%. Greater responses to P were obtained when the level was below 0.13%. Table 38.7 presents ranges of mineral composition in normal and deficient cacao leaves.

TABLE 38.7
Nutrient concentrations in normal and deficient cacao leaves (After Egbe and Omotoso, 1971)

	Criteria according to Loue (1961)			Criteria According to Murray (1967)		
Nutr.	Normal	Moderately Deficient	Severely Deficient	Normal	Low	Deficient
N	2.35-2.50	1.80-2.00	<1.80	>2.00	1.80-2.00	<1.80
P	>0.18	0.10-0.13	0.08-0.10	>0.20	0.13-0.20	<0.13
K	>1.20	1.00-1.20	<1.00	>2.00	1.20-2.00	<1.20
Ca				>0.40	0.30-0.40	<0.30
Mg				>0.45	0.20-0.45	<0.20

> = greater than
< = less than

Visual Deficiency Symptoms

Nutritional deficiencies symptoms can be considered as a complement to soil and tissue analyses. It is believed that deficiency symptoms appear almost in the last stages in which an element may be limiting either yields or plant growth. Factors such as nutritional imbalance, overliming, excess nutrient or lack of organic matter can induce some nutritional deficiencies. The same can happen with pest and disease attack. Under field conditions it is very difficult to find all the deficiencies and perhaps the most common are N, K, Zn, Fe and B; the deficiency of the last three elements generally occurs in alkaline soils or when sandy soils are limed, even with very low amounts of limestone.

Table 38.8 presents deficiency symptoms for cacao.

Table 38.8
Key to recognize mineral deficiency symptoms in cacao (After Alvim 1961 and 1964)

A. Plants markedly chlorotic

1. Leaves pale-green, reduced size, thicker than normal, often showing necrosis; necrosis advancing from the tip in later stages (Common in unshaded, weedy cacao). ... Nitrogen

2. Leaves pale-green, normal size and thickness, veins often paler than lamina (Uncommon in the field). ... Sulfur

3. New leaves only yellow, normal size, veins often remaining green; old leaves, green. (Common in badly-aerated soils lacking organic matter or highly alkaline). ... Iron

B. Chlorotic mottling between veins

1. Only old or mature leaves affected, necrotic areas between veins or marginal, often following chlorosis. (Common in acid soils and on nursery seedlings). ... Magnesium

2. Only young or recently hardened leaves showing chlorosis within inter-veinal areas and leaf margins, never alongside veins. (Uncommon, except in highly alkaline soil). ... Manganese

C. Leaves necrotic

1. Only old or mature leaves showing marginal scorch with sharp wavy division between necrotic and healthy tissue. (Common in highly-leached, acid, sandy soils). ... Potassium

2. Newly formed leaves showing necrotic areas ("islands") between veins, symmetrical on each side of mid-rib. Premature leaf shedding. (Uncommon in the field). ... Calcium

3. Mature leaves showing necrotic areas following chlorosis between veins. No premature leaf shedding. (See B.1). ... Magnesium

D. New leaves deformed

1. New leaves reduced in size, curved to spiral; lamina crinkled, hard, brittle; mid-rib and veins thicker than normal, often corky or cracked. (Occasional in leached acid soils, especially during the dry season, sometimes also in alkaline conditions.) — Boron

2. New leaves narrow, margins wavy, lamina sometimes sickle-shaped, chlorotic between secondary veins; old leaves with chlorotic spots alongside the mid-rib and main veins. (Common on sandy and alkaline soils). — Zinc

3. New leaves reduced in size, compressed near apex; secondary veins reduced in number, irregularly distributed; necrotic apex. (Uncommon in the field). — Copper

E. Absence of chlorosis, necrosis or leaf deformity

1. Reduced leaf canopy (premature shedding of lower leaves), occasional necrosis near leaf apices prior to shedding; sometimes leaves bronzy color. (Very common on infertile soils). — Phosphorus

2. New leaves narrow, translucent, faint chlorotic mottling in inter-veinal areas; marginal necrosis in older leaves. (Uncommon in the field.) — Molybdenum

SUPPLEMENTAL READING

ACQUAYE, D.K. 1964. Foliar analysis as a diagnostic technique in cacao nutrition. 1. Sampling procedures and analytical methods. Journal of Science of Food and Agriculture 12:855-863.

_____, SMITH, R.W.C. and LOCKARD, R.G. 1965. Potassium deficiency in unshaded Amazon cocoa (Theobroma cacao L.) in Ghana. Journal of Horticultural Science 40(2):100-108.

ALVIM, P. de T. 1961. Clave para los sintomas de deciencias en cacao. In Hardy, F., ed. Manual de Cacao. Turrialba, Costa Rica, Instituto Interamericano de Ciencia Agricolas. pp. 76-78.

ALVIM, P. de T. 1964. Como conhecer os sinais de fome do cacaueiro para aduba-lo satisfatoriamente. Cacau Atualidade (Brasil) 1(6):6-7.

BRADEAU, J. 1970. El Cacao. Barcelona, Blume. pp. 51-67.

BURRIDGE, J.C., LOCKARD, R.G. and ACQUAYE, D.K. 1964. The levels of nitrogen, phosphorus, potassium, calcium and magnesium in the leaves of cacao (Theobroma cacao L.) as affected by shade, fertilizer, irrigation and season. Annals of Botany, 28(11):401-417.

CABALA-ROSAND, P. et al. 1969. Deficiencias minerais e efeitos da adubacao na regiao cacaueira da Bahia. In Conferencia Internacional de Pesquisas em Cacau, 2a., Salvador/Itabuna, BA, Brasil. 1967. Memorias. Ilheus, BA, Brasil, CEPLAC. pp. 436-442.

CABALA-ROSAND, P., MIRANDA, E.R. de and PRADO, E.P. do. 1970. Efeito da remocao de sombra e da aplicacao de fertilizantes sobre a producao do cacaueiro da Bahia. Cacao (Costa Rica) 15(1):1-10.

_____, ALVIM, P. de T. and MIRANDA, E.R. de. 1974. Representacao da fertilidade do solo atraves de graficos poligonais. In Reuniao Brasileira de Fertilidade do Solo, 9a., Belo Horizonte, Brasil, 1974. Communicacoes da Equipe de Fertilidade do Centro de Pesquisas do Cacau. Ilheus, B, Brasil, CEPLAC. pp. 30-35.

_____, SANTANA, M.B.M. and MIRANDA, E.R. 1975. Fertilidade dos solos ocupados com cacaueiros no Sul da Bahia. Ilheus, Ba, Brasil CEPLAC/CEPEC. Boletim Tecnico no. 27, 31 p.

CABALA-ROSAND, P. et al. 1975. Exigencias nutricionais e fertilizacao do cacaueiro. Ilheus, Ba, Brasil. CEPLAC/CEPEC. Boletin Tecnico no. 30. 59 p.

_____, SANTANA, C.J.L. de and MIRANDA, E.R. de 1976. Respuestas del cacaotero al abonamiento en el Sur de Bahia, Brasil. Ilheus, Ba, Brasil. CEPLAC/ CEPEC. Boletim Tecnico no. 43. 24 p.

CENTRO DE PESQUISAS DO CACAU. 1978. Normas para utilizacao de fertilizantes e corretivos na regiao cacaueira da Bahia. Ilheus, Ba, Brasil. 74 p.

CUNNINGHAM, R.K. 1959. A review of the use of shade and fertilizer in the culture of cocoa. Tafo, Ghana. West African Cocoa Research Institute. Technical Bulletin no. 6. 15 p.

_____, and BURRIDGE, J.C. 1960. The growth of cacao (Theobroma cacao) with and without shade. Annals of Botany 24(96):258-262.

_____, and ARNOLD, P.W. 1962. The shade and fertilizer requirements of cacao (Theobroma cacao) in Ghana. Journal of Science of Food and Agriculture 13:213-221.

_____, 1964. Micronutrient deficiency in cacao in Ghana. Empire Journal of Experimental Agriculture 32(125):42-50.

EGBE, N.E. and OMOTOSO, T.I. 1971. Nutrition of cacao in Nigeria. In Cocoa Research Institute of Nigeria. Progress in Tree Crop Research in Nigeria (Cocoa, Kola and Coffee). Ibadan, Nigeria. pp. 78-95.

GEUS, J.G. de. 1967. Fertilizer Guide for Tropical and Subtropical Farming. Zurich, Centre d'Etude de l'Azote. 727 p.

JADIN, P. 1977. La fertilisation minerale des cacaoyers en Cote d'Ivoire a partir du "diagnostic sol". Cafe Cacao The 16(3):204-218.

KANAPATHY, K. 1976. Guide to Fertilizer Use in Peninsular Malaysia. Kuala Lumpur, Ministry of Agriculture and Rural Development. 160 p.

LOUE, A. 1961. Etude ces carences et des deficiences minerales sur la cacaoyer. Paris. Institut Francais du Cafe et du Cacao Bulletin no. 1. 52 p.

MASKELL, E.J., EVANS, H. and MURRAY, D.B. 1953. The symptoms of nutritional deficiencies in cacao produced in sand and water cultures. In St. Augustine. Trinidad. College of Tropical Agriculture. A Report on Cacao Research 1945/51. St. Augustine. pp. 53-64.

MURRAY, D.B. 1967. Leaf analysis applied to cocoa. Cocoa Growers Bulletin No. 9:25-35.

_____, and MALIPHANT, G.K. 1967. Problems in the use of leaf and tissue analysis in cacao. In Conference Internationale sur les Recherches Agronomiques Cacaoyeres, Abidjan, Cote d'Ivoire, 1965. Paris, IFFC. pp. 36-38.

SANTANA, M.B.M., CABALA-ROSAND, P. and MIRANDA, E.R. de. 1970. Nivel nutricional dos solos da Regiao Cacaueira da Bahia. Ilheus, BA, Brasil. CEPLAC/CEPEC. Boletim Tecnico no. 6. 25 p.

_____, CABALA-ROSAND, P. e MIRANDA, E.R. de 1973. Toxidez de aluminio em plantulas de cacau. Revista Theobroma (Brasil) 3(1):11-21.

SANTANA, M.B.M., CABALA-ROSAND, P. e MIRANDA, E.R. de. 1974. Efeito da concentracao de aluminio sobre o desenvolvimento de plantulas ce cacau e seringueira. In Reuniao Brasileira de Fertilidade do Solo, Belo Horizonte Brasil, 1974. Comunicacoes da Equipe de Fertilidade do Centro de Pesquisas do Cacau. Ilheus, Ba, Brasil, CEPLAC/CEPEC. pp. 44-48.

THONG, K.C. and NG, W.L. 1980. Growth and nutrient composition of monocrop cocoa plants on inland Malaysian soils. In International Conference on Cocoa and Coconuts, Kuala Lumpur, 1978. Proceedings. Kuala Lumpur, The Incorporated Society of Planters. pp. 262-286.

URQUHART, D.H. 1963. Cacao. Turrialba, Costa Rica, IICA, 322 p.

WESSEL, M. 1969. Cacao soil of Nigeria. In Conferencia Internacional de Pesquiss em Cacau, 2a., Salvador, Itabuna, B, Brasil, 1967. Memorias. Ilheus, Ba, Brasil, CEPLAC. pp. 417-430.

_____, 1970. Fertiliser experiments on farmers' cocoa in South Western Nigeria. Cocoa Growers' Bulletin No. 15, pp. 22-27.

_____, 1971. Fertiliser requirements of cacao (Theobroma cacao L.) in South-Western Nigeria. Amsterdam. Royal Tropical Institute. Communication no. 61. 104 p.

39
COFFEE

Jose Vicente-Chandler

There are about 15 billion coffee trees in the world planted on 10 million hectares yielding an average of about 600 kg of market coffee/ha, Annual production is about 4.4 million metric tons of Arabica coffee and 1.2 million tons of Robusta coffee. About 3.8 million tons are exported annually, 1.3 million to the USA. Discussions in this chapter are confined to Arabica coffee (Coffea arabica) since it is the most widely planted and studied species and produces the highest quality coffee.

The areas best suited to coffee production are generally at elevations of from 750 to 1,500 m above sea level with mean monthly temperatures ranging from 15 to 25°C and an annual rainfall from 1,600 to 2,400 mm. Although well established coffee trees can resist prolonged droughts, dry weather when the trees are bearing a large crop can cause considerable fruit drop, and also result in reduced yields the following year. Deep soils with bulk density of 1.0 to 1.2 with good structure and permeability and preferably with 3% or more of organic matter are essential to successful coffee production. Although coffee soils are often low in nutrients and are frequently acid, these deficiencies can be corrected by proper fertilization, liming and management.

CROP PHYSIOLOGY

Coffea arabica is a deciduous evergreen, and lost leaves are never replaced. Generally, several main roots penetrate to a depth of up to 1 m, and over 80% of the coffee roots are found in the upper 40 cm of soil, extending to over 1.5 m from the trunks. The greatest growth of coffee roots apparently occurs previous to the flush of top growth and flowering in the spring.

---

[1]Research Scientist, Agricultural Research Service, U.S. Department of Agriculture, Rio Piedras, Puerto Rico, 00927.

Coffee flowers several times during the spring, and the heaviest flowering often follows the first rains after a rather dry winter season. Consequently the crop ripens in stages throughout a period of several months, about 7 months after first flowering.

Fastest vegetative growth occurs at flowering time and later when the berries are developing. Slowest growth occurs after the harvest season when coffee can be almost dormant during the cool, shorter days of winter, particularly if the weather is dry, as is often the case.

The coffee crop is borne mainly on the branches produced during the previous year, and therefore growth during one growing season largely determines the size of the subsequent crop. Since much of the growth occurs during the season when the trees are also producing a crop of berries, heavily bearing trees generally cannot make vigorous vegetative growth, consequently the following crop will usually be smaller. Conversely, when a small crop is borne, the trees may produce much bearing wood, which can result in a heavy crop the following year, if climatic conditions are favorable. This explains much of the fluctuations in coffee yields that may occur from one year to another. Intensively managed coffee starts bearing heavily 3 to 4 years after planting and can produce high yields for 20 years or more.

Coffee grows well under shade but produces much more heavily when grown in full sunlight. This has been shown by experiments in Colombia, Costa Rica, Brazil, Guatemala and Puerto Rico where intensively managed coffee (nine varieties) produced about 40% more, about 2,000 kg of market coffee/ha, when grown in full sunlight than under shade trees. Trials in various countries have shown no difference in coffee quality attributable to shade, except that beans may be smaller in full sunlight as a result of the higher yields produced.

To cut costs, yields must be sharply increased by cultivating coffee intensively in full sunlight. However, only farmers disposed to carry out all the essential practices should attempt intensive coffee culture since omitting just one practice can jeopardize the entire effort. Coffee trees yielding little under dense shade can stand considerable mismanagement, whereas intensively managed plantations producing at close to maximum capacity will suffer severe damage if neglected.

NUTRIENT REQUIREMENTS

The fertilizer requirements of high-yielding coffee can be rationalized. Table 39.1 shows the approximate quantity of nutrients contained in 11,000 kg of coffee cherries yielding about 2,000 kg of market coffee, a production level often attained by intensively managed coffee plantations.

TABLE 39.1
Nutrient composition of coffee beans and processing byproducts

	Nitrogen	Phosphoric Acid (P_2O_5)	Potash (K_2O)
Hulls and pulp	17	4	27
Beans	45	8	38
Total	62	12	65

In addition to these nutrients which are removed from the plantation annually, about 50, 10 and 50 kg/ha of N, P_2O_5 and K_2O, respectively, are contained in the leaves and twigs grown yearly on the coffee trees. Thus, a total of 112, 22, and 115 kg/ha of N, P_2O_5 and K_2O are respectively taken up each year by high yielding coffee.

Experiments with many crops in the tropics show that only 40% of the N applied as fertilizer and 70% of the K and P are utilized by the crop, the remainder being lost, mostly by leaching.

Coffee is not likely to obtain more than 40 kg/ha each of N and K yearly from the relatively infertile soils on which it is usually grown. Such soils often have little P other than that remaining from previous fertilization.

From the above it can be concluded that high-yielding coffee will require annual applications of about 180, 30 and 110 kg/ha of N, P_2O_5 and K_2O, respectively. Since 2 metric tons of coffee pulp are required to supply the nutrients contained in 100 kg of 10-3-14 fertilizer, these nutrients can usually be supplied more cheaply as fertilizers except where labor is cheap and fertilizer very expensive.

Extensively managed, low-yielding coffee responds little or not at all to fertilization as shown by numerous experiments in India, Puerto Rico, Colombia and other countries. On the other hand, intensively managed, high-yielding coffee generally responds strongly to fertilization. In Hawaii, Dean and Beaumount (1938) showed a strong response to both N and K but not to P.

In Colombia, the application of different rates of 12-12-17-2 (NPKMg) fertilizer to intensively managed coffee growing in full sunlight had the following effect on coffee yields (Mestre-Mestre, 1972):

Fertilizer Applied (kg/ha yearly)	Yields of Market Coffee (kg/ha)
0	2,210
600	2,810
1,600	3,520
2,400	3,440

The application of 1,600 kg/ha of 12-12-17-2 was found to be close to optimum. This is equivalent to applying 192, 192, 272 and 32 kg/ha yearly of N, P_2O_5, K_2O, and M_gO, respectively. In view of the general lack of a strong response by coffee to P application, it is likely that only a fraction of the P applied was needed to produce these high yields.

Also in Colombia, intensively managed sun grown coffee responded over a 10-year period to applications of 120-240 kg of N/ha yearly at eight locations studied and to applications of K within this range at five locations, but did not respond to P at any location.

Numerous experiments indicate that intensively managed coffee growing in typical soils responds to applications of about 150-300 kg/ha yearly of N and K depending largely on yields; about 200 kg/ha of both N and K is required for yields of 2,000 kg of market coffee/ha. Response to P is rare in soils previously fertilized with this nutrient. On the other hand, P applications may be needed on previously unfertilized soils or those that fix P in unavailable form to plants.

SOIL TESTS

There is no adequate soil test for N. Available soil P values above 10-15 ppm (Bray #1) usually indicate sufficient P for producing high yields of coffee. Exchangeable K values above 300 ppm usually indicate adequate levels for high yields.

FOLIAR DIAGNOSIS

Leaf analysis may be used to diagnose nutrient deficiencies, although nutrient content of the leaves varies with season of the year, age and other factors.

The fourth pair of leaves from the tips of the branches is considered the best index of the nutritional status of coffee plants, better reflecting changes in composition of the leaves as well as the effect of fertilization and pruning. This pair of leaves also has the lowest coefficient of variability in both mineral and carbohydrate constituents. Flowering branches from the previous year's growth, preferably branches flowering for the first time, should be sampled. About 40 leaves constitute an adequate sample. Leaves should be taken from all sides of the plant at about 1.5 m height. Age of leaves, position of the plant, season of the year, variety and physiological status of the plant all affect leaf composition.

Table 39.2 presents foliar analysis levels for 12 nutrients in coffee leaves.

TABLE 39.2
Concentrations of nutrients in the fourth pair of coffee leaves and general nutrient status of the plant

	General Nutrient Status of the Plant		
	Low	Medium or Adequate	High
	percent of dry wt.		
Nitrogen	<2.5	2.5-3.0	>3
Phosphorus	< .10	.15	.18
Potassium	<1.5	2.0	2.5
Calcium	.6	1.0	1.4
Magnesium	.2	.4	.6
	parts per million (ppm)		
Sulphur	100	200	>200
Manganese	100	200	300 (>500 = toxic)
Iron	70	100	>100
Boron	30	60	>100 (can be toxic)
Copper	5	10	15
Zinc	5	10	20
Molybdenum	.1	.3	.5

Coffee yields may be almost nil when leaves contain less than 1% K, while high yields may result when the leaves contain 2% or more of K. Nitrogen concentrations in the leaves during the flowering season correlate well with coffee yields. At this time also, nutrient deficiencies can be corrected before the high demand for nutrients occurs.

NUTRIENT SOURCES

In general, although the cheapest source of nutrients should be used, those high in sodium should not be used and those containing chlorides should be used sparingly. Chlorine toxicity may result when leaves contain over 2,000 ppm Cl. It has been suggested that less than half of the K applied to coffee should be in the chloride form.

TIME AND PLACE OF APPLICATION

Coffee grows throughout the year, but maximum growth occurs after flowering in the spring and 4 to 6 months

later when the coffee berries are developing at a fast
rate. About half of the N and K in coffee berries accumulates over only a 2-month period. Slowest growth occurs immediately after the crop is harvested which generally coincides with the cooler, shorter, often drier days
of winter.

It is probably best to apply the fertilizer in three
equal applications, at first flowering, and 3 and 6
months later. Applying fertilizer in a band about 30 cm
from the trunk of the trees may give best results. In
Hawaii, applying 350 kg N/ha yearly in ten equal applications gave better results than when N was applied in only
two applications.

Manganese (Mn) Toxicity and Liming

The need for liming coffee plantations is related
both to soil acidity and to readily available Mn content,
both of which vary widely with soil type. Manganese
toxicity is not a problem on soils low in easily reducible Mn irrespective of their acidity, but is likely to be
a serious problem if acidity is high in soils with high
exchangeable or easily reducible Mn.

Foliar analysis should be used along with soil acidity and exchangeable and easily reducible Mn in determining the need for liming coffee. The critical level of Mn
in coffee leaves has not been determined, and values may
vary widely with season of the year and other factors.
However, leaf Mn values in excess of 1,000 ppm indicate a
severe Mn toxicity, and values of 500 to 1,000 ppm indicate a moderate Mn toxicity. Values of 100-300 ppm are
considered normal.

Low-yielding, shade-grown coffee generally does not
respond to liming. In Puerto Rico, over a 3-year period,
applications of ammonium sulphate, urea or ammonium nitrate at the rate of 240 kg of N/ha/yr decreased the pH
of an Ultisol by 0.7 units. Also, the effects of liming
over 4 years on yields of well fertilized sun-grown
coffee, and on the condition of two Ultisols, both containing almost no exchangeable Mn, were determined. Liming did not affect yields in either experiment although
it strikingly decreased the exchangeable aluminum content
and increased the pH of both soils (Table 39.3). Liming
increased the pH of the Alonso clay from 4 to 6.1 and
that of Los Guineos from 3.9 to 4.7. Exchangeable aluminum content of the Alonso clay was decreased from 3 to 0
me/100 g and that of Los Guineos from 3.8 to 1.3 me/100
g. Manganese content of the coffee leaves was not appreciably affected by liming, and ranged from about 100 to
300 ppm.

Similarly, Valencia and Bravo (1981) found that liming coffee trees growing in a soil with pH values as low
as 3.8 with 80% Al saturation of the cation exchange capacity (4.2 me/100 g of soil) did not affect yields even

TABLE 39.3
Effect of liming on coffee yields and acidity in the surface 15 cm of two Ultisols over a 4-year period (Abruna, et al., 1965)

Limestone Applied (ton/ha)	Yields of Market Coffee (kg/ha)	Soil pH	Exchangeable Aluminum in Soil (me/100 g)	Manganese in Coffee Leaves (ppm)	Exchangeable Calcium + Magnesium (me/100 g)	Saturation of the Effective Cation Exchange Capacity of the Soil with Aluminum (%)
			Alonso Clay			
0	1,880	4.0	3	290	1.7	55
16	1,870	6.1	0	235	22.6	0
			Los Guineos Clay			
0	2,510	3.9	3.8	145	1.8	60
12	2,810	4.7	1.3	110	4.7	20

when liming raised soil pH to 5.5 and exchangeable Al was reduced to 0.9 me/100 g of soil. This soil was also very low in exchangeable Mn as evidenced by leaf contents not exceeding 160 ppm.

These data show that high yields of coffee can be produced even on very acid soils provided that they contain little exchangeable or easily reducible Mn.

The situation is very different, however, with acid soils high in available Mn such as occur in many coffee growing areas. Under these conditions, coffee plantations can evidence alarming symptoms. In the initial stages, margins of the coffee leaves become deep-yellow, followed by complete yellowing of the young leaves. In later stages, older leaves drop off and bearing branches lose most of their berries. This is soon followed by severe dieback. Leaf analysis usually reveals normal contents of all nutrients except Mn, which varies from 1,000 to as high as 2,500 ppm compared with 100 to 200 ppm for normal leaves.

Two such severely affected plantings in Puerto Rico were treated with 8 mt of limestone/ha distributed under the coffee rows. Two years later yields of market coffee had increased from about 600 to over 2,000 kg/ha, the coffee leaves were deep green, and their Mn content had dropped from 1,500 to about 400 ppm. Similar results were obtained in Brazil where Moraes (1966) reported a 50% increase in coffee yields as a result of liming soils with pH values of 4 to 4.5.

MINOR ELEMENTS

Although symptoms of minor element deficiencies are rarely found in densely shaded, low-yielding coffee plantations, high-yielding plantations frequently display a severe deficiency of one or more minor elements, especially zinc, magnesium and boron.

Symptoms of sulphur deficiency, which are similar to those of severe N deficiency except that the leaves are mottled, can occur under some conditions. In Brazil, application of 35 kg/ha of S increased coffee yields by 90% and sulphate and S content of the leaves from 75 to 150 ppm (Freitas, et al., 1972). Sulphur deficiencies are not likely when potassium sulphate, single superphosphate or ammonium sulphate are used as fertilizers or when the soil contains more than about 100 kg/ha of sulphate sulphur.

Liming as recommended will provide the required Ca and also Mg if the limestone contains sufficient dolomite, as is often the case.

Thus, the probable minor element deficiencies are Fe, Mn, Zn, B, and Mg, if limestone containing little dolomite is used. Deficiency symptoms for these four elements, plus Mg, are as follows:

Iron (Fe). Chlorosis of the young leaves, the veins remaining green during the early stages. Eventually, the entire leaf may become yellow, but usually the size and shape are normal. Seedlings are particularly susceptible, but generally their growth rate is not reduced much. Symptoms of iron deficiency often appear during droughts and may disappear after the rains begin.

Zinc (Zn). Similar to those of iron, but in severe cases the young leaves are small, narrow, curled, and brittle, and the internodes are short. Loss of leaves and dieback of the branches often follow. Typical symptoms of a severe zinc deficiency have been corrected by applications of zinc sulfate.

Boron (B). Death of the growing tips, followed by the production of several secondary branches that give the twigs a broomlike appearance. The leaves are often cupped upward and the young ones have chlorotic areas near the tips and edges. Symptoms are usually particularly noticeable at the start of the rainy season, at flowering or after liming.

Manganese (Mn). Chlorosis of the leaves, but often lemon-yellow in contrast with pale-yellow or whitish color of leaves deficient in iron. Excessive quantities of Mn may be taken up by trees growing on acid soils. Under these conditions, uptake of iron is reduced and mottled chlorosis and yellowing of the leaf blades and loss of leaves follow. Trees treated with chelated iron recover rapidly, but liming is the most practical long-term solution to this problem.

Magnesium (Mg). Deficiency symptoms show up on the older leaves. Irregular chlorotic patches appear between the veins, progressing to necrotic areas near the tips. A green area often remains near the petiole.

OVERCOMING MINOR ELEMENT PROBLEMS

Heavy fertilized, intensively managed coffee growing in Los Guineos clay in Puerto Rico responded strongly to the application of 35 kg of Mg/ha (as magnesium sulphate) as shown in the following tabulation.

Mg Applied (kg/ha/yr)	Market Coffee Produced (kg/ha)	Average Mg Content of Leaves (%)
0	1,885	0.14
35	2,560	.30
70	2,780	.33

Iron must always be applied as a chelate to the soil, since it is usually rapidly fixed in the soil and is not absorbed through the coffee leaves in appreciable quantities. A few grams of chelate may be applied to seedlings to overcome deficiencies.

With the other minor elements, most rapid effects are obtained by applying the deficient minor element in a foliar spray. Suggested quantities of material to use per 1,000 liters of spray, enough to treat 1 ha of mature coffee, are as follows:

	kg
Manganese sulfate	3
Zinc sulfate	4
Borax	2
Calcium hydroxide (to neutralize any free sulfate)	4

Sticker-spreader (as recommended by manufacturer)

If mist or low-volume spray equipment is used, these materials should be mixed in about 150 liters of water. Frequency with which the spray is applied will depend on the severity of the deficiency; usually two applications a year during seasons of fast growth are enough.

Minor elements can also be applied to the soil as frits or as chelates. In Brazil the application of chelates of Cu, Fe, Mn, Zn, or of all combined, approximately tripled yields of coffee (Medcalf and Lott, 1956). There was no difference between the effect of the various chelates on coffee yields, and all chelates reduced Mn content of the leaves from 1,100 to 520 ppm. Thus, the beneficial effect of chelates could have been caused by reducing Mn toxicity.

Zinc may be applied to the soil at the rate of 10 kg/ha yearly. Magnesium may be applied as dolomite or as magnesium sulphate at the rate of 100 kg/ha yearly.

A deficiency of B may be corrected by applying borax (11% B) to the soil. About 14 g per tree yearly may be applied until the deficiency is corrected. Boron should be used only under the direction of a competent technician since it may be toxic to coffee if applied in excessive amounts.

ECONOMIC CONSIDERATIONS

Considering the high value of coffee (about US$200/ kg of market coffee) in relation to fertilizer ($25/100 kg) it appears generally economic to fertilize heavily providing all other factors (climate, proper pest and weed control, high-yielding varieties, proper pruning and high sunlight intensity) favor high yields. The application of 1 mt of fertilizer/ha worth about $250 to ensure the production of 2 mt of coffee worth over $4,000 is a good investment. When other factors limit yields, fertilizer requirements should be decreased proportionally.

SELECTED REFERENCES

ABRUNA, F., J. VICENTE-CHANDLER, L.A. BECERRA, and R. BOSQUE-LUGO, 1965. Effect of liming and fertilization on yields and foliar composition of high yielding new grown coffee. Journal of Agriculture, University of Puerto Rico, 49(4):413-28.

CARVAJAL, J.F. 1963. Leaf sampling of coffee for diagnostic purposes. Coffee and Cacao Technical Service, Turrialba, Costa Rica 5 (17):21-39.

CIBES, H., and G. SAMUELS, 1965. Mineral deficiency symptoms displayed by coffee trees grown under controlled conditions. Agriculture Experiment Station, University of Puerto Rico, Tech. Paper No. 14.

COOIL, B. and E.T. FUKUNAGA, 1959. Mineral nutrition: high fertilizer applications and their effects in coffee yields. In: Advances in Coffee Production Technology, B. Sacks and P. Sylvain (Editors). Coffee and Tea Industries, N.Y., N.Y.

DEAN, L.A., and J.H. BEAUMOUNT, 1938. Soil and fertilizer in relation to the yield, growth and composition of the coffee tree. American Society Horticultural Sciences 36:28-35.

FREITAS, L., F. PIMENTAL-GOMES, and W.L. LOTT, 1972. Effect of sulfur fertilizer on coffee. IRI Res. Inst., Bull. No.41, Rockefeller Plaza, N.Y., N.Y.

HUERTAS, A. 1963. Par de hojas representativas del estado nutricional del cafeto. Cenicafe 14(2): 111-126.

LOPEZ, A.M. 1967. Fertilizacion con cloruro de potasio y sulfato de potasio en plantaciones de cafe. Cenicafe 18(2):47-54.

MEDCALF, J.C., and W.L. LOTT, 1956. Metal chelates in coffee. IBEC Res.Inst. Bull. No. 11.

MESTRE-MESTRE, A. 1972. Determinacion de la rata optima de fertilizacion en plantaciones de cafe sin sombra. Cenicafe 28(2):51-60.

MORAES, F.R.P. 1966. Aumento da produtividade da cultura cafeeira. Efeito da fertilicao do solo sobre a producao do cafe. O. Agronomica Brasil 18 (5:6) 7-8.

PARRA, J. 1971. Correlaciones entre peso seco de cafe y analisis de suelo. Cenicafe 22(3):83-91.

URIBE-HENAO, A. and A. MESTRE-MESTRE, 1976. Efecto del nitrogeno, fosforo y potasio sobre la produccion de cafe. Cenicafe 27(4):158-173.

VALENCIA, G., and E. BRAVO, 1981. Influencia de encala-miento en la produccion de cafetales intensifi-cados. Cenicafe 32(1):3-14.

VICENTE-CHANDLER, J., F. ABRUNA, R. BOSQUE-LUGO, and S. SILVA, 1968. Intensive coffee culture in Puerto Rico. Agricultural Experiment Station, University of Puerto Rico, Bull. 211.

40
ALFALFA (Lucerne)

Dale Smith

Alfalfa (<u>Medicago</u> <u>sativa</u>) is a legume widely grown as feed for livestock, especially for dairy cows during milk production. It may be pastured, fed green as whole herbage or chopped, ensiled, or fed dry as loose hay, chopped hay, or in bales. The herbage also is dehydrated to be used in mixed rations of rabbits, poultry, swine, and horses, or as a leaf meal.

ECOLOGICAL ADAPTATION

Alfalfa grows best where winters are mild and summers are warm. In the upper latitudes where subfreezing temperatures prevail during winter, cold-hardy types have been developed. Maximum growth rate (kg/ha/day) of alfalfa occurs near 25 to 26° C (78-80°F), but it is grown in regions where summer temperatures seldom get above 22° C (e.g., Canada, Scandinavia) and in arid regions under irrigation where summer temperatures at times may exceed 54° C (e.g., Yuma Valley of Arizona and Imperial Valley of California, USA; Saudi Arabia and Bahrain). With cool temperatures, growth is slow and flowering delayed. At high temperatures, alfalfa may flower quickly with little yield (the "summer slump" period in the arid Southwest USA).

Alfalfa grows on a wide variety of soils if they have good surface and internal drainage. Alfalfa does not survive long in standing water and grows best on well-drained, deep loam or sandy loam soils. The soil pH should be near 7.0, but the crop is grown where the pH may be as high as 8.5 (e.g., irrigated lower valleys of New Mexico, Arizona and California). In Australia, alfalfa is grown on soils that are quite acid, but these soils are low in manganese (Mn), aluminum (Al), and iron (Fe). A primary reason for adding lime to many acid

[1]Adjunct Professor of Plant Sciences, University of Arizona, Tucson 85721, USA (and Professor Emeritus of Agronomy, University of Wisconsin, Madison).

soils is to raise the pH to about 7.0 to reduce the Mn, Al, and Fe in the soil solution, since these elements are toxic to alfalfa when in high concentration.

Alfalfa fits well into a rotation-cropping system with cultivated crops where alfalfa stands are kept for 2 or 3 years. Cultivated crops (e.g., maize, oats, sorghum, cotton) in the rotation benefit from the accumulated N and organic matter. On steep, erodable land where cultivated crops accelerate soil erosion, alfalfa is kept as long as stands are satisfactory to hold the soil.

MINERAL NUTRIENT REQUIREMENTS

About 11 million ha of alfalfa presently are being harvested for hay and silage in the USA, and an additional undetermined crop area is pastured. Top farm yields have been in the range of 12 to 16 mt/ha (dry matter) from three to four harvests annually in the North Lakes states of Minnesota, Wisconsin, Michigan, and New York, and in Ontario province of Canada; from four to five harvests in the Corn Belt states of Illinois, Indiana, Ohio and Pennsylvania; from 20 to 24 mt/ha from five to six harvests in the Great Plains states of Nebraska, Kansas, and Oklahoma; and from 28 to 32 mt/ha from seven to nine harvests in the irrigated, lower valleys of Arizona and California. Even so, average yields for the states usually are about one-half those of the top farmer yields.

AVERAGE YIELDS IN USA AND OTHER COUNTRIES

Average yield of alfalfa for the USA in 1982 was reported as 6.8 mt/ha of dry hay. Average yields in 1982 for the major regions of the USA were (mt/ha) Northcentral - 6.5, Northeast - 5.7, Southeast - 7.0, and the West - 8.0. Highest yield for individual states was 14.6 mt/ha for Arizona and 13.4 mt/ha for California; states where alfalfa is grown mostly under irrigation and is harvested nearly the entire year.

Yields for other countries are difficult to obtain; however, Table 40.1 presents the best information available.

PRESENT LEVELS OF MINERAL NUTRITION BY REGION

As farmers obtain higher yields and as the plant breeders release higher-yielding, persistent, pest-resistant cultivars, it will be necessary to increase fertilization, especially of P and K, if the stand is to survive and remain highly productive. In addition, it may become necessary to apply other elements, such as sulfur, magnesium, boron, iron, and zinc that may not have been required previously.

TABLE 40.1
Yields of alfalfa hay in selected countries for the years
noted

Country	Year	Metric Tons/Hectare
France	1967	5.40
Great Britain	1967-68	2.56 - 5.20
Belgium	1969	4.44 - 8.9
West Germany	1969	3.29 - 6.6
Australia	1968	3.85 - 7.8
Greece	1969	3.92 - 7.9
Hungary	1967	1.79 - 3.6
Romania	1968	1.45 - 2.9
Yugoslavia	1967	2.54 - 5.1
Australia	1966-67	2.53 - 5.1
Argentina	1968-69	3.23 - 6.6
Chile	1965	1.09 - 2.3
Peru	1965-66	5.30 -10.7
Brazil	1968	2.91 - 5.9

Adapted from: Bolton, J.L., B.P. Goplen, and H.
Baenizger. World distribution and historical development. p. 1-34. In: C.H. Hanson (ed.), Alfalfa Science
and Technology, Agronomy Monograph No. 15, American
Society Agronomy, Madison, Wisconsin, 1972.

In the Northcentral USA, P, K, S, and B are most often found to be deficient for alfalfa production. This
is also the case in the Northeast USA, but Mg and molybdenum (Mo) are deficient in some localities. In the
Great Plains, P and S are most often deficient but the
need for K is increasing. In the Pacific Northwest, P
and K are most often deficient. There may be a need for
Mg, and B sometimes occurs at toxic levels. Also, high
levels of sodium (Na) salts in the soil and irrigation
water hinder alfalfa production in some parts of the
Southwest. Alfalfa tolerates moderate salinity, and tolerance increases with plant age if soil fertility needs
are met. Alfalfa is now becoming an economic crop in the
Southeast, and the nutrient problems of these acid soils
are being assessed.

MINERAL NUTRIENTS NEEDS FOR ACCEPTABLE YIELDS

Soil test values sought. Soil analysis results are
the best guide to determine whether nutrients need to be
applied. Soil analysis should not include just the usual
P and K, but should include an entire nutrient profile.
Soil samples can be taken any time of the year when the
soil is not frozen, or too dry or too wet. An ideal time
would be before the crop season begins or after it has
been completed.

Some knowledge of the properties of the soil is
necessary in order to decide how much of a nutrient (fertilizer) should be applied. A soil test suggests the
amount of a nutrient available to alfalfa at the time of
the test, but it does not show the soil's capacity to release nutrients (supplying-power), or in reverse, the
soil's capacity to tie-up (fix) added nutrients. Some
soils have a high supplying-power (e.g. for K), while
others may have little or none. The less the supplying-
power of the soil, the greater the importance of fertilizer applications to establish the stand and to maintain
yields thereafter. Thus, the amount of a nutrient that
will be available during a growing season will depend on
the amount already in the soil solution, the amount released and/or fixed by the soil, and the amount applied
as fertilizer. Alfalfa is a heavy user of soil nutrients. A 20 mt/ha yield (dry matter) of alfalfa during a
season will remove these amounts of nutrients (kg/ha);
K-550, Ca-390, Mg-66, P-58, S-55, B-0.05, Mn-0.5, Zn-0.4,
Cu-0.2, and Mo-0.01. These amounts must be supplied by
the soil, or fertilizers must be added. Wisconsin recommendations suggest that by the end of the crop season and
prior to applying top-dressed fertilizer for production
next year, soil tests should not be below 55 kg/ha for
available P, 380 - 440 kg/ha for exchangeable K, 44 kg/ha
for sulfate-sulfur (SO_4-S), and 3 kg/ha for B. Higher
levels in the soil may be required with low soil temperatures, with very dry or wet soil conditions, poor drainage, or root damage from diseases and insects.

Plant tissue values sought. A soil analysis indicates what nutrients are in the soil, but does not necessarily indicate the amounts of nutrients that plants take
up from soil. Plants can indicate best what is available
from the soil. There is a need for greater use of tissue
analysis, not just of P and K, but for all of the major
and micronutrients. A deficiency of one nutrient may
limit the effectiveness of the others. Tissue analyses
of the entire elemental profile (10 to 12 elements) can
be of use, at the least, as a diagnostic tool to determine whether the crop is obtaining sufficient nutrients
for maximum growth and yield. Low nutrient levels in the
plant also may be due to low soil temperatures, low soil
moisture, poor drainage, or root damage, even though adequate amounts of nutrients are available in the soil.

Tissue analyses should be made soon after the stand
is well established in the seeding year, and in succeeding crop years. It would be best to sample tissues during the first or second cuttings in the spring, because
these are high-yielding harvests with high nutrient demands. Sampling should be done when soil moisture is
adequate and not when it is very wet or very dry.

Suggested minimum sufficiency tissue values are giv-
en in Table 40.2. As higher yields are attained, higher

tissue nutrient concentrations may be needed. Using the
total herbage (entire shoot above stubble) for tissue
analysis allows the results also to be used to evaluate
feeding value of the herbage for livestock. This is not
possible when only the top 12 cm or the middle section of
the shoots are sampled.

TABLE 40.2
Suggested minimum sufficiency levels for nutrients needed
for maximum growth and yield in alfalfa herbage at first
flower of the first or second harvests

Nutrient	Percent (dry weight)	Nutrient	Parts per million (dry weight)
	%		ppm
Nitrogen (N)	2.50	Manganese (Mn)	25
Phosphorus (P)	0.26	Iron (Fe)	30
Potassium (K)	2.50	Boron (B)	25
Calcium (Ca)	1.80	Copper (Cu)	11
Magnesium (Mg)	0.30	Zinc (Zn)	20
Sulfur (S)	0.25	Molybdenum (Mo)	0.5

Visual symptoms as indicators of deficiencies. Additional support to the identification of a specific
problem may be supplied through the use of visual signs.
A key to plant nutrient deficiency symptoms can be found
in Chapter 3.

IMPROVED CULTIVARS OF ALFALFA

There appears to be little, if any, difference in
the concentrations (dry weight) of elements contained in
the herbage of alfalfa strains and cultivars now available to the farmer, when grown under the same soil and
climate conditions and sampled at the same stage of
growth. The amount of nutrients removed from the soil is
related primarily to total dry matter production; the
higher the yield the larger the amount of nutrients removed from the soil.

ROLE OF NUTRIENTS IN GROWING ALFALFA

Barnyard manure is sometimes top-dressed on alfalfa
stands, but to prevent burning by N it should be applied
when plants are dormant in late autumn or early spring,
or just after cutting. Cow manure is most widely available and averages about 13 kg N/mt, 3.4 kg P/mt, and 10
kg K/mt, as well as some micronutrients. The N, however,
is of little value to properly nodulated alfalfa and will
promote the invasion of grasses and weeds.

Commercial fertilizers generally are used to supply
the needed nutrients. In addition to Ca, supplied by
lime, P and K are the nutrients needed most often. Phosphorus and K usually are applied as mixed fertilizers.
Sulfur is supplied when the potassium carrier is K_2SO_4,
or when Mg is needed and applied as $MgSO_4$. Sulfur can be
applied prior to seeding as granular elemental S or in
the sulfate (SO_4) form. If top-dressed, S should be in
the sulfate form to be effective that year. Boron can be
added to the soil by top-dressing borax or using borated
fertilizers. Fertilizers also are available that supply
all of the micronutrients. Sometimes micronutrient applications do not boost yield (e.g., Cu, Mo), but help in
other ways like N-fixation and N utilization by the
plant.

Lime usually is considered a soil amendment since it
is applied primarily to correct soil acidity. Calcium
and sometimes Mg (in dolomitic lime) are added when liming soils, but Ca deficiency in alfalfa is rare under
field conditions.

EFFECTIVE USE OF FERTILIZERS AND AMENDMENTS

Liming. Where the soil pH is below 6.5, lime usually is applied to bring pH to an optimum range of 6.8-7.0
for alfalfa. For soils below pH 6.0, lime should be
worked into the soil several months before seeding to allow time to neutralize the acidity. Speed of neutralization is dependent to a large part on soil moisture, adequate mixing into the soil, and the fineness of grind and
quality of the lime. The finer the grind, the more rapidly limestone is dissolved in the soil solution.

Soil acidity per se is not toxic to alfalfa, but it
does influence the availability of soil nutrients. As
soil acidity increases, concentrations of Al, Fe, Mn, and
B increase in the soil solution. Excessive Mn can be
toxic. In contrast, P and Mo decrease in availability as
soil acidity increases; the P is tied up as Al and Fe
phosphates. Overliming decreases the availability of P
and B.

Nitrogen. It is accepted generally that alfalfa
will not respond to N fertilization when the plants are
well-nodulated with N-fixing, symbiotic bacteria. The
result of applying N to alfalfa is that N-fixation by the
symbiotic bacteria in the root nodules will be reduced
proportional to the amount of N fertilizer applied. With
a new seeding, it may be 6 weeks before effective nodules
are formed on the seedling plant roots. Some N applied
to the seedbed before seeding may help during this short
period. In the Southeast USA, applying 10 to 20 kg N/ha
at seeding is sometimes beneficial.

Phosphorus, potassium, and other nutrients. The initial amounts of nutrients needed at time of seeding must be at·least sufficient to satisfy the crop during its first year of growth. These amounts can be estimated by having soil samples analyzed well in advance of seedbed preparation. Thus, fertilizer needs will be based on a "test," not a "guess." Fertilizers containing the needed nutrients usually are applied broadcast and worked into the soil before seeding. High rates of most fertilizer materials should not be placed with the seed, since the germinating seed may be damaged. Where soils are too acid for alfalfa (below pH 6.0), response to the nutrients applied may not be fully realized unless the soil acidity has been corrected by liming.

Top dressing established stands, especially with P and K, is usually necessary to replace nutrients removed from soil in the hay, silage, or pasturage. Phosphorus and K can be top-dressed almost any time of the year, provided the foliage is dry so that burning will not occur. It is desirable, however, when top-dressing only once a season, to top-dress so that the nutrients are available prior to the highest-yielding cuttings (usually the spring harvest). In northern areas where alfalfa must develop cold resistance to survive the winter, top-dressing in early autumn will allow a high concentration of nutrients in the soil solution during the autumn months when alfalfa is hardening for winter. In particular, K appears to enhance the development of cold- and drought-hardiness. Excessive uptake of a nutrient, such as K, can be reduced by adding the annual amount in split applications. Split applications during the growing season also may be necessary for nutrients that may be fixed in the soil, such as P and Mg.

NUTRIENT MOVEMENT IN SOIL, AND ROOT ABSORPTION

Potassium and P top-dressed on alfalfa are relatively immobile in a mineral soil and therefore remain relatively close to the point of application. They remain mostly in the top 7.5 to 15 cm of soil, even 3 to 4 years after application. Phosphorus is less mobile than K, and therefore P should be mixed into the top 15 to 25 cm of soil during seedbed preparation. Maximum efficiency can be obtained by top-dressing K. Potassium generally is applied·as either KCl or K_2SO_4. The chloride (Cl^-) ion moves rapidly down through the soil at about the same speed as one would expect for nitrate (NO_3). The sulfate-sulfur (SO_4-S) moves more slowly and is intermediate between K and Cl^- in its movement through the soil. Nutrients (at least K) appear to be absorbed mostly by roots near the surface of the soil, primarily by roots feeding in the top 20-24 cm of soil. Thus, alfalfa cultivars with heavy surface-rooting will be more efficient in obtaining top-dressed nutrients, especially the rela-

tively immobile P and K. The taproot of the alfalfa is
not so important for nutrient absorption as it is for obtaining water from deep soil depths. Where alfalfa is
not irrigated, it sometimes must grow during dry periods
of the year on moisture obtained from the subsoil.

RELATION OF WEED CONTROL, INSECT PESTS, AND DISEASES TO
NUTRIENT NEEDS

Annual weeds are seldom a problem in thick, vigorous
and healthy alfalfa stands. They may invade stands during establishment because there is little competition, or
later if the alfalfa is thinned by diseases, insects,
winterkilling, standing water, or low soil fertility.
Perennial weeds may invade during seedling establishment
and then survive the harvest system used thereafter.
Weeds use soil nutrients and many accumulate higher concentrations of nutrients than alfalfa. Diseases and insects that injure the feeding roots and crowns of alfalfa
will limit nutrient absorption from the soil.

CROP MATURATION AND HARVEST

With disease-tolerant cultivars and proper use of
insecticides and fertilizers, it is now possible in most
regions of the USA to cut alfalfa at late bud to first
flower (approximately 1/10 bloom) to obtain the highest
quality forage. However, a delay to 25-50% bloom may be
necessary in regions of very high temperature in order to
maintain stands. Highest yields per ha of protein, energy, P, Mg, and other constituents needed in animal feeding occur at about the first-flower stage. Most of the
yield increase after first flower is not from new leaves,
but from stem elongation and stem lignification. Cutting
at bud to first flower at each harvest will remove nutrients from the soil at a rapid rate, so that one or more
soil nutrients may rapidly become limiting if not replaced by fertilization. Harvesting at early maturity also
will conserve soil moisture for the next harvest.

41
RED CLOVER

D. E. Peaslee and N. L. Taylor

Red clover (<u>Trifolium</u> <u>pratense</u>) is used for hay, pasture, and soil improvement throughout the temperate regions of the world. Its use extends into the sub-tropics on upland soils and at high elevations.

Red clover has been estimated to fix to 280 kg N/ha by means of rhizobia and has high value as a soil-improving crop, often being used for that purpose alone. It is also valued for its mid-summer production in fairly humid areas to provide hay and pasture for livestock when cool-season grasses are largely dormant. Because of high digestibility and protein, it adds feeding value for livestock and increases yields of meat, milk and fiber over that produced by grasses sown alone.

Red clover is the most widely grown of the true clovers. Grown alone and with grasses it constitutes the most important hay crop in cooler climates of humid regions. It is grown on about 5.5 to 6.5 million ha in the USA and Canada. In Europe it is a major crop in the Scandinavian countries, England, Scotland, Wales, and Ireland and extends across the continent into the USSR and Japan, either as a cultivated or wild species. It is also grown, although less extensively, in the southern hemisphere, particularly in South Africa, Chile, New Zealand and Australia.

ECOLOGICAL ADAPTATION

Red clover is best adapted where summer temperatures are moderately cool to warm, and where adequate rainfall is available throughout the grazing season. For example, in the USA this includes the humid region of the east and midwest extending westward into eastern North and South Dakota, Nebraska and Kansas. It extends north into Ontario and Quebec and south into Tennessee and North

[1]Professors, Department of Agronomy, University of Kentucky, Lexington, Kentucky 40506, USA.

Carolina. It is used as a winter annual in southeastern USA. It is also grown in much of the Pacific northwest, primarily under irrigation for seed production.

Fertile, well-drained soils of high moisture holding capacity are best for red clover. Loams, silt loams, and even fairly heavy textured soils are better suited than light sandy or gravelly soils. The red clover tap root is much branched, particularly in older stands, but most of it is concentrated in the top 30 cm of soil. Red clover is sensitive to poor subsurface drainage, and plants may be injured during winter months on soils that are poorly drained when precipitation exceeds evapotranspiration.

Red clover, a short-lived perennial, transitionally is used in short rotations in farming systems consisting of a cultivated crop such as maize or tobacco, followed by small grain (wheat, barley, oats) sown in the fall, and red clover is sown in late winter or early spring in partially disturbed grass sods. Due to the short life of the clover stand, pastures usually are renovated every third year. In regions with hot, dry summers, red clover is usually sown alone in early fall, and grazed the next spring until the stand is depleted in early summer. Plantings also may be made in early fall if the stand is depleted in early summer. Plantings also may be made in early fall in the Pacific Coast states (USA) for seed production in the next year. In the second and third years, it is usually harvested for hay, silage, or is grazed by livestock.

Traditionally, red clover fits well in beef cattle enterprises where animals are fed to a yearling age before shipment elsewhere for fattening on grain. However, red clover also may be used as a source of feed for dairy animals, sheep, goats and horses.

MINERAL NUTRIENT REQUIREMENTS

Under optimum soil nutrient levels, potential yields of red clover are far greater than average yields commonly obtained. An estimation of potential may be seen from Table 41.1 which gives hay yields from a single stand of "Kenstar" red clover sown at Lexington, Kentucky on a well-fertilized soil.

Yields were not measured in 1971, the year of sowing, or in 1975 when the stand had declined to the point that hay harvesting was not economically feasible. Obviously, higher yields may be obtained in longer seasons with irrigation, where rainfall is deficient for optimum plant growth.

Average yields of red clover hay in the USA, on the other hand, are much lower, averaging about 4.5 to 5.5 mt/ha. Figures from other countries are not available but may be estimated to be even less. Although many of the factors involved in low yields are sociological rather than agronomic, yield increases could be attained with proper fertilization.

TABLE 41.1
Hay yield in 1972 to 1974 from a stand of "Kenstar" red clover sown March 30, 1971 at Lexington, Kentucky

| Year | Cutting (Crop) (metric tons per hectare) | | | |
	1st cutting	2nd cutting	3rd cutting	Total
1972	4.48	3.51	2.33	10.32
1973	1.42	3.73	3.23	8.38
1974	1.59	2.01	2.51	6.11
TOTAL	7.49	9.25	8.07	24.81

Levels of nutrition that are currently maintained in the crop in various regions of the world are difficult to ascertain. The best indicators would be plant and soil analyses summaries but these are not generally available on a worldwide basis. In lieu of such data, Table 41.2 shows levels of several nutrients that have been reported in the literature during recent years. Most of the information is from research experiments and may not typically represent nutritional status of farm-grown red clover.

When nutrient supplies in soils are reasonably balanced, regardless of whether the absolute supply is moderately high or moderately low, dry matter production will tend to adjust upward or downward in response to nutrient supply. This compensatory response tends to move nutrient concentrations in plant tissues toward a central norm that is characteristic for each nutrient. The compensation effect is somewhat evident in Table 41.2, where despite large differences in soils and climatic zones, most nutrient concentrations are in a relatively narrow range. Potassium, Fe, and Mo all had approximately a two-fold range in concentration.

Soil Tests

Soil test levels for red clover production generally are in the medium to medium high range for P and K. Test levels are affected by soil test methods which include a chemical extractant, soil-solution ratio, and length of extraction time. In Kentucky, for 95% of maximum yield, using Bray's chemical extraction solution [1 (0.05 $\underline{N}$ HCl + 0.25 $\underline{N}$ NH_4F)], extractable P should be 20 ppm or more when extracted for 5 minutes at a 1:10 soil:solution ratio. Potassium as extracted by a suitable solution [1 $\underline{N}$ NH_4OAc (pH 7, 5 minutes at a 1:5 ratio)] should be 80 ppm

TABLE 41.2
Concentrations of mineral nutrients in shoots of red clover plants (initial bloom stage) growing in various regions of temperate climates of the world

	Major Nutrients					Minor Nutrients					
	—percent by weight—					—parts per million—					
	N	P	K	Ca	Mg	Fe	Mn	Zn	Cu	Mo	B
USA, Wisconsin	2.76	0.27	3.47	1.07	0.40	104	52	21	9	—	18
Scotland	—	0.37	3.80	1.48	0.28	—	—	—	—	—	—
Poland	—	0.21	1.36	1.68	0.46	230	38	55	9	.09	—
USA, West Virginia*	2.80	0.26	1.82	1.45	0.30	77	48	26	13	2.20	22
USA, Kentucky	—	0.38	2.16	1.91	—	162	94	43	10	—	36
England	3.80	0.29	2.15	2.50	0.24	68	49	21	7	0.38	32

* Survey of 28 samples.

or more for about 95% of maximum yield. (See Table 41.3
for examples of responses). These are base levels for
silt loam soils as tested by certain methods.

TABLE 41.3
Effect of Bray's extractable phosphorus, and NH₄OAc extractable potassium on mean annual yields of red clover-
timothy hay in Kentucky

Extractable P Parts Per Million	Relative Yield of Crop	Extractable K Parts Per Million	Relative Yield of Crop
5	60%	50	70%
10	85%	60	75%
15	90%	70	87%
20	95%	80	94%
25	98%	90	98%

Optimum soil pH levels influence available Mn, Al,
Mo, and P levels in soils, but growth and fewest disease
problems occur when the soil pH is between 6.4 and 6.8
(Tables 41.4 and 41.5). Red clover is especially sensitive to Mn toxicity, but usually this nutrient decreases
markedly in availability when soil pH is decreased below
5.7. The fact that increased productivity occurs at pH
values greater than 5.7 suggests that important effects
on other nutrients such as Mo or P may be occurring. Nodulation is usually adequate when soil pH is greater than
6.0.

Appropriate soil test levels should be maintained by
topdressing fields with fertilizers to correct deficiencies. Maintenance additions of fertilizers should not be
made solely on the basis of nutrient removal in the harvested crop but rather on the basis of stabilizing total
soil fertility.

TABLE 41.4
Soil acidity and red clover production in 8 years of crop
rotation in Ohio

Soil pH	Relative yield[1]
4.7	12
5.0	21
5.7	53
6.8	98
7.5	100

[1]Expressed as $\frac{yield}{maximum\ yield\ obtained}$ x 100

Note: best yield non-acid soils

452

TABLE 41.5
Relative yields and boron in red clover plant tissue in the greenhouse, as influenced by soil pH and boron treatments

| Soil pH | Red Clover Top Growth Applied Boron, mg/kg Soil | | |
	0	0.5	2.0
	Tops	Tops	Tops
5.3	0.54 (22)*	0.68 (44)	0.60 (114)
5.8	0.74 (21)	0.86 (27)	0.78 (59)
6.3	0.62 (20)	0.93 (26)	0.90 (50)
6.8	0.52 (20)	0.98 (22)	0.94 (40)

* Boron in tops is given in parentheses ( ), expressed in parts per million
Canada data

Tissue Analysis

Interpretation of plant tissue analyses data requires a well-calibrated set of tests to indicate deficiencies, preferably when information is interpreted for regional climate, soil, and plant varietal situations. Importance of using a carefully selected stage of plant maturity in collecting plant samples for analyses is illustrated in Table 41.6. Concentrations decreased 40, 25, and 36% for N, P, and K, respectively as red clover progressed from vegetative to seed development stages.

Critical ranges for several nutrients in red clover are reported in Table 41.7, but values for S and micronutrients were taken from alfalfa. Comparative studies of some critical nutrient levels for red clover, alfalfa and white clover showed similar behavior among the species. Therefore, critical ranges for micronutrients in Table 41.7 should serve as a useful guide but not as a diagnostic basis until verified for red clover.

Visual Symptoms

Diagnosis of nutrient deficiencies or toxicities based on visual symptoms is difficult, unless stresses are severe. For P and K, yields usually will be decreased appreciably before visual symptoms are common in the field. Different species of legumes generally follow similar patterns of deficiency symptoms which are as follows:

Nitrogen (N). Slow growth, with pale green rather than yellow leaves starting at the base of the plant. Leaf chlorosis is normally general over the entire leaf or leaflet.

TABLE 41.6
Concentrations of nutrients in red clover shoots at various stages of maturity in West Virginia

| | Major Mineral Nutrients | | | | | Minor Mineral Nutrients | | | | | |
| | N | P | K | Ca | Mg | Mn | Fe | B | Zn | Cu | Mo |
	—percent by weight—					—parts per million—					
Vegetative	5.0	0.36	2.5	1.4	.29	40	106	22	27	12	3.1
Bud	4.1	0.32	2.4	1.6	.38	51	80	22	25	11	3.2
1/4 Bloom	4.0	0.31	2.1	1.6	.28	50	106	26	26	11	2.8
Full Bloom	3.3	0.27	1.5	1.6	.29	42	86	22	23	11	3.1
Seed	3.0	0.27	1.6	1.3	.33	50	74	19	23	10	3.3

As reported by Barton S. Baker and R. L. Reid, 1977. Mineral concentration of forage species grown in central West Virginia on various soil series. Bulletin 657. West Virginia University Agriculture and Forest Experiment Station, Morgantown.

TABLE 41.7
Ranges and mean values of nutrient concentrations in red clover shoots just prior to flowering, expressed in parts per million (ppm)

| Nutrient | Major Mineral Nutrients | | | | | Minor Mineral Nutrients | | | | |
	P	K	Ca	Mg	S	Fe	Mn	Zn	Cu	B
	—percent by weight—					—parts per million—				
Critical Range	0.18–0.25	1.59–2.0	1.0–1.8	0.18–0.24	0.20–0.22(†)	—	20–30(†)	11–20(†)	7–10(†)	10–20(†)
Mean (§)	0.26	1.82	1.45	0.30	—	77	48	26	13	22

(†) Critical ranges are for alfalfa shoot material at early bloom (in lieu of red clover) as reported by W. E. Martin and J. E. Matocha. "Plant analysis as an aid in fertilization of forage crops", In Soil Testing and Analysis, 1973. Revised edition. L. M. Walsh and J. D. Beaton, Editors.

(§) As reported by Barton S. Baker and R. L. Reid, 1977. Bulletin 657. West Virginia University and Forest Experiment Station, Morgantown.

Phosphorus (P). Usually there will be no symptoms of P deficiency except slow or stunted growth.

Potassium (K). Interveinal yellow mottling around leaflet margin which is more prominent near tip or midvein. Chlorotic areas soon go necrotic (dead), giving a flecked appearance. Actual deficiency is usually quite severe before leaf symptoms appear.

Boron (B). Reddish-purple discolorations on margins of leaflets, gradually spreading to include the entire leaf. Symptoms are more prominent in younger plant parts. Some tendency toward rosetting of main branch.

IMPROVED VARIETIES

Recently, varieties of red clover have been bred for greater persistence and yield. Two such varieties are "Arlington" developed by the USDA in cooperation with the Wisconsin Agricultural Experiment Station; and "Kenstar" developed by the Kentucky Agricultural Experiment Station in cooperation with USDA. The tetraploid varieties such as "Hungarpoly" bred in Europe are other examples. These varieties possess higher yielding ability which may be expressed only at higher levels of fertility and management. According to a West Virginia study, a crop of red clover yielding 8 mt/ha of hay removed from the soil 196 kg N/ha, 26 kg P/ha, 247 kg K/ha and 18 kg Mg/ha in one year. Newer varieties are expected to need higher amounts of lime to maintain soil pH, to supply Ca and Mg, and to make available the other essential mineral nutrients, as well as to promote the growth of soil microorganisms and N-fixing bacteria in root nodules. As yield levels are increased, typical of those shown in Table 41.1, higher levels of mineral nutrients other than N must be applied. Consequently, nutrient requirements for new varieties and types will have to be worked out as yield potentials are established.

FERTILIZING RED CLOVER

Adjusting and maintaining soils at pH 6.4 to 6.8 (very mildly acid) for most soils in the temperate zones are basic to efficient production of red clover. Potential toxicities of Al, Cu, Mn, Fe, Zn and B can be avoided in this pH range, and yet utilization of soil P and Mo is generally efficient. Since most regions where red clover is grown tend to have acid to moderately acid soils, liming must be practiced. It is important to decrease soil acidity in as much of the root zone as is practical by incorporating and mixing liming materials with the soil. This will encourage normal root development, thereby aiding nutrient and water uptake from soils.

Animal manures are economical, effective sources of mineral nutrients for crops and are especially useful for maintaining soil fertility. For correcting deficiencies of major nutrients in red clover, there are two features of animal manures that are detrimental, relatively high N content compared to that of P and K, and the overall low contents of mineral nutrients. Correcting P or K deficiencies with manure alone would require such high rates that excess N for legumes could cause problems. Assuming nominal contents of 4.5 kg, 1 kg, and 4 kg of N, P, and K respectively, per ton of liquid/solid animal manure, probably no more than 5 mt/ha of manure should be applied without concern for the detrimental effect of N on the longevity or health of red clover stands.

In comparison to manures, commercial fertilizers are often more costly sources of nutrients, but in overcoming deficiencies and in bringing low fertility land into legume production there are several advantages. Fertilizers provide nearly total flexibility in which major or minor nutrients are applied and where they are applied. Many soils tend to immobilize P and many of the micronutrients. Potassium generally is immobilized to a lesser degree. Placing fertilizer in bands or zones greatly lessens immobilization, thereby enhancing fertilizer availability and lowering the amount of fertilizer required to reach a given production level. Red clover yields were 59% and 100% of maximum for P fertilizer in broadcast and banded applications, respectively. Applications of fertilizer in bands near the row at seeding require special equipment. Under low fertility conditions where banding is most beneficial, the fertilizer must be close enough to seeds to provide early nutrition, but not close enough to allow fertilizer salts to accumulate near the seed and kill seedlings. Two to three cm is a practical distance to place bands of fertilizer near the seeded drill row.

Normally in temperate climates, fertilizer topdressing (surface applications) for maintenance should be made after a final harvest but prior to winter dormancy. In this way crowns and roots will benefit from increased nutrient availability over winter and during rapid spring growth.

BENEFITS FROM ADEQUATE NUTRIENT SUPPLY

Red clover-grass mixture response to fertilizers was evaluated in a Kentucky (USA) experiment containing several levels of residual fertilizer P. Since untreated plots tested very low in P, P response was excellent, and clover production appeared to increase more than that of grass. Production costs and net profits were calculated from response curves to soil test P (Table 41.3), and approximately 15 units of fertilizer P were required to change soil test P by one unit. Residual buildup in the soil was distributed over a 5-year period. Cost factors are based on those used by the Department of Agricultural Economics, University of Kentucky. Production costs were quite high as indicated in Table 41.8, but even at the nominal hay price of $50/mt and discounting any improvements in hay quality, the last increment of fertilizer paid for itself and returned over 40% on the investment. Under conditions of this response situation and costs involved, if additional capital for fertilizer was available, excellent returns on investment could have been realized by operating at a slightly higher fertilizer application.

458

TABLE 41.8

Cost and net return analysis of red clover-grass response to phosphorus on a low phosphorus soil, Kentucky, USA

Annual Phosphate Fertilization Rate	Annual Hay Yields	Phosphate Cost (†)	Fixed and Variable Costs (§)	Total Costs	Crop Value (¶)	Net Return
—kg/ha of P_2O_5—	—mt/ha—	—$/ha—	—$/ha—	—$/ha—	—$/ha—	—$/ha—
0	8.40	0	263.40	263.40	420.00	156.60
104	12.59	34.46	283.43	317.88	629.50	311.62
208	13.71	68.94	289.35	358.29	685.50	327.21

† Phosphate cost calculated for indicated annual applications at $0.331/kg.

§ Fixed costs included seed and lime plus interest, depreciation and taxes on all capital investment for buildings, land, and equipment. Variable machinery and labor costs were calculated as being proportional to increased yields.

¶ Hay value calculated at $50/mt.

42
LESPEDEZA

M. S. Offutt

Lespedezas are potentially important legumes for pasture, hay, seed and soil improvement in the temperate zones, and are of greatest importance in the southeastern part of the USA. All are warm season plants and grow well only during the summer months.

Approximately 140 species of lespedeza have been identified. The three species of economic importance in the USA are the two annual species, Korean (Lespedeza stipulacea) and common striate (L. striata), and one perennial species, sericea (L. cuneata), all of which were introduced into the USA from eastern Asia.

ECOLOGICAL ADAPTATION

The practical limits of the three lespedeza species in the USA extend from central Texas, Oklahoma and Kansas eastward to the Atlantic Coast. Other regions of the world with similar climates are suited for lespedeza. Korean varieties are best adapted to the northern two-thirds of this region. Most striate varieties require a longer growing season and, as a result, are better suited for the southern part of the region. Sericea varieties grow well over most of the area.

Daylength and temperature influence vegetative growth, flowering, and seed maturity in the annual lespedezas. A short-day plant, lespedeza remains vegetative when daylengths are longer than their critical daylength; and flower and set seed when daylengths become shorter than the critical daylength. The critical daylength for floral initiation is longer for early maturing varieties than it is for those which mature later. In the northern hemisphere, when a variety is grown at a more southern latitude, floral initiation and seed maturity occur earlier because daylengths are always shorter to the south between March 21 and September 21. Temperatures are

[1]Formerly Professor of Agronomy, University of Arkansas, Fayettsville, Arkansas 72701, USA (now deceased).

460

higher in the more southern latitudes, and this tends to reduce the number of days between flowering and seed maturity. Rate of vegetative growth is retarded when seedlings of a variety emerge before the daylengths in the spring are equal to or greater than the critical daylengths for that variety. Careful selection of annual lespedeza varieties for any given area is essential therefore, if the highest yields of forage and seed are to be obtained.

General Soil Adaptation

The lespedezas will grow on almost any soil, except those which are very sandy. They do well on the sandy loam soils of the Coastal Plain, the clays of the Piedmont, and the limestone soils of Virginia, Tennessee, and Kentucky, and will grow well on soils too acid to grow clover or alfalfa. They will not, however, tolerate high-lime or extremely wet soils. Like most other crops, lespedeza does better on good land and makes its best growth in fertile bottomland soil.

Appropriate Place in Farming Systems

The annual lespedezas are easy to establish, produce relatively high yields of forage and seed on less fertile soils, and will renew stands each year through volunteer reseeding with only minimal management. These characteristics have been largely responsible for the wide acceptance and use of the annual lespedezas on livestock farms.

A succession of small grain and annual lespedeza may be maintained for many years if the small grain is not fertilized too heavily with N. Lespedeza normally shatters enough seed to ensure a heavy volunteer stand the next spring even when harvested for seed. Small grain may be planted in the late fall or early spring before the volunteer lespedeza seeds germinate. After the lespedeza seed has matured in the early fall, the seedbed for the small grain can be prepared with a field cultivator, or by disking, to keep the lespedeza seed on or near the soil surface to ensure a volunteer stand the following spring.

A combination of winter wheat and Korean lespedeza can provide the key to a balanced pasture and forage system on many livestock farms. Both crops can be grown on the same land each year, and good returns can be obtained even on soils of marginal fertility. Winter wheat normally provides earlier spring grazing than most permanent pastures. When permanent pastures are ready for grazing, livestock can be removed and the wheat used later for silage or grain. The lespedeza can then be used for summer pasture, hay, or seed as the need arises. If all of the lespedeza is not needed for any one use, the area can be divided with an electric fence anytime during the season, allowing any surplus growth to be used as desired. Most years these two crops are available for grazing when permanent pastures generally are lowest in production, thereby providing a more uniform carrying capacity of livestock feed throughout the year. Full production is obtained from this combination the first year, and no decision on utilization is required until the need is determined.

General farming methods usually are not feasible on hilly, rocky, brush-covered land. Where these conditions exist, brush may be controlled with aerial applications of herbicides, and the area seeded with annual lespedeza and an adapted grass in the fall or early the following spring. The lespedeza will add to the forage quality and quantity and, in addition, will also provide some N for use by associated grasses.

Annual lespedeza contributes significantly to pastures and meadows without special management. Less competition is encountered when lespedeza is seeded with bunchgrasses, especially if the seeding rate of the grass is reduced to ensure moderately open stands of grass. In well-sodded vigorous grasses, moderately close grazing or cutting early in the spring for hay is needed to maintain the annual lespedeza for use later in the season.

Sericea (perennial lespedeza) is an excellent soil-building legume. It is especially valuable in this respect on badly depleted soils where it is difficult to establish other legumes. Since it is leafier and less woody than other perennial lespedeza, sericea also has a place as a pasture, hay, and seed crop. For the highest quality forage, sericea should be grazed or cut for hay when the plants are tender and growing vigorously.

MINERAL NUTRIENT REQUIREMENTS

Potential Yields When Nutrient Needs Are Fully Met

With favorable soil, moisture, and fertility conditions, yields of 8 mt of hay and 900 kg/ha of seed have been obtained from the annual lespedezas. Also, such yields are within the potential of sericea under ideal conditions.

Soil Testing

Desirable soil test values for lespedeza are shown in Table 42.1.

Plant Tissue Values

Mineral nutrient concentrations in plant tissues considered adequate for optimum plant growth and animal nutrition are presented in Table 42.2.

TABLE 42.1
Desirable soil test values for the production of
lespedeza

Soil pH	6.5 to 6.8 (mildly acid)
Organic matter, in percent of total soil weight	2.0 to 3.0 percent
Extractable phosphorus (P)	80 to 100 kg/ha
Exchangeable potassium (K)	175 to 250 kg/ha
Exchangeable calcium (Ca)	1,000 to 2,000 kg/ha
Exchangeable magnesium (Mg)	100 to 150 kg/ha

TABLE 42.2
Plant tissue nutrient values considered adequate for
optimum plant growth, and for animal nutrition

Nutrient Element	Concentrations within Plant Tissue[a]		
	Percent		
Nitrogen (N)	2.75	to	3.00
Phosphorus (P)	0.18	to	0.25
Potassium (K)	0.90	to	1.20
Calcium (Ca)	1.00	to	1.30
Magnesium (Mg)	0.20	to	0.25

[a]The plant tissue values given are on a dry weight basis
of plant material harvested in the early bloom (hay)
stage of maturity.

Visual Symptoms

In many areas where lespedeza is grown, close observation of the different plant species present often will
provide important information on nutrient deficiencies in
the soil. If broomsedge grass (Andropogon virginicus)
plants are found in any quantity, the soil can be expected to be quite acid, and low in P. Sheep sorrell (Rumex
acetosella) is considered as an indicator of acid soil,
and poorjoe (Diodia teres) is regarded as an indicator of
a generally low nutrient level. Annual lespedeza becomes
chlorotic (yellowish) rather early in its life cycle on
soils with a high free lime content; this free calcium
carbonate in the soil depresses the availability of Fe to
lespedeza. If this condition is severe, the plants may
die before flowering.

Improved Varieties in Relation to Nutrient Requirements

In general, Korean varieties are better adapted to
alkaline (high pH) soils than striate, whereas striate
varieties are more tolerant of acid soils than Korean.
Sericea varieties also can be grown successfully on the

more acid soils, even where the pH may be as low as 5.0. Newer varieties of all three species have been developed and their performance has been evaluated for the most part on soils with higher nutrient levels than were used in the past. It seems likely, therefore, that most of the newer varieties may require higher soil nutrient levels than older varieties.

Correcting Deficiencies

When a soil test indicates a deficiency of exchangeable Ca and/or a pH below 6.5, agricultural ground limestone should be applied to correct the deficiency. If the soil test indicates that lime is needed and that exchangeable Mg also is low, then dolomitic limestone should be considered.

It usually is not practical to apply N to pure stands of lespedeza because it tends only to increase the competitiveness of any weeds present. If properly inoculated with the most efficient strains of rhizobia, lespedeza will fix atmospheric N in a form and quantity to supply all of its own N needs. If, on the other hand, the lespedeza is being grown in mixtures with one or more adapted grasses, nutrient elements in fertilizers, including N, would be utilized much more efficiently and effectively.

Soil P deficiencies are usually corrected by adding superphosphate (0-20-0 analysis) manufactured by treating rock phosphate with sulfuric acid. Superphosphate also will contain enough S to correct most S deficiencies. Rain and snow also may add a significant amount of S to the soil, especially near large industrial centers.

Muriate of potash (KCl) fertilizers are satisfactory materials to correct K deficiencies.

Iron chelates appear to be the most promising materials to correct Fe deficiency. These materials may be applied directly to the soil, or better still, sprayed on the lespedeza foliage as soon as any chlorosis is observed. Also, since Korean lespedeza is better adapted to high lime conditions, only varieties of that species should be seeded on high pH soils.

Effective Use of Fertilizers and Amendments

If the soil pH is 5.5 or lower, 3 mt/ha of agricultural limestone should be applied and mixed with the top 10-15 cm of soil, before seeding. If this is not feasible, over a period of years surface applications have been shown to be nearly as effective as soil incorporation. Higher rates will be required where the exchangeable Ca is less than 1,000 kg/ha, or if pH is 5.5 or lower, and the soil has a high clay content. In contrast, lower rates will be needed if the exchangeable Ca is greater than 1,000 kg/ha, the pH is above 5.5, and the soil has a relatively high sand content. If lime is needed, it should be applied as soon as possible, because it often takes 6 months or more before the soil pH is changed significantly.

The annual lespedezas respond favorably to lime and mixed fertilizers on deficient soils. Data from 34 tests in 8 states (USA) have been summarized in Table 42.3 to show the response of annual lespedeza to soil treatments with lime, P and K.

TABLE 42.3
Response of annual lespedezas to soil treatments[a]

Soil treatment	Hay Yield	Relative Hay Yield
	kg/hectare	Percent
None	1,557	100
Lime	2,587	166
Phosphorus (P)	2,095	135
Potassium (K)	1,763	113
Lime + phosphorus	3,032	195
Lime + phosphorus + potassium	3,162	232

[a]Average response of annual lespedezas to soil treatments in 34 tests conducted in 8 states of the upper part of the USA lespedeza region during the period 1930 to 1950.

For the annuals, liming to pH 6.0 or 6.5 (mildly acid) and applying 40 to 80 kg P_2O_5/ha have given successful production. If applied along with adequate lime and 40 to 80 kg P_2O_5/ha, K at the rate of 40 to 80 kg K_2O/ha often gives a marked response. Sericea lespedeza tolerates acid soils down to a pH of about 5.0 and gives little response to lime.

Relation of Time of Planting to Mineral Nutrient Supply

Volunteer stands of the annuals, Korean and striate, emerge as soon as the soil warms to 68°F in the spring and stays at that temperature or higher for a few days. This soil temperature normally is reached sometime between late February and early April. When establishing new stands, seeding is most frequently done by drilling or broadcasting the seed between February 15 and April 15 (in north temperate region). If lime is needed it should be applied the previous year and mixed with the soil during seedbed preparation. Phosphorus and K fertilizers usually are applied in early spring either before establishing new stands or before volunteer stands have emerged and started to grow.

If either N fertilizer or animal manure is to be applied to lespedeza growing in mixture with one or more species of grass, then the application should be timed to favor the grass. The best time to apply N, therefore, will depend on the growth habit and temperature requirements of the grass(es) involved. Usually it is best to apply N to cool season grasses in early spring and to warm season grasses in late spring or early summer.

Crop Maturation and Harvest in Relation to Mineral Nutrition

Soil amendments do not have much effect on maturity and harvest dates of lespedeza, except on soils very low in P. On these soils, application of P fertilizers hastens maturity and allows somewhat earlier seed harvest. For the most part, however, because they are short-day plants, seed maturity in the lespedezas is related more to existing daylengths and temperatures than to soil amendments.

BENEFITS FROM ADEQUATE MINERAL NUTRITION

Crop Yields

Yields of annual lespedezas have been increased from about 2000 kg of hay and 200 kg/ha of seed to about 6000 kg of hay and 750 kg/ha of seed by providing adequate mineral nutrition. Yield increases of the same general magnitude also have been realized from sericea lespedeza by applying adequate amounts of P and K.

43
ORCHARDGRASS

D. D. Wolf and H. E. White

Orchardgrass (<u>Dactylis</u> <u>glomerata</u>) is known also as cocksfoot in English-speaking countries outside the USA and is a name that describes the flower head. Used only for livestock feed, it is found throughout the world wherever favorable growing conditions are present. In the northeastern USA, orchardgrass is adapted between 35 and 42 N latitude. It also does well in other areas at high elevations where water is adequate. It is not extremely winter hardy, becomes dormant with drought stress, and begins growth early in the spring with almost one half of the total seasonal yield occurring before early June.

ECOLOGICAL ADAPTATION

The fibrous root system develops primarily in the upper meter of well-aerated soils and can very efficiently remove water and nutrients. As compared with other similar grasses, relative growth is better on somewhat infertile soils; however, with adequate nutrients and water, it will respond to productive soils. Its bunch type growth habit forms an open sod providing good protection against erosion. Properly managed stands will remain productive for seven or more years. Orchardgrass is limited primarily to cultivated fields and is used in rotations for hay and pasture.

Pure stands are seldom recommended unless intended for seed production since legumes provide N and increase feed quality. When intended for pasture, 5 kg/ha of red clover and 1 kg/ha of ladino clover should be included with 14 kg/ha of high quality seed of well adapted orchardgrass. As clover disappears from the stand, overseeding in later winter will be successful if competition is controlled during early spring growth. When intended for hay production only, red clover at 6 kg/ha or alfalfa

[1]Department of Agronomy, Agricultural Experiment Station, Virginia Polytechnic Institute and State University, Blacksburg, Virginia 24061, USA.

468

at 11 kg/ha can be included. Variety selection is based primarily on earliness of maturity, yielding ability, and persistence. Primary differences between varieties concern the date of flowering during spring growth.

SOIL TESTS

Nitrogen, P, and K are the only major elements that might be needed from fertilizers in most cultivated agricultural soils (Table 43.1). Orchardgrass is a very efficient competitor for K. With low available soil K, alfalfa in a mixture will show deficiencies without affecting grass growth. Adjusting the upper 15 cm of soil to pH 5.7-7.0 will eliminate or minimize deficiencies of minor nutrients in most agricultural soils.

TABLE 43.1
Soil test results and fertilizer recommendations for orchardgrass production on a moderately productive soil

	Soil Test Results		Fertilizer Needed *	
Class	P	K	P	K
	ppm		kg/ha	
Low	0- 6	0- 38	35-35	80-120
Medium	6-18	38- 88	20-35	50- 80
High	18-55	88-155	15-20	30- 50
Very High	55-up	155- up	0	0

*Nitrogen needs should be governed by the percent of clover, the need for productivity, and other management factors (range = 132 to 176 kg/ha in split application).

Data from Virginia Extension recommendations.

PLANT TISSUE COMPOSITION

Nutrient composition depends largely on plant maturity, soil fertility levels, and soil characteristics. Generally, major nutrient concentrations decrease as the plant develops stems and seed heads while most minor nutrients remain unchanged (Table 43.2). Composition of vegetative growth in early spring is quite similar to regrowth during summer. One should resist a "shot gun" application or possible excessive amounts of nutrients. An orchardgrass pasture fertilized with high N plus micronutrients (Co, Mo, Cu, Zn, S) resulted in thyroid enlargement of newborn lambs, low lamb birthweights, and low liver Cu (subclinically deficient) as compared to high N treatments without micronutrients. Before a broad recommendation for micronutrient fertilization of orchardgrass is made, the possible soil-plant-animal interactions need to be known.

TABLE 43.2
Nutrient composition of orchardgrass sampled during vegetative (Veg), early bloom (EB), and early seed (ES) development in spring and at 2 regrowths (45 to 60 cm height) during summer*

| Nutrient Element | First Growth | | | Regrowth | |
| | Veg | EB | ES | 1st | 2nd |
			%		
N	3.77	1.71	1.40	3.00	–
P	0.85	0.34	0.27	0.44	0.56
K	3.86	2.91	2.16	3.04	3.31
Ca	0.39	0.27	0.29	0.45	0.58
Mg	0.12	0.11	0.11	0.13	0.16
Na	0.02	0.01	0.01	0.02	0.01
Si	0.30	0.25	0.37	0.42	0.41
			ppm		
Mn	157	158	173	170	183
Fe	308	93	235	210	168
B	5	4	3	4	6
Cu	18	19	18	15	15
Zn	47	40	37	44	38
Al	273	88	129	296	243
Sr	11	6	5	12	10
Mo	0.6	0.3	0.3	0.4	0.4
Co	0.6	0.4	0.8	0.3	0.5
Ba	16	16	24	21	22

*Data from R. L. Reid et al. 1970. Bulletin 589T. West Virginia Agricultural Experiment Station.

PRODUCING ORCHARDGRASS

The first growth of the season develops stems and seed heads during a reproductive stage. Maximum dry matter accumulates during early seed formation; however, nutritive value for livestock declines due to low digestibility and decreased protein of the stem portion (Table 43.3). Highest protein yields occur before seed heads are visible, highest digestible nutrients coincide with early flowering, and highest dry matter is accumulated when in early seed formation stage. First hay harvest should be made in an early head development stage with some yield sacrifice and some compromise of quality to obtain acceptable feeding value. Regrowth after this first hay harvest consists only of leafy herbage of fairly uniform quality. When used for pasture, the highest productivity and stand longevity occur with short rotation grazing periods. Nitrogen at 70 to 100 kg/ha is needed for pure stands before first spring growth to produce about 45% of the seasonal

TABLE 43.3
First and aftermath harvest yields, digestibility (IVDM), and protein of orchardgrass when first harvest was taken at several maturities. Aftermath harvest taken whenever regrowth was 30 to 45 cm height*

First Harvest Date Maturity	Yield			IVDM		Protein		First Harvest	
	First	After	Total	First	After	First	After	IVDM	Protein
24 Apr Prejoint	2860	5020	7880	81	71	24.2	21.4	2320	690
5 May Early Head	3560	5700	8260	72	70	18.8	17.4	2560	670
21 May Early Bloom	5460	4880	10,340	62	73	10.7	18.8	3380	580
4 June Early Seed	5100	4070	9170	52	69	8.8	18.5	2650	450

*Data taken from J. B. Washko et al. 1967. Bulletin 557T. West Virginia Agricultural Experiment Station.

yields. Additional applications of 35 to 50 kg/ha after each hay harvest will be economical during years with favorable weather conditions. Since low yields occur without some applied N, income may show net losses without some N fertilizer. In Virginia, each kg of N added up to 110 kg/ha returned 36 kg of saleable hay during a 9-month growing season.

SUMMARY

Orchardgrass, when grown under favorable environmental conditions and managed properly, is a valuable source of livestock feed. This grass can be utilized for grazing, hay, or silage from stands expected to persist for 6 to 8 years without reseeding. A legume should normally be grown in a mixture to provide N and improve feed quality, especially protein.

44
TALL FESCUE

S. R. Wilkinson

Tall fescue (<u>Festuca</u> <u>arundinacea</u>) is a perennial pasture and forage grass with the following attributes: (i) excellent seed yields (up to 1,000 kg/ha), (ii) good to excellent herbage yields (18,000 kg/ha), (iii) general ease of establishment from seed, (iv) excellent persistence, (v) adaptation to a wide range of soil and environmental conditions, (vi) excellent compatibility and tolerance of various management practices including grazing, (vii) comparatively long grazing seasons where adapted, and (viii) an apparent low susceptibility to serious disease and insect pests which affect its persistence and productivity. As expected from these attributes, tall fescue may provide excellent carrying capacity for grazing. However, grazing animal performance has often been below that obtained from other cool season perennial grasses such as orchardgrass.

Tall fescue's adaptation ranges from cool, dry regions to wet, warm regions and is related primarily to temperature, rainfall, and elevation, and much less affected by soil factors. Tall fescue is well adapted in the transition zone between warm and cool season forage species. Tall fescue is native to Britain and is well adapted to much of Europe, New Zealand, and parts of Australia and Africa. There are about 12 to 14 million ha in the USA. Its many uses, in addition to grazing and forage, include stabilization of land subject to erosion, turf, and cover on athletic and recreation fields. The combination of multiple uses and wide adaptability suggests that generalizations regarding nutrient requirements must take into account specific soils and climates.

[1]Contribution from Southern Piedmont Conservation Research Center, USDA, ARS, Watkinsville, GA 30677 in cooperation with the University of Georgia Agricultural Experiment Stations.

[2]Soil Scientist, USDA, ARS, Southern Piedmont Conservation Research Center, P.O. Box 555, Watkinsville, GA 30677, USA.

NUTRIENT REQUIREMENTS

   Nitrogen (N).   Figure 44.1 illustrates N response
curves obtained from a variety of locations under rainfed
and irrigated conditions.   Yield responses to N were
found at all locations.   The N fertilizer requirement for
80% of maximum yield varied from 150 kilograms per hectare (kg/ha) in the North Georgia Mountains to 410 kg/ha
in the Southern Piedmont of Georgia.   Tall fescue grown
in more humid regions with long growing seasons has a
higher N requirement than where soil N supplies are high,
or the growing season is short.   Under irrigation in Texas, about 356 kg N/ha was required to attain 80% of maximum yield.   Up to about 200 kg N/ha, its N use efficiency
in producing dry matter yield can be excellent.
   Nitrogen fertilization may be used to modify seasonal growth patterns.   Frequent split N applications will
help distribute the forage yield over the season, but
will not override the tendency of tall fescue to produce
high yields in spring and autumn.   Heavy N fertilization
in the spring and autumn enhances total yields.   When N
rates are greater than 50 kg N/ha, applications should be
split, with more frequent splits at the higher rates to
maximize N use efficiency, increase forage production and
optimize seasonal distribution of yield.
   Nitrogen sufficiency levels in forage (total top
growth) are in the range of 2.8-3.4% N (on a dry matter
basis), and the grass can accumulate nitrate nitrogen
(NO_3-N) when the percent N in forage increases beyond
3%.   The NO_3-N levels in tall fescue associated with
high, near maximum growth rates appear to be in the range
of 700-1000 parts per million (ppm).   When growth is limited by drought, high temperature (above 29°C), low sunlight, or when N fertilizer or soil N levels are excessive in relation to growth conditions, tall fescue has
the capability of accumulating NO_3-N levels exceeding
1%.   In general, NO_3-N accumulation in field grown plants
whose growth is not severely restricted by environmental
conditions is less than 2,000 ppm at fertilization rates
up to 400 kg N/ha per year.

   Nitrogen (N) sources.   Neither ammonium nor nitrate
N sources appear to have any clear-cut superiority in
producing tall fescue yields.   Application of urea or
liquid N fertilizers with high urea contents may result
in losses by N volatilization, and consequently lower N
use efficiency.   Such losses are dependent on weather
conditions, and are minimum when urea does not reside for
long periods on surfaces of the grass or soil.

   Response to waste materials containing N.   Tall fescue has been found to be tolerant of waste applications
providing several times its N requirement; however, herbage quality problems may occur at these heavy rates.   For

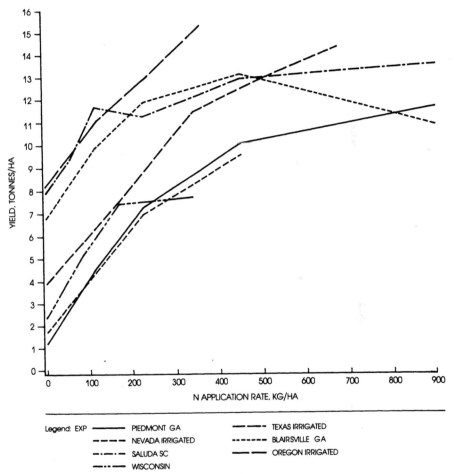

Figure 44.1   Dry matter response of tall fescue to nitrogen fertilization at locations in South Carolina, Georgia, Nevada, Wisconsin, Oregon and Texas, USA.

example, broiler litter application rates of about 22 mt/ha per year resulted in potentially toxic levels of NO_3-N, high levels of alkaloid, an increased incidence of grass tetany, and the occurrence in beef cows of fat necrosis (a form of tall fescue toxicosis) in Georgia.

Disposal rates of liquid sewage containing potentially toxic elements such as cadmium, lead and zinc may also pose a problem because of adherence of the sludge to plant leaf surfaces, raising the potential for ingestion by grazing animals of potentially dangerous levels of these elements. Once sludge has dried on the leaves of tall fescue, it is very difficult to remove by washing or by rainfall. Liquid sewage sludge should be applied to swards immediately after cutting and before regrowth occurs to reduce contamination of the forage.

Stand persistence under high N fertilization. Loss of stand associated with high N fertilization may be expressed as dead patches and/or thinned stands which decrease yields in subsequent cuttings. In general, when stand deterioration occurs, the larger the yield of the preceding cut the greater the stand deterioration. Consequently, the larger the N rate, and the longer the growth interval, the greater the potential for stand loss. Apparent causes include raising the growing points of the grass above the point of defoliation and the lack of developed growing points for regrowth, exhaustion of carbohydrate reserves, and reduced root mass. Increasing rates of N fertilization coupled with increased frequency of close cutting (5 cm) will increase tiller numbers, and may enhance weed populations. Tall fescue can be encouraged by lower rates of N and higher cutting heights.

Productivity in association with forage legumes or other grasses. Tall fescue-legume associations have potential dry matter production levels equivalent to that obtained with 200 kg N/ha. Compatible legumes include ladino clover (Trifolium repens), red clover (Trifolium pratense), and birdsfoot trefoil (Lotus corniculatus). In many regions annual legumes such as arrowleaf clover and crimson clover have potential as companion crops. Legume persistence is reduced by applying N and by allowing tall fescue to shade the legume, therefore close grazing or frequent close clipping are necessary to maintain the legume/tall fescue association. This may be difficult to manage during rapid growth periods of the grass. It may be necessary to interseed clovers every 2 or 3 years to maintain a favorable clover-grass balance. Interseeding with the warm season Kobe lespedeza may provide additional summer grazing in warmer areas.

Where both cool season and warm season grasses are adapted, tall fescue-bermuda grass (Cynodon dactylon) or tall fescue-bahiagrass (Paspalum notatum) associations have been established and/or maintained. These associations can be managed to increase the length of grazing season and improve the overall distribution of yield. Tall fescue has persisted in central Alabama in association with bahiagrass when fertilized during the winter with 224 kg N/ha and when a minimum summer stubble height of 10 cm has been maintained. Applying N fertilizer during the summer and close clipping will accelerate the shift from tall fescue to bahiagrass. Tall fescue can also be interseeded into Coastal bermudagrass in the Georgia Piedmont, and maintained by applying N fertilizers during winter and maintaining 10 cm stubble height during summer. During summer, applying N fertilizer and close clipping will cause a shift from tall fescue to Coastal bermudagrass.

Other uses of tall fescue. Tall fescue grown for seed production should be fertilized earlier (December through March) than for forage production. Levels of N fertilization recommended for seed production are from 37 to 100 kg N/ha. Post-seed-harvest fields managed for forage or grazing should be fertilized and grazed or cut to encourage fall tiller development.

Tall fescue may be "stockpiled" to provide winter forage of sufficient quality to meet nutritional needs of many classes of grazing livestock in the mid-latitudes of the USA. Nitrogen fertilization increases the amount of stockpiled tall fescue, and its nutritional quality, as measured by in-vitro dry matter digestibility, total non-structural carbohydrates (digestible) and mineral constituents. A general recommendation for mid-latitude regions of the USA would be to fertilize in late August or early September with moderate rates of N (100 kg N/ha), and allow the forage growth to accumulate for grazing after frost.

Phosphorus (P). Many mine spoil areas, eroded areas and subsoils exposed by earth moving are so P deficient that P applications of at least 56 kg/ha are necessary for early growth and winter survival of all seedlings.

Severe P deficiencies are apparently rare in established tall fescue because of residual P from previous fertilization and the ability of the grass to obtain slowly available soil P. Also, most N fertilization levels are relatively low, less than 150 kg/ha, and thus P removal under grazing is very low. Response to P fertilization of Alto tall fescue was less than 19% when sodium bicarbonate extractable P (by soil analysis) was less than 8.2 ppm, but was 80% or higher when levels of bicarbonate soluble P were 2 to 5 ppm. Fifty percent of maximum yield was obtained when tissue P levels were 0.19% P, and for maximum yields, 0.35% P content in tissues was needed.

In terms of plant tissue diagnosis, less than 0.24% P may be considered deficient; 0.24-0.32% critical; 0.32-0.40% adequate; and more than 0.45% P, high.

In humid regions, P deficiency is considered unlikely when the level of soil P from a dilute double acid soil extraction is more than 22 kg P_2O_5 per ha (10 ppm P). When legumes are grown in association with tall fescue, the association must receive P at the level required by the legume. Soil P requirements for clover in tall fescue pastures result in an increased P need at a medium soil P availability (increased from 25 kg P/ha to 40 kg P/ha under Georgia conditions). The importance of adequate P and K for legumes in tall fescue is demonstrated in Table 44.1. Annual applications of P and K to a tall fescue/ladino clover association on Tilsit silt loam soil (a fine silty, mixed, mesic Typic Fragiudult) in Kentucky increased clover content, total herbage production and total N in the tall fescue and clover over the 4-year period. An indirect benefit of improved P nutrition of grass-legume associations may be increased N fixation by the legume. Undesirable broomsedge invasion in many southern tall fescue pastures may be the result of low soil P and K, as well as selective grazing by livestock.

TABLE 44.1
Effect of P and K fertilizer on percent clover, yield, and N removal of tall fescue/ladino clover associations in Kentucky (Tilsit silt loam soil) (1955-58 average) (Wilkinson and Mays, 1979)

Fertility Treatment N-P-K	Clover Content (Average for 12 Harvests over 4-Year Period)	Total Herbage Production	Total N/ha Obtained over 4-Years from		
			Fescue	Clover	Total
kg/ha/year	%		------ kg/ha -----		
0 - 0 - 0	27.9	4,260	289	198	489
0 -98 -130	38.8	7,399	497	472	969
Increase from P & K	10.9	3,139	208	274	480

Potassium (K). Soil test levels for K should be medium to high, depending on the yield and desired productivity level. Tall fescue can use effectively non-exchangeable soil K. The amounts and rates of release of soil K depend on the past fertilization (residual K), soil type, and in particular the types and amounts of clay in the soil.

Deficient, critical, adequate, high, and excessive amounts of K in tall fescue are <2.2% (deficient); 2.5% (critical); and 2.8% (adequate); 3.0 to 3.5% (high); and greater than 4.0% (excessive), respectively. At least 2.0% K appears necessary for near maximum growth of the "Kentucky-31" strain. Concentrations of K less than 1.5% in the forage have been associated with moderate to severe deficiency. Tall fescue leaves affected by K deficiency grow more slowly and turn brown along margins and at the leaf tip. Potassium-deficient stands suffer winter injury.

Soils with relatively low available K, such as coarse textured soils, can be depleted rapidly by heavily N-fertilized tall fescue. Yields from 7,000 to 12,000 kg/ha having a minimum K concentration of 2.0% will remove from 140 kg K/ha to 240 kg K/ha, respectively, in harvested forage. Soils with low cation exchange capacities and low release rates of residual and non-exchangeable K (e.g. sandy soils), will require frequent, relatively small applications of K, whereas soils with high exchange capacity and high release rates of residual and nonexchangeable K may be fertilized less frequently or will require only maintenance K applications. In Oklahoma, yields of tall fescue grown on Taloka loam soil (a fine, mixed, thermic Mollic Albaqualf) fertilized at 717 kg N/ha/yr increased from 4,500 to 8,200 kg/ha by fertilizing annually with 79 kg P/ha and 149 kg K/ha. There was a complete tall fescue stand loss after 2 years of fertilization with 717 kg N/ha without P and K fertilization. Potassium deficiency may lead to stand losses wherever N fertilization is adequate for high yields but K fertilization is not.

Higher soil K levels are needed to maintain legume-grass swards than for tall fescue alone, because legumes are usually less competitive for K, particularly under conditions of low K availability in the soil.

Potassium requirements are lower under grazing than for harvested forage. For example, the removal of 500 kg animal liveweight gain constitutes about a 0.9 kg K/ha loss, whereas removal of 8,000 kg forage containing 2.0% K removes 160 kg K/ha per year. Seventy to 90% of the K ingested by grazing animals may be returned to the soil in the urine. Fertilization to maintain a medium soil K test based on annual sampling is the recommended approach under grazing to assure that the growth potential of tall fescue is not limited by K deficiency.

Lime, Calcium (Ca) and Magnesium (Mg). Lime requirement relates to assuring an optimum soil pH, and since liming is normally accomplished with calcitic or dolomitic limestone, liming will normally meet the necessary Ca or Mg requirements. The lower soil pH limit for growth of tall fescue is in the range of 4.5-4.7 (strongly acid), and the upper soil pH limit for growth and adaptability of tall fescue is 9.5 (strongly alkaline). Tall fescue is moderately susceptible to aluminum (Al) and manganese (Mn) toxicity. Tall fescue is more susceptible to Al injury on acid mine spoils than is creeping lovegrass (_Eragrostis curvula_).

In spite of tall fescue's tolerance to a wide range of soil pH, a 6 to 7 pH range is desirable because legumes have a higher soil pH requirement, P is more soluble at higher pH, and N mineralization rates from plant and animal residues may be low in extremely acid soils. Consequently, N requirements may be increased. The actual amounts of lime required to establish and maintain pH between 6 and 7 are best determined by soil tests correlated for specific soils and locations.

Critical levels of Ca and Mg for tall fescue have not been determined. However, based on results from similar grasses, tall fescue in vegetative growth may be deficient in Ca and Mg if concentrations are lower than 0.2 and 0.1%, respectively, as shown by plant analysis.

Many tall fescue pastures are used for cow-calf operations. In temperate climates where tall fescue is grown, hypomagnesemic (deficient magnesium) grass tetany may be an economic problem for cattle producers. This problem can be brought into perspective when it is recognized that "safe" Mg levels in herbage for prevention of grass tetany are considered to be in the range of 0.20-0.25% Mg, or roughly double that required to assure a non-limiting concentration for growth. In practice, Mg concentrations in tall fescue vary with season of year, stage of maturity, available Mg and K supplies in the soil, the balance between K, Ca, and Mg in the soil, soil textures and mineral content, soil temperature and soil moisture. Fertilization of tall fescue with ammonium nitrate may increase Mg concentrations in forage if K fertilization is low. If the soil K supply is high, uneconomically high rates of Mg will be required to overcome K/Mg antagonisms. Dolomitic limestone applications are not effective under these high K conditions. Corrective fertilization practices include minimizing luxury consumption of K by the fescue plants and increasing supplies of available Mg. Magnesium oxide, kieserite, and dolomite are sources recommended for acid soils, while for alkaline soils sulfate or chloride sources appear to be suitable for correction of Mg deficiency. Epsom salts are effective but expensive. Kainite can also be used for correcting Mg deficiencies for plant growth, but may be less effective in elevating Mg concentrations to "safe" levels from a tetany point of view.

New tall fescue strains which have a higher Ca and Mg content and lower mineral ratios (K/Ca + MgO) offer promise as a means of minimizing or alleviating grass tetany. Tall fescue breeding efforts suggest that there is germplasm potential for new strains with increased tolerance to poor soil and improved nutrient use efficiency.

Sulfur (S). The only yield increases in tall fescue from S fertilization were reported in the state of Washington; 56 kg S/ha more than doubled the average yield of second, third, and fourth cuttings of Alta and "Fawn" strains of tall fescue, and halved the N:S ratio in the herbage. The lack of response to S fertilization of tall fescue relates to the widespread contribution of S in rainfall, the ability of soils to absorb atmospheric S and the ability of tall fescue roots to exploit subsoil S reserves. Based on results from other cool-season perennial grasses, an adequate level of S for tall fescue would be in the range of 0.25-0.30% of plant dry weight. In Kentucky, cellulose digestion by microorganisms in the animal rumen (stomach) was stimulated by the addition of 0.05 or 0.15% S to tall fescue which contained 0.2% S. Nitrogen/sulfur ratios in forage diets may need to be about 10 to 1 for maximum quality.

Micronutrients. There are no documented field cases of a tall fescue growth response to miconutrients. However, tall fescue may accumulate unessential silica in quantities slightly greater than 4% of the dry weight. Such accumulations may have potentially negative effects on forage digestibility. Liming acid soils may decrease soluble silica in soil solution (over the pH range of 5.0 to 6.8). Nitrogen fertilization may also decrease silica content by improving water-use efficiency, and the increased tall fescue yield, particularly if harvested frequently, will likely contain a lower silica concentration.

UTILIZATION OF INCREASED PRODUCTION FROM FERTILIZER

The high forage yield potential of tall fescue (18 mt/ha) which can be obtained with adequate N, P, and K fertilization is of little value unless it is converted to animal products (meat, milk, wool). Lechtenberg et al. (1976) found that tall fescue fertilized with 112 kg N/ha in southern Indiana produced lower calf gains, cow gains, and cow conception rates than orchardgrass fertilized with 112 kg N/ha, or tall fescue-ladino and red clover not fertilized with N (Table 44.2). Poor grazing animal performance has also been reported in Missouri, Illinois, North Carolina, and Georgia.

Legume-tall fescue associations improve animal performance. However, legume adaptability and persistence problems exist over much of the region of tall fescue's adaptation, and producers often have to rely on N fertilizers to sustain pasture production. Research at the Black Belt Substation of the Alabama Agricultural Experiment Station has shown rather dramatically increased utilization of pasture, hay, or seed of tall fescue on black montmorillonite soils, when the grass was free of infestation with an endophytic fungus (Table 44.3).

TABLE 44.2
Comparison of performance of cows and calves grazing orchardgrass and tall fescue pastures fertilized with 112 kg N/ha/yr and tall fescue-ladino and red clover pastures in Southern Indiana (3-yr average) (Lechtenberg et. al. 1976)

Pasture Treatment	% Legumes	Calf Gain	Cow Gains	Cow Conception
		kg/day	kg/day	%
Orchardgrass	-	0.81	0.26	90
Tall fescue	-	0.58	0.01	71
Tall fescue-legume	29.4	0.83	0.26	92

Average daily gains on pasture and hay and seed yields were reduced 56%, 12%, and 79% when pasture, hay, and seed, respectively, were infected with this fungus. The average daily gains for fungus-free tall fescue in this study are comparable to those obtained from other cool season grasses.

If this fungus is shown to be the cause of the erratic and/or poor performance often experienced in cattle that graze tall fescue, and if tall fescue pastures free of this fungus can be economically established and maintained, then improved animal production from the fertilized pastures is expected.

TABLE 44.3
Steer performance on tall fescue pasture, hay, or seed free of fungus or with the fungus present (Schmidt, et.al. 1981)

Type of Fall Fescue Material Fed	Free of Fungus	Fungus Present	% Decrease from Fungus
Pasture, Average Daily Gain, kg/day	0.67	0.30	56
Pasture, Beef Gain, kg/ha	442	235	57
Hay, Average Daily Consumption, kg/day	4.8	4.2	12
Hay, Average Daily Gain, kg/day	0.66	0.29	57
Seed, Average Daily Consumption, kg/day	6.40	4.1	35
Seed, Average Daily Gain, kg/day	0.96	0.20	79

SUPPLEMENTAL READING

LECHTENBERG, V.L., W.H. SMITH, and D.C. PETRIG. 1976. Pasture performance of beef cows and calves grazing orchardgrass, tall fescue, and tall fescue-legume herbage. In Hill Lands. Proceedings of an International Symposium, J. Luchok, J.D. Cawthorn, and M.J. Breslin (eds.). pp. 738-742. West Virginia Univ. Press, Morgantown, West Virginia.

MATCHES, A.G., 1979. Management. In Tall Fescue. Buckner, Robert C. and Lowell P. Bush (eds.) pp. 171-200. Agronomy Monograph No. 20, American Society of Agronomy, Crop Science Society, Soil Science Society of America, Madison, Wisconsin.

SCHMIDT, S.P., C.S. HOVELAND, C.C. KING, E.M. CLARK, N.D. DAVIS, D.M. BALL, L.A. SMITH, H.W. GRIMES, AND J.L. HOLLIMAN. 1981. Fungus limits fescue pasture gains. Highlights of Agricultural Research 28:3 Alabama Agricultural Experiment Station, Auburn, Alabama.

WILKINSON, S.R. and D.A. MAYS. 1979. Mineral Nutrition. In Tall Fescue. Buckner, Robert C. and Lowell P. Bush (eds.) pp. 41-74. Agronomy Monograph No. 20, American Society of Agronomy, Crop Science Society Soil Science Society of America, Madison, Wisconsin.

45
BERMUDAGRASS (Stargrass)

Glenn W. Burton

Bermudagrass, <u>Cynodon</u> <u>dactylon</u>, is the most important pasture, lawn and soil conserving grass in the tropics and subtropics. Fifty years ago when it was the
"worst weed" in the South's cotton fields in the USA, it
was also recognized as an excellent pasture grass by the
few cattlemen who used it. Grazing and feeding experiments over the past 30 years have confirmed the opinions
of those early graziers.

Bermudagrass's sod-forming properties and ability to
tolerate continuous, close defoliation make it an excellent turf grass. These traits, plus its toughness make
it the best species for football and other athletic
fields. No other tropical grass is so well-suited for
golf greens, fairways, and tees.

The seed that has spread bermudagrass around the
world and the stolons and rhizomes that make it a weed
are characteristics that make it well-suited for soil
conservation waterways, dams, and so on. The soil it has
saved, even when it was a weed in cotton fields, may have
been worth more than the labor expended in its control.

ECOLOGICIAL ADAPTATION

Bermudagrass is adapted to a wide range of climatic
conditions and may be found in arid as well as humid regions throughout the tropics and subtropics. Although
cold beyond 35°N and S latitudes generally limits its
distribution, genotypes capable of surviving throughout
the USA and in Berlin, Germany have been found. Bermudagrass starts growth at 25°C and grows best with temperatures ranging from 35° to 45°C.

---

[1]Research Geneticist, Agricultural Research, Science and
Education Administration, U.S. Department of Agriculture, and the University of Georgia, College of Agriculture Experiment Stations, Coastal Plain Station, Agronomy
Department, Tifton, Georgia 31793, USA.

Bermudagrass has an unusually wide soil adaptation. It grows well between pH 5 and 8 (moderately acid to mildly alkaline). Soil textures ranging from clay to sand grow good bermudagrass if they are well-drained and adequately fertilized. Although bermudagrass will survive on infertile soils, it responds unusually well to increases in soil fertility. Bermudagrass has greater salt tolerance than most plants.

On the farm, bermudagrass is more important as a pasture plant for livestock. Improved hybrids such as "Coastal" grow tall enough to be excellent, easily-cured hay plants. Bermudagrass is a valuable soil conservation tool for stabilizing dams, terraces, and waterways. It can also make a beautiful home lawn.

IMPROVED VARIETIES

Coastal bermudagrass was the first hybrid of Cynodon dactylon to be produced and released. It has been planted vegetatively on more than 4 million hectares across the southern USA. In much of the South, it is the principal hay plant as well as the most important pasture crop. Nutrient requirements are similar to those described for common bermudagrass; however, coastal bermudagrass yields nearly twice as much as common bermudagrass with the same fertilization and management and hence provides a much higher return to money invested in fertilizer. Other named bermudagrass hybrids believed to have similar nutrient requirements to Coastal are Suwannee, Midland, Coastcross-1, Alicia, Callie, and Tifton 44.

NUTRIENT REQUIREMENTS

The potential yield of bermudagrass is determined by genotype, available water, soil fertility, number of days with temperatures above 25°C, and management. Coastal bermudagrass at Tifton, Georgia with approximately 220 frost-free days, adequate water and nutrients and a 6-week cutting schedule will produce up to 25 mt/ha of hay (16% H_2O content). Coastal will yield nearly twice as much as common bermudagrass.

In the southern USA where common bermudagrass is fertilized very little in pastures, it probably yields 2 to 3 mt/ha of dry forage but is utilized for grazing. On more fertile soils, this production will double or treble. In most of the world, common bermudagrass yields will be determined by the natural fertility of the soil and probably will average the equivalent of 2 to 5 mt/ha of hay. Hybrids like Coastal and Tifton 44 usually receive more fertilizer and will probably yield 10 to 15 mt/ha.

In the southern USA Coastal will receive fertilizer to provide 250 kg/ha of N, 75 kg/ha of P_2O_5 and 25 kg/ha of K_2O for hay, and probably less than half as much for grazing. Common bermudagrass will receive less. In the rest of the world, fertilizer applied will usually be much less.

The mineral nutrient requirements for bermudagrass will be determined by the N rates used and the presence or absence of legumes. Because of the deep roots and strong feeding power of bermudagrass, P, K, Ca, Mg, and S are usually adequate even with "low" soil test readings, if bermuda is grown alone without N applications. If legumes are grown with bermudagrass, medium to high soil test levels or applications of P, K, Ca, Mg, and S will be required. If P is supplied from 20% super phosphate, the Ca and S in the super phosphate will satisfy the needs of bermudagrass for those elements. Using dolomitic limestone should also satisfy the Ca as well as the Mg needs. Nitrogen in the form of ammonium sulfate or ammonium thiosulfate can be used to satisfy the sulfur required by bermudagrass.

Coastal bermudagrass growing on an infertile, deep sand has failed to respond to minor element treatments. Research has shown that Coastal hay cut at 6-week intervals must contain at least 1.0% K and 0.14% P on a dry basis for optimum yields. Applying enough P and K to raise the content of these elements appreciably failed to increase yields enough to pay for the added fertilizer. However, grass cut more frequently will probably need a higher K level in the forage.

Either P or K deficiencies weaken stands, reduce yields, and increase its susceptibility to winter injury. In addition, K deficiency symptoms induce an abundance of tiny, brown spots 1 mm or less in diameter on the leaves.

ROLE OF NUTRIENT REQUIREMENTS IN THE CULTURE OF BERMUDAGRASS

Poultry manure is an excellent fertilizer material for coastal bermudagrass. Heavy applications up to 25 mt/ha have been tolerated by the grass and have produced high yields of high protein forage. Other animal manures are also effective. Some of the N in manures breaks down to form ammonia that will be lost to the air if the manures are left on the top of the ground. Rain or irrigation to carry the soluble nitrogenous materials into the soil will reduce these losses.

Effluent from sewage disposal plants is an excellent fertilizer for bermudagrass if the sewage does not carry large amounts of metallic elements such as copper, cadmium and zinc. Sewage from industrial plants may contain such elements.

Surface applications of all fertilizer elements except ammonia and urea nitrogen have been as effective as fertilizer placements in the soil. Urease, an enzyme produced by microorganisms and present in all soil and forage (except for a few days after fire has removed the stubble) causes urea to decompose and give off ammonia gas until the urea is washed into the soil by rain or irrigation. Average losses of 20% of urea nitrogen applied to mowed stubble of Coastal bermudagrass have been recorded. Anhydrous ammonia injected into the soil has been an effective nitrogen source for Coastal bermuda.

In regions where bermudagrass is killed back to the ground by frost or severe drought, removing the dead stubble with fire shortly before new growth begins is an excellent practice. Fire destroys insect eggs, disease spores, and weeds, allows the soil to warm up faster, makes the grass start growth earlier, and over the years has increased Coastal bermuda hay yields by 1 mt/ha/year.

Fertilizer should be applied to bermudagrass so it will be available to stimulate growth when temperature and water favor it. Applying the first fertilizer about 1 week after the average last 0°C temperature in the spring will enable the grass to make the most efficient use of water that is available.

Bermudagrass, like all other plants, will take up more fertilizer than needed for economic growth when it is in abundant supply. To minimize this luxury consumption of such elements as P and K, these elements should be applied in a 4-1-2 or 3-1-2 ratio of $N-P_2O_5$ and K_2O whenever N is applied for hay production. On pastures where most of the minerals applied are returned to the soil in the urine and manure, less P and K in relation to the N applied will usually suffice. A single annual application of fertilizer in the spring will produce high protein grass early and low protein grass late in the season. Splitting the annual application into three or more applications distributed through the spring and summer will give a more uniform protein content of hay or forage, and will usually give a more uniform seasonal distribution of the forage.

MEASUREMENT OF BENEFITS FROM ADEQUATE MINERAL NUTRITION

Adequate mineral nutrition maintains good stands and maximizes cold tolerance. Nitrogen is the fertilizer element that usually has the greatest effect on yield and protein content of bermudagrass. Table 45.1 shows how N and cutting frequency affected the yield and protein content of Coastal bermudagrass at Tifton, Georgia in one 220-day, frost-free growing season with adequate rainfall. These N response data will enable one to calculate the cost of growing hay and protein at varying costs of land, fertilizer, and lime. For example, allow $100/ha/yr for land rent, $1.05 for a kg of N plus P_2O_5 and K_2O to make a 4-1-2 ratio, and 1.8 kg of lime/kg of N to neutralize the acid residue from the N (1981 prices at Tifton, Georgia). If rainfall and other growing conditions in 1981 at Tifton are average, yields of hay and protein in Table 45.1 may be expected.

TABLE 45.1
Effect of nitrogen and cutting frequency on the annual
yield, protein content and yield (dry weight basis) of
Coastal bermudagrass hay at Tifton, Georgia

Nitrogen (N) kg/ha	Hay (mt/ha) when cut every a/		Protein (%) when cut every		Total protein, kg/ha
	4 wks	6wks	4wks	6wks	
0	2.7	3.1	9.1	7.6	200
110	10.0	12.8	11.2	7.8	900
330	17.7	21.7	15.2	11.3	2220
660	21.7	26.9	17.0	13.8	3190

a P and K adequate, 16% moisture hay.

Costs of growing coastal bermudagrass to the hay
stage (4 to 6 weeks) with annual fertilizer applications
of 0 to 660 kg/ha of N plus P and K and lime calculated
from Table 45.1 and 1981 prices are presented in Table
45.2. These data indicate that the 110 kg N/ha fertilization treatment gives the lowest cost ($16.84) per ton of
hay. The lowest cost per kg of protein ($0.20) was ob-
tained with the 330 kg N/ha treatment. Protein in soybean meal is currently selling for $0.80 per kg. Because
cutting on a 6-week interval increases the yield of hay,
it reduces the cost per ton. It does not affect protein
yield and cost, but does reduce the digestibility as well
as the protein content of the hay (Table 45.1).

UTILIZATION OF INCREASED PRODUCTION

Coastal bermudagrass hay is generally sold at local
farmers' markets without being graded. Enterprising hay
growers who produce a quality product create their own
market with dairy and beef cattle farmers who are willing
to pay a premium for increased protein content and higher
digestibility.
Much of the coastal bermuda hay is produced by cattlemen to feed their own livestock. For their benefit
and for those who wish to produce a quality product, the
following information is provided.

NUTRIENT REQUIREMENTS FOR CATTLE

The nutrient requirements for different classes of
livestock vary greatly. To maintain her weight, a dry
pregnant mature cow weighing 450 kg requires only 6.8 kg
dry matter containing 5.9% protein in a ration that is
50% digestible (Table 45.3). Most any kind of Coastal
bermuda hay will satisfy this need without supplement.
This same cow producing milk for her calf will require
9.9 kg dry matter containing 9.2% protein in a 57% digestible hay ration to maintain her body weight. Coastal

TABLE 45.2
Cost of growing Coastal bermudagrass hay at Tifton, Georgia in 1981

N kg/ha	Land + fertilizer cost[a]	Hay cut every				Total Protein	
		4 weeks		6 weeks			
		yield mt/ha	Cost $/mt	yield mt/ha	Cost $/mt	yield kg/ha	Cost $/ha
0	$100.00	2.7	37.04	3.1	32.26	200	0.50
110	215.50	10.0	21.55	12.8	16.84	900	0.24
330	446.50	17.7	20.58	21.7	20.58	2220	0.20
660	793.00	26.9	29.48	26.9	29.48	3190	0.25

[a]Land rent = $100/ha, N = 58¢/kg, P and K to make a 4-1-2 ratio of N-P₂O₅-K₂O = 44.5¢, and lime to neutralize the acid residue from the N = 2.5¢ (US dollars).

491

bermudagrass hay cut at 6-week intervals from a field
fertilized with 330 kg N/ha/yr or hay cut at 4-week
intervals from a field fertilized with 110 kg N/ha/yr
should be adequate for such cows without supplement
feeds.

Table 45.3 shows that growing heifers (200 kg) need
a ration containing at least 10% protein. Because they
can eat only about 4.5 to 5.0 kg of dry matter per day,
the only way they can satisfy their body requirements and
gain more than 0.25 kg per day is to eat a more highly
digestible ration. To make gains of 0.50 and 0.75 kg per
day, their ration must be 63 to 69% digestible. Older
heifers (300 kg) with a greater feed capacity can make
0.50 kg gain per day with a 57% digestible ration, but
must have a 63% digestible ration to gain 0.75 kg per
day. The nutrient requirements for most of these rations
can be met with Coastal bermudagrass (hay, silage, or
pellets) without expensive supplements.

TABLE 45.3
Daily nutrient requirements for various classes of
livestock

Average Daily Gain kg	Dry Matter Required/ Day kg	Protein Required/ Day %	Total Digestible Nutrients in Ration %
Dry pregnant mature cow (450 kg)			
0.0	6.8	5.9	50
Cow (450 kg) nursing calf			
0.0	9.9	9.2	57
Growing heifers (200 kg)			
0.0	3.3	7.8	57
.25	4.6	10.0	57
.50	5.0	11.1	63
.75	5.4	11.1	69
Growing heifers (300 kg)			
0.0	4.5	7.8	57
.25	6.2	8.9	57
.50	8.2	10.0	57
.75	8.6	11.1	63

[a]National Academy of Sciences, 1970, Nutrient requirements for cattle, Fourth Revised Edition.

46
STYLOSANTHES SPECIES

Jose G. Salinas and Ramon Gauldron

Many tropical forage legumes have been identified as tolerant to acid soil constraints, and many of these have their center of origin in acid soil regions. This suggests that adaptation to soil constraints is part of the evolutionary process. Among these forage legumes, the genus Stylosanthes is widely found in tropical areas of the world, although it is most numerous in tropical America. This chapter is concerned mainly with ecological adaptation and nutrient requirements of some Stylosanthes species that are considered promising for tropical pastures.

ECOLOGICAL ADAPTATION

Stylosanthes contains approximately 39 species which are widely distributed within the latitudes 40°N and 40° S. This broad zone includes different climatic regions such as the humid tropics, seasonal rainforests, seasonal rain subtropics, piedmonts, savannas, deciduous forests and dry tropics. However, most Stylosanthes species are distributed in seasonal tropical and subtropical environments. Many have also been introduced into tropical areas for use as cover crops and pastures. Many species and ecotypes are of special interest for tropical regions because they are adapted to a wide range of climatic and soil conditions.

Most Stylosanthes species grow from sea level to about 2000 meters with a wide range in annual precipitation. For instance, many species are well established in Madagascar in arid regions with just 350 mm of mean annual rainfall, in Australia from 900 mm to 4000 mm, in Brazil from 1000 to 1700 mm, and Colombia between 1000 and 2500 mm mean annual rainfall.

---
[1]Soil Scientist and Research Assistant, respectively, Tropical Pastures Program, Centro Internacional de Agricultura Tropical (CIAT), Apartado Aereo 6713, Cali, Colombia.

494

Temperature has been indicated as an essential factor for adaptation of most Stylosanthes species, since temperature fluctuations can exert a modifying influence through delaying or accelerating flower elongation. Significant reductions in plant height, dry matter production, root growth, and nodule number have been found below 23°C and above 35°C. In addition, frosts may prevent seed formation in high altitudes. Both flowering time and seed set are important in the adaptation of introduced plants. Photoperiod is considered the main controlling factor of flowering in Stylosanthes.

Stylosanthes lines vary widely among and within species with respect to adaptation to different environments. For instance, S. humilis is an annual, well adapted to areas with short summer growing seasons, whereas S. guianensis is highly valued because it is perennial. There are ecotypes of S. guianensis which vary in frost resistance, date of flowering, and growth habit. Ecotypes of S. capitata differ in many morphological characters including size and stage of inflorescence, growth habit, and flowering time.

Because of such morphological characteristics associated with physiological variability, many Stylosanthes species have been introduced successfully into diverse environments. Most Stylosanthes species have a reputation as forage legumes that grow well under adverse soil conditions characterized by low base status, low phosphorus availability, and strong acidity.

Very acid soil conditions and resulting high levels of exchangeable aluminum are toxic to most cultivated crops. In general, soil adaptation of Stylosanthes species and ecotypes is closely related to their centers of origin. Also, some morphological and physiological characteristics of Stylosanthes plants may explain the broad adaptation to different soil conditions. Among these characteristics are mechanisms of adaptation to soil water stress including, a deep root system, stomatal closing, shedding of leaves, and leaf movement to avoid excessive evapotranspiration.

In general, most Stylosanthes species grow on well-drained sands, sandy loams, and clay loam soils. However, some species such as S. capitata prefer infertile sandy soils.

DIAGNOSIS OF MINERAL DISORDERS

In general, determination of minerals that are deficient or toxic for leguminous pastures can be made by soil analysis, plant tissue analysis, and by observing foliar symptoms of mineral disorders. Diagnosis of mineral disorders in Stylosanthes species is not always clear due to their wide range of adaptability. In addition, diagnosis of mineral disorders in legumes is somewhat more complex than in non-legumes because of the legume symbiosis with rhizobia. Consequently, diagnosis of mineral disorders will depend largely on pooling information from soil tests, plant tissue analysis, and/or visual symptoms of mineral disorders.

Soil Analysis

Although the main purposes for soil tests relate to the availability of nutrients, the analytical techniques are empirical in nature. Therefore, it is necessary to calibrate soil test values against actual plant responses before they can be used to predict soil nutrient status or to define a soil as having low native fertility. The actual techniques used in soil analyses vary greatly between laboratories, but details of the more commonly used ones for tropical soils can be found in the handbook, "Analytical methods for acid soils and tropical plants" (Salinas and Garcia, 1979).

Based on these analytical procedures, data on tentative critical levels of soil parameters for Stylosanthes are summarized in Table 46.1. These values are not absolute but they can be used as a general guide. Calibrations may differ between soil types and Stylosanthes species, consequently, soil test values should not be used solely to predict the amount of fertilizer needed to correct nutrient deficiencies. In addition, when using low rates of fertilizers, conventional soil test extraction procedures often do not reflect the amount of fertilizer applied. Therefore, it is difficult to make fertilizer recommendations based only on soil tests.

Plant Tissue Analysis

Critical concentrations in the plant tissue for several Stylosanthes species have been determined for some nutrients (Tables 46.2, 46.3, 46.4 and 46.5). Nutrient levels in Stylosanthes are being used more frequently as a diagnostic index due to limitations in soil tests for acid tropical soils under low input management. In addition, a critical concentration from plant analyses may be used over a wide range of soil types and environmental conditions. In general, the critical values shown in Tables 46.3, 46.4 and 46.5 indicate that plant tissues of Stylosanthes species are quite low in most nutrients; this may also indicate lower internal and external nutrient requirements as compared with other tropical pasture legumes.

Foliar Symptoms

Symptoms of nutrient deficiencies in Stylosanthes are not always clear since symptoms vary among species, soil fertility situations, and environmental conditions.

TABLE 46.1
Tentative levels of soil parameters for Stylosanthes species adapted to tropical acid soils

Soil Parameter	Low	Level Critical	High	Method of Analysis*	Reference
pH	< 4.5	4.5–5.5	> 5.5	1:1 water-soil ratio	Salinas & Garcia, 1979
Al-Saturation (Al toxicity) (%)	< 70	70–90	> 90	(Al/Al+Ca+Mg) x 100	CIAT, 1980
Ca-Saturation (%)	< 10	10–30	> 30	(Ca/Al+Ca+Mg) x 100	CIAT, 1980
Mg-Saturation (%)	< 5	5–15	> 15	(Mg/Al+Ca+Mg) x 100	CIAT, 1980
P (ppm)	< 3	3–5	> 5	Bray-II-extract	CIAT, 1981
K (meq/100g soil)	< 0.05	0.05–0.10	> 0.10	Bray-II-extract	Salinas & Garcia, 1979
S (ppm)	< 10	10–15	> 15	0.008M Calcium Phosph.	CIAT, 1981
Zn (ppm)	< 0.5	0.5–0.8	> 0.8	Mehlich- 2 extract	Salinas & Garcia, 1979
Cu (ppm)	< 0.1	0.1–0.4	> 0.4	DTPA Extract	Bruce, 1978
B (ppm)	< 0.3	0.3–0.5	> 0.5	Hot water-Extract	Salinas & Garcia, 1979
Mn-toxicity (ppm)	< 20	20–50	> 50	1 $\underline{N}$ KCl	Salinas & Sanz, 1981

* 1N KCl extractant for Al, Ca and Mg.
$\underline{Bray}$-II = 0.1N HCl + 0.03 N NH_4F
Mehlich -2 = 0.05 N HCl + 0.025N H₂SO₄
DTPA = 0.005M DIPA + 0.01 M CaCl₂ + 0.1 M Triethanolamine
< = less than
> = greater than

TABLE 46.2
Estimated calcium and lime requirements for several Stylosanthes species and ecotypes

Species	Ecotype	Calcium Rate	Lime Rate (CaCO$_3$ equiv.)	Source
		kg/ha		
S. guianensis	La Libertad	50	150	Spain et al., 1975
S. guianensis	-	250	800	Carvalho, 1978
S. capitata	CIAT-1315	50	150	CIAT, 1982
S. capitata	CIAT-1318	100	300	CIAT, 1982
S. capitata	CIAT-1405	200	600	CIAT, 1982
S. capitata	CIAT-1419	300	1000	CIAT, 1982
S. capitata	CIAT-1899	300	1000	CIAT, 1982
S. macrocephala	CIAT-1643	200	600	CIAT, 1982
S. humilis	CIAT-118	150	500	CIAT, 1980
S. hamata	CIAT-174	1200[1]	4000	Grof et al., 1979

[1] Ca rate also associated with a decrease of Al toxicity in the soil.

498

TABLE 46.3
Ranges of critical calcium and magnesium concentrations in the plant tissue
of _Stylosanthes_ species

Element	Species	Plant Tissue	Range of Critical Element Concentration (%)	Reference
Calcium	S. capitata	plant tops	1.00 ± 0.30[1]	CIAT, 1982
	S. macrocephala	plant tops	0.70 ± 0.15[1]	CIAT, 1982
	S. guianensis	plant tops	0.85 ± 0.20	CIAT, 1981
	S. humilis	plant tops	2.00 ± 0.40	Andrew and Hegarty, 1969
	Mean		1.10 ± 0.30	
Magnesium	S. capitata	plant tops	0.25 ± 0.04	CIAT, 1982
	S. guianensis	plant tops	0.30 ± 0.03	CIAT, 1981
	S. macrocephala	plant tops	0.18 ± 0.02	Flores, 1982
	S. humilis	plant tops	0.35 ± 0.06	Andrew and Robins, 1969
	Mean		0.27 ± 0.04	

[1]Ranges correspond to values obtained with different ecotypes and associated with 80% of maximum yield at 8 weeks of plant growth.

However, foliar symptoms of mineral disorders in several Stylosanthes species have been obtained in nutrient solution cultures, and many of the mineral deficiency symptoms have been observed under field conditions. Detailed descriptions and color photographs of mineral deficiencies and toxicities have been published (CIAT,1982). A brief summary follows:

Deficiencies

Nitrogen (N). Uniform chlorosis in the whole plant from the beginning; under severe conditions lower leaves with red pigmentation and border necrosis.

Phosphorus (P). Reduced plant growth, dark bluish green leaf color, thick leaves, and dropping of lower leaves.

Potassium (K). Reduced plant growth, small and chlorotic upper leaves, marginal yellowing and border necrosis of lower leaves.

Calcium (Ca). Apical chlorosis and marginal yellowing of younger leaves, malformation of newly expanded leaflets, abundance of new regrowth which turns brown quickly, dark green color of lower leaves.

Magnesium (Mg). Marked interveinal chlorosis on the lower and mid-positioned leaves, under severe conditions yellowing of young leaves and border necrosis on lower leaves.

Sulfur (S). General yellowing of upper leaves, which soon spreads throughout the plant, similar to N-deficiency; under severe conditions, lower leaves develop marginal necrosis, become yellow in color, and fall off.

Zinc (Zn). Interveinal chlorosis in top leaves associated with a slight degree of bronzing of the young emerging leaves; under severe conditions leaflets curved epinastically.

Boron (B). Fast necrosis of young leaves, short petioles, irregular growth of young leaves, lateral leaflets show unequal size and are malformed; root growth restricted, dark in color, and suppressed lateral root development.

Iron (Fe). Uniform chlorosis of upper leaves, which become white under severe deficiency; at this stage marginal necrosis in young leaves.

Toxicities

Aluminium (Al). Reduced root growth, root tips inhibited, necrosis and severe stunting and thickening of lateral roots; yellowing of older leaves under severe toxicity.

Manganese (Mn). Reduced top growth and chlorosis on newly expanded leaves similar to Fe-deficiency symptoms, irregular chlorosis on intermediate and lower leaves; under severe conditions, irregular brown spots near the main veins in lower leaves.

MINERAL NUTRIENT REQUIREMENTS

During the last years in many tropical areas attention has been focused on low-input management technology, based on three main principles: (i) adaptation of plants to soil constraints, rather than elimination of all soil constraints; (ii) maximization of production through efficient use of existing resources; and (iii) advantageous use of the favorable attributes of acid, infertile soils. Hence, the genus Stylosanthes has received particular attention in tropical areas because of the adaptability of several species and ecotypes to acid soils with low native soil fertility.

It is important to realize that soil fertility is a relative matter and depends on the scale used to measure it. For maize, most acid soils would be judged as infertile, but for adapted species such as Stylosanthes capitata such acid soils may appear quite fertile. Consequently the nutrient requirements of Stylosanthes will largely depend on the soil-plant relationships existing in a specific ecosystem. For instance, the more acid the soil and the more limited the availability of plant nutrients, the more restricted the range of plant adaptation to the ecosystem. Nutrient requirements for tropical pastures generally involve two plant growth stages, establishment and maintenance. This paper refers to the nutritional requirements of establishment which in most cases covers the first year of plant growth. Nutrient requirements for Stylosanthes species covering the maintenance period are minimal and still need substantial investigation.

Lime, Calcium and Magnesium

Aluminum is the predominant cation in the exchange complex in most acid soils and is frequently a growth limiting factor, being particularly severe at soil pH 5.0 or below. Most species of Stylosanthes are generally considered tolerant to acid, infertile soils and specifically to Al toxicity. When acid soils are limed, adequate yields (80% of maximum) have been attained at

150, 300 and 600 kg $CaCO_3$ equiv./ha with several ecotypes of Stylosanthes capitata. This indicates that generally lime can be applied in small amounts, mainly to supply Ca as a nutrient.

The importance of Ca in legume growth and soil fertility in temperate regions is well known, but evidence in tropical areas shows that adequate growth of some tropical legumes may be obtained under conditions of low Ca status in the soil (Munns and Fox, 1977). Many Stylosanthes species appear to be quite efficient in extracting Ca from the soil. Table 46.2 summarizes some of the experience about external requirements of Ca, and lime applications within species and ecotypes of Stylosanthes during the first year of establishment.

Basic slag or rock phosphate may also meet the low external Ca requirement. On the other hand, depressed yields have been observed in several Stylosanthes species at lime rates over 1 ton/ha. In these cases the yield decreases are probably related to some nutritional imbalance, or to the lack of adaptation of these species to the modified soil condition.

Despite the low Mg content of many acid tropical soils, few field responses to Mg have been observed on pasture legumes including Stylosanthes species. The lack of response may be due to the presence of Mg in many liming materials such as dolomitic lime. Unless dolomitic lime is used, a recommended rate for many Stylosanthes species growing in acid tropical soils is about 12 kg Mg/ha.

Critical concentrations for Ca and Mg in plant tissue have been determined for several Stylosanthes species (Table 46.3). These values are not absolute but are merely a guide in the interpretation of plant tissue analysis. In addition, the critical levels are considered as a range of values rather than unique ones, since they can fluctuate with the species and/or ecotypes used, soil texture and moisture, time of sampling, soil fertility, and interaction with other nutrients.

Nitrogen, Phosphorus and Potassium

Nitrogen values reported for tropical legumes vary between 0.9% and 6.0%, with a mean N percentage of 2.8 for all legumes. Values presented in Table 46.4 for Stylosanthes species fluctuate around this mean value but are considered high in terms of forage quality. This explains the interest in this legume, both to increase forage quality and as a source of N for associated grasses in the vast tropical savannas.

Table 46.4 also shows ranges of critical P and K concentrations in plant tissue of Stylosanthes species. Although differences in response to P within the genus Stylosanthes exist, many Stylosanthes species grow in soil with low available P (3-5 ppm P-Bray II). This has

TABLE 46.4
Ranges of critical N, P, and K concentrations in plant tissue of Stylosanthes species

Element	Species	Plant Tissue	Range of Critical Element Concentration (%)	Reference
Nitrogen	S. capitata	leaves	2.40 ± 0.5	CIAT, 1982
	S. guianensis	plant tops	1.90 ± 0.3	Blunt & Humphreys, 1970
	S. hamata	plant tops	2.60 ± 0.3	Aitken, 1979
	S. humilis	plant tops	3.40 ± 0.6	Andrew & Robins, 1969
	Mean		2.60 ± 0.5	
Phosphorus	S. capitata	plant tops	0.14 ± 0.04[1]	CIAT, 1982
	S. macrocephala	plant tops	0.12 ± 0.03[1]	CIAT, 1982
	S. guianensis	plant tops	0.25 ± 0.05[2]	Jones, 1974
	S. hamata	apical tissue	0.27 ± 0.03[2]	Wilapon et al., 1979
	S. humilis	plant tops	0.27 ± 0.03[2]	Jones, 1974
	Mean		0.21 ± 0.04	
Potassium	S. capitata	plant tops	1.08 ± 0.13[1]	CIAT, 1982
	S. macrocephala	plant tops	1.05 ± 0.15[1]	CIAT, 1982
	S. guianensis	plant tops	0.82 ± 0.03[2]	Brolman and Sonoda, 1975
	S. hamata	apical tissue	0.70 ± 0.02[2]	Aitken, 1979
	S. humilis	plant tops	0.60 ± 0.02[3]	Andrew & Robins, 1969
	Mean		0.85 ± 0.07	

[1] Ranges correspond to values obtained with different ecotypes and associated with 80% of maximum yield at 8 weeks of plant growth.

[2] Ranges associated with 90% of maximum yield at 8 weeks of plant growth.

[3] Preflowering stage.

been attributed in part to a low internal P requirement. In addition, <u>Stylosanthes</u> species may possess inherent low P uptake characteristics which, combined with their low initial growth rate, enable them to survive and produce under conditions of low P supply.

Between 20-40 kg P/ha was required to obtain 80% of maximum dry matter yield of several <u>Stylosanthes</u> species on acid soils in Tropical America, with available P of 1 to 4 ppm (Bray II extraction)(Sanchez and Salinas, 1981). Phosphorus is generally applied as single or triple superphosphate, basic slag, or ground phosphate rock. There has been considerable interest in the use of the last two P sources on tropical soils, because of the low cost and the possibility of using local phosphate sources.

Critical K concentrations in plant tissue of <u>Stylosanthes</u> species show a mean value of 0.85% $\pm$ 0.07. This is a low internal requirement, and symptoms of K deficiency among <u>Stylosanthes</u> species were not observed within this range of K concentrations. However, when <u>Stylosanthes</u> species grow in grass-legume mixtures, their external K requirements seem to increase in order to maintain the internal K content near the critical value. This has been attributed to: (i) a poorer competitive ability of the legume than grasses for soil K, (ii) an advantage of grasses in K uptake in competition with legumes, and (iii) a lower cation exchange capacity of legume roots as compared with grasses.

The K requirement for adequate establishment of <u>Stylosanthes</u> species in tropical soils with exchangeable K of 0.05 to 0.01 meq/100g ranges from 20-30 kg K/ha (Coelho and Blue, 1979; CIAT, 1981, 1982). When land is cleared by slash and burn systems in forest lands, K applications may be omitted initially (Kerridge, 1978).

Sulfur and Micronutrients

Table 46.5 shows ranges of S, Zn and Cu concentrations in plant tissues of <u>Stylosanthes</u> species. These critical concentrations represent values obtained for plant tops harvested at 8 weeks of plant growth and are considered acceptable for pasture establishment. These concentrations must be evaluated further for maintenance levels under grazing.

Critical S concentration of <u>Stylosanthes</u> species growing on tropical soils with available S less than 10 ppm (calcium phosphate extraction) can be reached by applying about 20 kg S/ha. Common fertilizer sources of S are sulfur flower and gypsum, but simple superphosphate, ammonium sulfate, sulfur-coated urea, and rock phosphates with partial acidulation with sulfuric acid can also supply S for plant use.

TABLE 46.5

Ranges of critical sulfur, zinc and copper concentrations in plant tissue of Stylosanthes species

Element	Species	Plant Tissue	Range of Critical Element Concentration (%)	Reference
Sulphur	S. capitata	plant tops	0.16 ± 0.02[1]	CIAT, 1981
	S. guianensis	plant tops	0.13 ± 0.04	Miller and Jones, 1977
	S. hamata	plant tissue	0.11 ± 0.02[2]	Aitken, 1979
	S. humilis	plant tops	0.14 ± 0.02[2]	Andrew, 1977
	Mean		0.13 ± 0.02	
			(ppm)	
Zinc	S. capitata	plant tops	25 ± 5	CIAT, 1982
	S. guianensis	plant tops	20 ± 5	Winter & Jones, 1977
	S. hamata	plant tops	15 ± 5	Bruce, 1978
	S. humilis	plant tops	21 ± 2	Crack, 1971
	Mean		20 ± 4	
Copper	S. capitata	plant tops	8.0 ± 2.0[1]	CIAT, 1982
	S. guianensis	plant tops	3.5 ± 0.5	Winter & Jones, 1977
	S. hamata	plant tops	4.5 ± 0.5	Bruce, 1978
	S. humilis	plant tops	4.0 ± 1.0	Webb, 1975
	Mean		5.0 + 1.0	

[1]Ranges correspond to values obtained with different ecotypes and associated with 80% of maximum yield at 8 weeks of plant growth.

[2]Ranges associated with 90% of maximum yield at 8 weeks of plant growth.

Zinc sulphate and copper sulphate have been the most commonly used sources of Zn and Cu. Recommended rates are 3 kg Zn/ha and 2 kg Cu/ha. Residual effects of Cu and Zn in tropical soils appear to be considerable, and reapplications may be made about every 4 years. More research is required for Zn and Cu to compare a range of soils, particularly acid sandy soils.

SELECTED REFERENCES

AITKEN, R.L. 1979. Apical tissue analysis for determining the sulphur status of Stylosanthes hamata cv. verano. pp.83-87. In: Annual Report 1979, Pasture Improvement Project, Khon Kaen University, Faculty of Agriculture, Thailand.

ANDREW, C.S. 1977. The effect of sulphur on the growth, sulphur and nitrogen concentration, and critical sulphur concentrations of some tropical and temperate legumes. Australian Journal of Agricultural Research 28: 807-820.

ANDREW, C.S. and M.P. HEGARTY. 1969. Comparative responses to manganese excess of 8 tropical and 4 temperate pasture legume species. Australian Journal of Agricultural Research 29L: 687-696.

ANDREW, C.S. and M.F. ROBINS. 1969. The effect of potassium on the growth and chemical composition of some tropical and temperate pasture legumes. I. Growth and critical percentages of potassium. Australian Journal of Agricultural Research 20: 999-1007.

BLUNT, C.G. and L. R. HUMPHREYS. 1970. Phosphate response of mixed swards at Mount Cotton, Southeastern Queensland. Australian Journal of Experimental Agricultural and Animal Husbandry 10: 431-443.

BROLMAN, J.B. and R.M. SONODA. 1975. Differential responses of three Stylosanthes guianensis varieties to three levels of potassium. Tropical Agriculture (Trinidad) 52: 139-142.

BRUCE, R.C. 1978. A review of the trace element nutrition of tropical pasture legumes in Northern Australia. Tropical Grasslands 12: 170-182.

CARVALHO, M. de, 1978. A comparative study of the responses of six Stylosanthes species to acid soil factors with particular reference to aluminum. Ph.D. Thesis, Department of Agriculture, University of Queensland, Australia, 298p.

Centro Internacional de Agricultura Tropical (CIAT). 1980, 1981, 1982. Tropical Pasture Program. Annual Report 1979, 1980, 1981. Cali, Colombia.

COELHO, R.W. and W.G. BLUE. 1979. Potassium nutrition of five species of the tropical legume Stylosanthes in an Aeric Haplaquod. Proceeding, Soil Crop Science Society Florida 38: 90-93.

CRACK, B.J. 1971. Studies on some neutral red duplex soils in Northeastern Queensland. 2. Glasshouse assessment of plant nutrient status. Australian Journal of Experimental Agriculture and Animal Husbandry 11: 336-342.

FLORES, A.J. 1982. A preliminary agronomic evaluation of fifty-two accessions of Stylosanthes macrocephala under acid soil conditions. M.S. Thesis, New Mexico State University, Las Cruces.

GROF, B., R. SCHULTZE-KRAFT and F. MUILLER. 1979. Stylosanthes capitata Vog., some agronomic attributes and resistance to anthracnose (Colletotrichum gloesporioides Penz.) Tropical Grassland 13: 28-37.

JONES, R.K. 1974. Nutrient requirements for the establishment of improved pasture. pp. 17-33. In: Proceeding of the seminar on potential to increase beef production in tropical America. CIAT, Cali, Colombia.

KERRIDGE, P.C. 1978. Fertilization of acid tropical soils in relation to pasture legumes. pp. 395-415 In, C.S. Andrew and E.J. Kamprath (ed.), Mineral Nutrition of Legumes in Tropical and Subtropical Soils. CSIRO, Melbourne, Australia.

MILLER, C.P. and R.K. JONES. 1977. Nutrient requirements of Stylosanthes guianensis pastures on a enchrozem in North Queensland. Australian Journal of Experimental Agriculture and Animal Husbandry 17: 607-613.

MUNNS, D.N. and R.L. FOX. 1977. Comparative lime requirements of tropical and temperate legumes. Plant Soil 46: 533-548.

SALINAS, J.G. and R. GARCIA. 1979. Metodos Analiticos para suelos acidos y plantas. Centro Internacional de Agricultura Tropical (CIAT), Cali, Colombia. 54p.

SALINAS, J.G. and J.I. SANZ. 1981. Sintomas de deficiencias de macronutrimentos y nutrimentos secundarios en pastos tropicales. CIAT, Cali, Colombia. 28p.

SANCHEZ, P.A. and J.G. SALINAS. 1981. Low-input technology for managing Oxisols and Ultisols in Tropical America. Advances in Agronomy 34: 279-406.

SPAIN, J.M., C.A. FRANCIS, R.H. HOWELER, and F. CALVO. 1975. Differential species and varietal tolerance to soil acidity in tropical crops and pastures. pp. 308-329. In: E. Bornemisza and A. Alvarado (ed.s), Soil Management in Tropical America. North Carolina State University, Raleigh.

WEBB, A.A., 1975. Studies on major soils of the Forayth Granite. 2. Glasshouse nutrient assessment. Queensland Journal of Agricultural and Animal Science 32: 19-26.

WILAPON, N., R.L. AITKEN, and J.D. HUGHES. 1979. The use of apical tissue analysis to determine the phosphorus status of Stylosanthes hamata cv. verano. pp. 107-111. In, Annual Report 1979, Pasture Improvement Project, Khon Khaen University, Faculty of Agriculture, Thailand.

WINTER, W.H. and R.K. JONES, 1977. Nutrient responses on a yellow earth soil in northern Cape York Peninsula. Tropical Grassland 11: 247-255.

47
GLYCINE

P. C. Kerridge

Glycine (<u>Neonotonia wightii</u>) is a perennial pasture legume of African origin that also occurs naturally in tropical Asia. It can be grown in sub-tropical areas of medium rainfall on soils of medium to high fertility. It was previously known as <u>Glycine wightii</u> and <u>Glycine javanica</u> and is known commonly as glycine and perennial soybean. Where it is adapted, glycine is a successful pasture legume because it combines well with grasses, grows and seeds vigorously once established, and fixes large amounts of nitrogen. It is palatable and non-toxic and can be conserved as hay or silage. Glycine pastures require good management to maintain a good nutrient supply and to prevent overgrazing. It is grown commercially in Australia, East Africa, North, Central and South America and the Caribbean area.

ECOLOGICAL ADAPTATION

Glycine is a summer growing perennial adapted to a subtropical climate. Some of the most successful areas of adaptation have been tropical highlands where temperature extremes are moderated by altitude, but it will grow 30° north or south of the equator at sea level. High temperatures (35°C) can limit growth while cool night or soil temperatures (16°C) slow growth which ceases at about 13°C. Frost causes shedding of leaves but it is more frost tolerant than <u>Centrosema pubescens</u> and most <u>Stylosanthes</u> species. It is a short day plant.

Best growth is observed in areas of 1200 to 1500 mm rainfall evenly distributed during the summer months. However, certain cultivars will grow where the annual rainfall is only 600 mm, and on well-drained soils good growth has been observed at 2000 mm. It is only moderately drought tolerant and is not tolerant of flooding.

[1]Division of Tropical Crops and Pastures, Commonwealth Scientific and Industrial Research Organization (CSIRO), Brisbane, Australia.

Glycine grows best in freely drained fertile soils derived from basic parent material where the pH is 5.7-6.0 or greater. These include the deep red soils (ultisols) developed on basalt, alluvial soils, and soils developed on serpentine. It is not well adapted to acid soils or soils with poor drainage. It is only slightly tolerant of saline conditions. Glycine is more restricted by soil type than most other tropical or subtropical pasture legumes.

Glycine develops a deep persistent taproot which gives some tolerance to drought, frost, fire and heavy grazing. The long trailing stems root sparsely at the nodes or can climb and form a canopy over tall grass.

NUTRIENT REQUIREMENTS

Nutrients play a major role in influencing the proportion of legume in a pasture. The nutrient requirement for legumes will usually be higher than for grasses because of special requirements for symbiotic N fixation and a lower ability to compete for nutrients. Thus except for N, when the nutrient requirements for glycine are satisfied there will be adequate nutrients for grass growth also.

One limitation to more widespread use of glycine has been its slow establishment, which has often been associated with slowness to nodulate. In turn, slow nodulation appears to be associated with both a build-up of sufficient rhizobia to effect nodule initiation and an inherent characteristic of the plant not to nodulate freely. Liming or lime pelleting have improved nodulation and seedling yields on acid soils (< pH 5.5). Lime pelleting will have the greatest effect where low numbers of rhizobia are present on the seed at planting, particularly when dry periods follow sowing. Lime pelleting is also advisable where seed is sown in contact with superphosphate.

On soils where glycine is adapted, the most common nutrient responses, other than N, are to phosphorus (P), sulfur (S), molybdenum (Mo), potassium (K) and lime. Aluminum (Al) and Mn toxicities may occur on acid soils (< pH 5.5). Positive responses have also been recorded to zinc (Zn).

Glycine has a greater requirement for P than Centrosema and Stylosanthes species. Glycine is very sensitive to Mo deficiency and is less efficient than most tropical species in extracting Mo from the soil. Liming may be required also to overcome Mn and Al toxicity. Potassium and S will often not be required on many of the more fertile soils to which glycine is adapted.

DIAGNOSIS OF NUTRIENT REQUIREMENTS

Nutrient deficiency can be diagnosed by one or more of the following approaches - visual symptoms, plant and soil chemical analysis and by field test strips.

Visual Symptoms

Foliar symptoms are particularly useful in the diagnosis of N, S, Mo and K deficiency and Mn toxicity. Symptoms and toxicities have been described and illustrated.

Deficiencies

Nitrogen (N). Uniform chlorosis of the leaves, starting with the lower leaves.

Sulfur (S). Uniform chlorosis of the leaves but starting with the upper leaves.

Molybdenum (Mo). Same as for nitrogen deficiency when the plant is dependent on symbiotic N.

Potassium (K). Interveinal, then general chlorosis of lower leaves, with brown necrotic areas developing as severity increases; may be interveinal necrotic spotting where deficiency develops rapidly; basal portion of leaf remains darker green.

Phosphorus (P). Dark green to bluish green small leaves, severely reduced plant growth.

Calcium (Ca). Broad interveinal chlorosis of basal portion of the middle leaves, subsequently becoming necrotic; tips die back; roots dark brown, thickened and with little branching.

Magnesium (Mg). Broad interveinal chlorosis of lower leaves followed by large necrotic area, margin and tips curved downwards.

Copper (Cu). Wilting and pale green color of younger expanded leaves, most pronounced at midday, with tip necrosis as severity increases.

Zinc (Zn). Interveinal chlorosis of young growth, with intense necrotic spotting; leaves become thick, brittle and curved, with puckering of interveinal tissue.

Manganese (Mn). Upper leaves chlorotic and developing mottled appearance, puckered and curved downwards; in severe deficiency leaves have brown necrotic spotting at emergence.

Iron (Fe). Intense chlorosis of young upper leaves, which become almost white, leaves thin but not deformed.

Boron (B). Dark green tips, leaves thick and may be malformed; root growth poor with brown tips.

Toxicity

Manganese (Mn). Mottled, interveinal chlorosis on young expanded leaves, developing into irregular necrotic spots near mid-rib and main veins, surface puckering of leaf.

Boron (B). Marginal and tip chlorosis, then necrosis on older leaves. Distinguishable from K and Mg deficiencies by being confined to margin edges.

Aluminum (Al). Reduced top growth; short, thick roots, little lateral development.

Growth is often severely inhibited by the time foliar symptoms appear but as they often first appear in only a small area of a pasture they give a warning of incipient deficiency (or toxicity) in the paddock as a whole.

Plant Chemical Analysis

Critical concentration values or ranges have been established for several nutrients in pot culture based on concentrations in the shoot dry matter at the pre-flowering stage of growth.

Nutrient	Critical Nutrient Concentrations		
	Symptoms Present	Sufficiency	Excess
P	0.15%	0.22-0.25%	
S	0.07%	0.17 %	
K	0.60%	0.80 %	
Mo		<0.02 ppm	
Mn	30 ppm		560 ppm

Under favorable conditions for growth, an N concentration of less than 3.0% suggests some other nutrient or factor is affecting nodulation or nitrogen fixation. An index for one nutrient will only be useful where all other nutrients are adequate.

Equivalent values for field samples might be obtained from whole shoot samples containing 5 to 6 fully expanded leaves. Samples need to be collected when growth is not inhibited by other factors such as moisture and temperature. The surface soil should have been continuously moist for 2 to 3 weeks prior to sampling.

Plant analysis can also be used in diagnosis by comparison of the nutrient concentrations of tissue from an area of poor growth with that from an adjacent area where plant growth is vigorous.

Soil Chemical Analysis

Soil tests have not been developed specifically for glycine, but critical values can be inferred from research on other legume pastures adapted to a similar climate. Values will vary with soil type and should only be used as a guide.

Minimum Desirable Range for Soil Tests

pH	5.5-6.0
P	20-25 ppm (0.5 N Na HCO_3 or 0.01 N H_2SO_4 extracts)
K	0.15-0.20 me per 100 g soil
SO_4-S	4-6 ppm (0.01 M Ca $(H_2PO_4)_2$ extract weighted over profile)
Zn	0.2-0.5 ppm)
Cu	0.1-0.3 ppm) DTPA extract
Mn	1 ppm)

Field Plots

For diagnostic purposes an omission design is useful. Nutrients are omitted, one at a time, from a complete nutrient mixture and the resultant mixtures applied in strips across a pasture. Replication may be used or else each omission treatment is placed between two complete treatments.

For confirmation of a suspected deficiency, made by visual observation or plant and soil analysis, test strips of an addition of that nutrient alone may be compared with nil treatments on either side. Visual symptoms of deficiencies can also be confirmed by spraying a soluble salt of the nutrient onto a portion of the foliage showing the symptoms.

Diagnosis in Practice

A combination of these techniques needs to be employed in diagnosis. For example, suppose a plant is exhibiting general chlorosis. This might be due to N, Mo, S or Fe deficiency. Iron chlorosis would not be expected on soils of pH < 7.0. However, it could be checked by spraying $FeSO_4 \cdot 2H_2O$ or Fe-EDTA onto the leaves and watching for greening over several days where the spray wets the leaf. Sulphur deficiency, as for Fe, is usually more severe on younger portions of the plant. It could be confirmed by lack of response to N fertilizer and plant analysis. Nitrogen deficiency, which may be due to nodulation failure, nodulation with ineffective rhizobia, or some limitation on N fixation (e.g. other nutrient or waterlogging), could be checked by application of mineral N fertilizer. Molybdenum deficient plants are usually well nodulated and will respond rapidly to a spray application of a soluble Mo salt.

Response to liming may be due to overcoming toxicities of Al or Mn or deficiencies of Ca or Mo. These effects can be sorted out with separate applications of Mo, low rate of Ca and a heavy rate of P.

Deficiencies at levels where visual symptoms are absent are more difficult to diagnose. Diagnosis by soil or plant analysis may need to be confirmed by fertilizer application. But the typically large variations in legume composition may make it difficult to detect a response. A uniformity harvest before nutrient application and the use of covariance in statistical analysis may help in overcoming this variation. For Mo, greater precision can be obtained by comparing the change in tissue N concentration before and after Mo application.

Another problem in diagnosis is obtaining a representative sample of plant or soil material. Spatial variability can be large due to natural soil variation, uneven fertilizer application, and return of dung and urine. This can be reduced by taking a large number of samples or by sampling from fixed locations where an area is being monitored regularly. For field plots it is advisable to use long strips and evaluate yield differences by estimation.

Improved Varieties

Most cultivars of glycine have been selected in relation to climatic rather than edaphic factors. These include frost, drought and length of growing season. However, there are a few examples of differential response by cultivars to edaphic factors. On the Atherton Tableland, Queensland, the tetraploid variety Clarence and selection CPI 13300 were most persistent and gave higher yields than the diploid varieties Tinaroo and Cooper on a range of moderately acid soils. The tetraploid cultivars were more tolerant to aluminum than diploid cultivars when grown in nutrient solution. Another tetraploid variety, Malawi, has also shown better persistence and given higher yields where soil pH is below 6. A Brazilian selection IRI No. 1 (SPI) shows some tolerance to high soil Mn levels.

ROLE OF NUTRIENTS IN MANAGEMENT

Slow nodulation has been a major problem during establishment. Where this is due to acid soil conditions, the problem can be alleviated by ensuring the seed is inoculated with sufficient rhizobia and lime pelleting, or in more acid conditions by liming prior to sowing. Poor legume vigor at establishment may also be due to grass competition. This can occur under conditions of high available soil N where the grass responds more rapidly than the legume to correction of P deficiency. Grasses may also compete strongly due to their greater tolerance of acid soil conditions. Judicious grazing is required to prevent shading of glycine seedlings during establishment.

Soluble phosphate fertilizers will be necessary for establishment on P deficient soils. Rock phosphates will be less effective on the soils of higher pH and Ca status to which glycine is adapted than the more acid soils on which Stylosanthes and Centrosema will be grown. In Queensland the rates of P application on responsive soils with a moderately high capacity to absorb P have been 45 kg/ha at establishment and 15-30 kg/ha for maintenance. Phosphorus responses to 30-50 kg/ha have also been obtained in Cuba and Brazil.

Potassium deficiency is usually corrected by 50 kg K/ha. Maintenance requirement should be met by 25 kg K/ha in grazed pasture but will be higher where the pasture is cut for hay or silage.

Molybdenum deficiency can be corrected at establishment by the application of 100 g Mo/ha as molybdenum trioxide in the seed pellet coating, by the spray application of a soluble Mo salt, or by mixing Mo compounds in a basal fertilizer. Problems have arisen with the latter method due to inability to ensure even mixing. Frequency of application will depend on the ability of the soil to absorb Mo. On soils with a high capacity to absorb Mo this may need to be every 2 years while on lower absorbing soils it will be every 4 to 5 years. Soils that absorb P strongly also absorb Mo strongly.

There is no information specific to glycine on S requirement. However, it is expected that an initial deficiency would be corrected by 15-20 kg S/ha, with annual applications of 6-12 kg S/ha being adequate for maintenance in grazed pastures. Adequate S would be supplied in single superphosphate where this is used as the source of P. Otherwise gypsum or "flowers of sulphur" can be used.

Maintenance fertilizers can be applied at any time of the year. However, if leaching losses are likely to occur, then K and S application should be confined to the season when growth occurs.

Nitrogen fertilizer has been used strategically to increase grass production in glycine pastures during cooler months or to maximize the use of irrigation during dry periods. The proportion of legume has decreased during periods of prolonged N application (6 months) but glycine has the capacity to persist and increase once N application is stopped. From four to six consecutive 50 kg N/ha applications have been used under irrigation while one to two applications would suffice under dryland conditions.

Nitrogen should be used cautiously on glycine pastures. Unless the extra production can be utilized immediately, its value will be reduced due to a decline in quality as the grass matures. Also N usage may cover up nutrient deficiencies in the legume and lead to its disappearance. This could occur where there is poor nodulation and N fixation due to acid soil conditions and where deficiencies appear rapidly, e.g. with K where a hay or silage crop is removed; and with Mo after native soil N is depleted.

Herbaceous weeds may be a problem in a newly established pasture but do not persist in older pastures where fertilization is adequate and grazing not too severe. Diseases and pests may affect production but where stands are vigorous they survive attack. In Australia, the Amnemus weevil larvae attack the roots and can cause considerable damage where climatic and soil conditions are marginal for glycine. A beneficial effect of nutrition was observed in an Mo deficient soil following an Amnemus attack as glycine disappeared in the absence of Mo but recovered where Mo had been applied.

MEASUREMENT OF BENEFITS FROM ADEQUATE NUTRITION

Glycine does not persist under low fertility conditions. Where fertility is adequate or corrected by fertilizers, the presence of glycine affects forage quality directly through its high protein content and indirectly through N transfer to the associated grasses.

Correctly fertilized stands of glycine have produced from 4 to 12 tons dry matter (DM) per hectare depending on climate and management, with the young leaves having an N content of 2.9 to 3.5%. There is an accompanying increase in quantity and quality of the companion grasses. The lower yields are more common in Australia, possibly because most areas where glycine is grown are marginal for the plant in terms of edaphic and climate requirements. The higher yields have been reported from the Caribbean area and Brazil. Nevertheless, glycine has been successful in restricted areas in Australia, e.g. its use has resulted in a two to three fold increase in productivity in the Kairi area of the Atherton Tableland, Queensland.

UTILIZATION OF INCREASED PRODUCTION

Most glycine pastures are used for dairy rather than beef production. This is because the relatively long growing seasons and moderate climate in areas where glycine is adapted are also suitable for European dairy breeds.

There has been considerable research on dairy cattle nutrition using glycine pastures at Kairi. High milk production per animal has been achieved at a stocking rate of two cows per ha without supplementation. But, under heavier stocking as the amount of feed is reduced and diet selection is restricted, there is a reduction in milk yield per animal and a response to supplementation. In one experiment, where an increase in stocking rate from two to four cows per ha decreased the average pasture on offer from 2700 to 1100 kg DM per ha, the milk yield decreased from 12.7 to 9.6 kg/cow/day. In a further experiment, where very heavy stocking resulted in a rapid decrease in available pasture from 3600 to 1800 kg DM per ha and glycine content from 49 to 24%, milk production declined from 13.0 to 6.5 kg/cow/day. Increased carrying capacity may result in higher production per hectare but has to be assessed in relation to the cost of concentrates, effects on length of lactation, and likely harmful effect on glycine persistence. The Na contents on these pastures is low, 0.01 to 0.02% for glycine and 0.08 to 0.10% for panic grass, and increases in milk production have been obtained by supplementing cows with Na. Low Na concentrations have also been recorded on soils derived from limestone in the Caribbean area.

Productivity will decline during the cooler and drier months due to a decrease in the quantity and quality of pasture available. Hay and silage have been made successfully from glycine pastures to help overcome periods of feed shortage. However, on the Atherton Tableland the strategic use of N with or without irrigation and concentrate supplements are now more commonly used to overcome periods of feed shortage than conservation.

SELECTED REFERENCES

General

ANDREW, C.S. and KAMPRATH, E.J. 1978. Mineral Nutrition of Legumes in Tropical and Sub-Tropical Soils. CSIRO, Melbourne, Australia.

ANDREW, C.S. and PIETERS, W.H.J. 1976. Foliar symptoms of mineral disorders in Glycine wightii. Tech. Paper No. 18. Division of Tropical Crops and Pastures, CSIRO, Brisbane.

FUNES, F. 1979. Los Pastos en Cuba. Tomo 1. Production. Ministerio de la Agricultura, Calle 30y Avenida 7ª, Playa, Ciudad de la Habana, Cuba.

RODRIGUES, L.R. de A., PEDREIRA, J.V.S., MATTOS, H.B. de 1975. Adaptacao ecologica de algumas plantas forrageiras. Zootecnia 13: 201-18.

SHAW, N.H. and BRYAN, W.W. 1976. Tropical Pasture Research Principles and Methods. Comm. Bureau of Pastures and Field Crops. Bull. 51. CAB, Hurley, England.

SKERMAN, P.J. 1977. Tropical Forage Legumes. FAO Plant Production and Protection Series No. 2. FAO, Rome.

Specific

ANDREW, C.S. 1976. Effect of Ca, pH and nitrogen on the growth and chemical composition of some tropical and temperate pasture legumes. I. Nodulation and growth. Australian Journal of Agricultural Research 27: 611-23.

COWAN, R.T. and DAVISON, T.M. 1978. Feeding maize to maintain milk yields during a short period of low pasture availability. Australian Journal of Experimental Agriculture Animal Husbandry 18: 325-28.

GARTNER, J.A., FERGUSON, J.E., WALKER, R.W., and GOWARD, E.A. 1974. Evaluating perennial grass/legume swards on the Atherton Tableland in North Queensland. Qld. Journal of Agriculture Animal Science 31: 1-17.

JOHANSEN, C., KERRIDGE, P.C., MARKLEY, K.E., LUCK, P.E., COOK, B.G., LOWE, K.F., and OSTROWSKI, H. 1978. Growth and molybdenum response of tropical legume/ grass swards at six sites in south-eastern Queensland over a five year period. Tropical Agronomy Tech. Mem. No. 10. CSIRO, Brisbane.

KERRIDGE, P.C., COOK, B.G., and EVERETT, M.L. 1977. Application of molybdenum trioxide in the seed pellet for sub-tropical pasture legumes. Tropical Grassland 7:228-9,32.

48
CENTROSEMA

Albert E. Kretschmer, Jr., and George H. Snyder

Centrosema belongs to the tribe Phaseoleae of the subfamily Papilionoideae, of Leguminosae. It is a well-known, primarily tropical, Western Hemisphere legume genus with some 35 species.

Most Centrosema species can be characterized as perennial and viney although a few are subshrubs and annuals. Main centers of diversity are Brazil and Central America. C. virginianum, a perennial and probably the most polymorphic species, has a natural range from New Jersey (about 40°N latitude) and Virginia west to Kentucky and Arkansas, south through Mexico, and the Bahama Islands into northern Argentina (about 35°S latitude). Centrosema was originally known as a wet tropical genus but more recently C. pascuorum and C. rotundifolium were collected from tropical arid areas.

Centro (C. pubescens) although native to tropical America is naturalized in tropical Africa, India, and Asia. It has been used successfully as a cover crop in rubber in Malaysia for 50 to 60 years, fixing 150 to 200 kg N/ha.

Common centro (a "short day" species) was first introduced into Australia from Java in 1930. It has been used as a herbaceous forage legume for grazing in mixtures with grasses for 30 to 40 years. In 1971, 'Belalto' formerly thought to be C. pubescens, but recently classified as C. schiedianum (a closely allied species) was released by Australia for grazing. A Brazilian C. pubescens cultivar "Deodora" is also available and in use in Brazil.

Common and Belalto centros have been tried or used commercially as forage in mixtures with grass in many parts of the tropics.

---

[1]Professor of Agronomy, University of Florida, P.O. Box 248, Ft. Pierce, FL 33452, and Professor of Soil Chemistry, University of Florida, P.O. Drawer A, Belle Glade, FL 33430.

ECOLOGICAL ADAPTATION

Common centro is adapted to the humid tropics and subtropics. Its growth rate is drastically reduced when night temperatures fall below about 15C, and maximum growth occurs when day/night temperatures are above 30/20C. Belalto centro may be more productive than common centro at temperatures below those for maximum growth of common centro.

Centro is less tolerant to shading than greenleaf desmodium (Desmodium intortum) and more tolerant than Siratro (Macroptilium atropurpureum).

Common and Belalto centro are adapted to soils having moderate to good drainage, physical characteristics, and chemical fertility. Although both require about 1000 mm or more of annual rainfall, they are able to become perennial during droughts of about three months. Neither can withstand prolonged flooding.

PRODUCTIVITY AND NUTRITIONAL QUALITY

When effectively nodulated, N fixation by centro generally is 75 to 280 kg/ha, but can be as high as 520 kg/ha/year. When appropriate Rhizobium bacteria are not present in the soil, common centro seeds should be inoculated with commercial strains that cause effective N fixation. In areas where centro is native, inoculant may not be needed although its low cost warrants its use under most circumstances. There are profuse and sparse nodulating types of common centro, and the characteristic is heritable. The sparse types can be inoculated satisfactorily with other Rhizobium strains. A commercial "cowpea" type inoculant may be effective. Lack of effective inoculation results in lower dry matter production and lower plant N concentrations.

Annual dry matter production of commercial centro and centro-grass mixtures has been determined in many areas of the world. Some of the yields in mt/ha/yr are as follows: Florida - 13.3 in various grass mixtures; Belize - 3.4 centro only; Ghana - 7.6 centro only (9 months); 27.9 with pangola (9 months), 30.4 with Andropogen gayanus (9 months), Venezuela - 9.4 with pangola (9 months), Costa Rica - 19.1 with guineagrass, 13.8 with African stargrass, 14.5 with pangola and 17.8 with jaraguagrass; Hawaii - 12.3 centro only, 10.3 with pangola, 10.5 with napiergrass; Nigeria - 10.6 with signalgrass; and Solomon Islands - centro component 0.1 to 4.8 t/ha depending on companion grass. Other ecotypes of C. pubescens may yield more than common centro.

Because flowering and seed maturation in <u>C</u>. <u>pubes-</u><u>cens</u> occur over a period of several weeks or months, and seed pods shatter when mature, seed yields may be low when conventional harvesting methods are used. Even so, 150 kg/ha have been harvested by direct heading in Australia, and hand-picked yields can be 300 to 600 kg/ha and as high as 950 kg/ha.

Most of the animal response data with centro have been obtained in Australia and Africa. In the establishment phase, centro is slow and thus may be damaged by heavy grazing, and even afterwards must be managed to permit periodic regrowth. Selected comparisons of grass alone and grass-centro mixtures on beef gains show the following respective values: (kg liveweight gain per ha) 67-134 (guineagrass, Australia, 6 months); 374-460 and 440-610 (guineagrass, Australia, annual), 349-410 (pangola, Brazil annual). About one animal unit (400 kg) per ha probably is an optimum fixed stocking rate, while in managed pastures stocking rates should be varied to permit periods of regrowth of centro. Realistic animal gains would be in the range of 0.4 to 1.0 kg/head/day.

Centro is intermediate in palatability but equal to most other tropical legumes. Tannins or oxalate may contribute to its lower palatability.

FERTILIZER REQUIREMENTS

In many soils, P is the nutrient most limiting for centro establishment and production. Application rates of about 75 kg P/ha often are used for centro establishment in soils suspected of being low in available P. However, the generalization that often is valid for temperate legumes, that high rates of P will favor legume growth relative to grasses, may not always hold for centro-tropical grass combinations.

Information is very limited for determining proper maintenance rates of P. It is believed that most tropical legumes have higher P requirements during the establishment and early growth stages than in later stages. We believe it is better to use one heavy application of P at the time of establishment to assure a good stand of vigorously growing centro, and then omit annual maintenance fertilizations instead of metering out the same total rate over a period of years. Considerable work is needed to identify optimum P rates for centro maintenance in various grass combinations under varying management systems. Annual applications of 10-20 kg P/ha from rock phosphate or superphosphate sometimes are used for centro-grass pastures, but in most cases information is lacking to verify the usefulness of this practice.

Several attempts have been made to reduce P requirements for legume establishment. There may be an advantage in banding seed with P, as opposed to broadcasting. Band placement appears to benefit the legume more than it does an associated grass. With banding it should be possible to reduce P application rates; however banding may lead to other problems. For example, root growth may be restricted to the band area, rendering the seedlings more susceptible to drought or other stresses. Later a broadcast application of P may be needed to promote legume growth throughout the pasture.

Relative to other legumes, centro appears to be especially dependent on vesicular-arbuscular mycorrhiza for P uptake in P deficient soils. Apparently mycorrhiza activity reduces the threshold level required for P uptake by centro, and the improved ability to utilize soil P may result both from more complete exploration of soil by virtue of the beneficial fungi which spread beyond the depletion zone and exploit P in the area explored by the hyphae. Perhaps it is because time is required for the mycorrhiza association to develop that higher P rates are needed at time of establishment than when centro has been established for some time.

Liming acid soils frequently has been beneficial for centro, but there is no clear consensus as to the reason for this enhancement or the degree to which acid soils should be limed. In very acid, infertile soils a small amount of lime (100-500 kg/ha) may be required to supply adequate Ca.

Many tropical legumes, including centro, survive in acid soils not suited for temperate species. Small amounts of lime usually are adequate for maintaining tropical legumes in pastures. When maximum centro growth is desired, only experimentation under the particular conditions encountered is needed to determine the optimum lime requirement. In most cases where liming is used for centro establishment and maintenance, initial application rates have been about 500-1000 kg/ha. Efforts are being made to develop Centrosema with greater acid soil tolerance. Some Centrosema species appear far more sensitive to soil acidity than C. pubescens.

When centro growth is stimulated by optimum P and lime additions, growth may be limited by other nutrients. Frequently Mo additions have been beneficial for tropical legumes, and for centro this response generally is associated with enhanced Rhizobium activity. Chlorosis associated with Zn deficiency has been reported in centro in the Brazilian Cerrado and Amazonian areas. Sulfur responses also are commonly reported. If 20% superphosphate is used to satisfy P needs, adequate S probably will be supplied, otherwise, rates of 10-30 kg S/ha may be needed. Deficiencies of K, Mg, Cu, and B also have been reported for centro. In Australia centro is considered intermediate in response to Cu deficiency. No Mn or Co deficiencies of tropical legumes have been reported but Mn excess could be encountered. Under high pH conditions, such as in soils derived from limestone, Fe deficiency of centro has been observed in the Solomon Islands and in the Bahama Islands.

SOIL TESTING

The prediction of nutrient deficiencies by soil
testing would be ideal, but as yet this is not possible
for the variety of nutrients and soils involved. A critical soil P level (Bray II) of about 3 ppm was found for
5 tropical legumes (centro not included) in the same
soil, a Colombian Oxisol. A microbiological test for Mo
deficiency successfully predicted Mo deficient soils for
centro growth in Brazil (Franco, et al., 1978). Standley
et al. (1981) in Australia recommended that Cu should be
applied when soil extractable Cu was below 0.2 ppm, that
no response would be expected above 4 ppm Cu, and plant
toxicity might occur when extractable Cu was above 10
ppm. Zinc should be applied when extractable Zn is below
0.3 ppm, but above 0.8 ppm growth responses to Zn applications would not be expected. With Mn, response may occur when soil values are below 1 ppm while extractable Mn
over 50 ppm indicates possible toxic levels. Clearly
much more work is needed before soil analyses can be used
routinely for determining fertilizer requirements of
centro.

TISSUE ANALYSIS

Tissue analyses offer a somewhat more universal
method for determining nutrient deficiencies in centro
and for guiding maintenance fertilizations (Table 48.1).
Even with tissue analysis a considerable range of values
is reported for healthy and deficient centro by various
workers. Part of the differences in Table 48.1 results
from the portion of the plant used for analysis (compare
Shorrocks' critical values for leaves and other complete
plant values for K or Cu) and the season when sampling
occurred (compare wet and dry season critical values by
Grof). Also critical values can be modified by stage of
plant growth. For example, Oliveira et al. (1978) determined the macronutrient contents of centro leaves and
stems at 21 day intervals, from 21 to 147 days. Highest
contents in leaves and stems, respectively, occurred at
the following harvests: N, 84 days (leaves) - 147 days
(stems); P, 126-126; K, 126-84; Ca, 21-21; Mg, 105-21;
and S, 42-21 days. Nutrient imbalances and grass competition also affect critical values. The authors suggest
the following minimum values for centro foliage above
which no deficiency would be expected: P-0.16%, K-1.0%,
Ca-0.5%, Mg-0.2% and S-0.2%; and Cu-4 ppm, Mn-10 ppm, Zn-
16 ppm, B-4 ppm and Mo-0.17 ppm.

TABLE 48.1
Range and critical concentrations of nutrient elements in the foliage (dry matter) of Centrosema pubescens

Author	Type Study	Range						Critical Concentration[2]				
		N	P	K	Ca	Mg	S	P	K	Ca	Mg	S
Shorrocks, 1964[2] [3]	P	2.7–5.0	0.35–0.50	1.3–2.0	0.6–1.0	0.30–0.46	0.50–0.70	0.12–0.19	0.55–0.94	0.1–0.2	0.07–0.18	0.09–0.18
Steel and Humphreys, 1974	P	1.6–2.1	0.12–0.18	–	–	–	–	0.16	–	–	–	–
Andrew and Robins, 1969b	P	–	–	0.4–1.9	–	–	–	–	0.75	–	–	–
Andrew and Robins, 1969a	P	2.8–3.7	–	1.9–2.1	0.7–0.9	0.24–0.32	–	–	–	–	–	–
Andrew and Hegarty, 1969	C	–	–	–	1.7–2.0	–	–	–	–	–	–	–
Jones and Quagliato, 1973	P	–	–	–	–	–	0.12–0.22	–	–	–	–	0.18
Andrews and Norris, 1961	P	1.9–3.5	–	–	0.4–1.7	–	–	–	–	–	–	–
Andrew and Pieters, 1970b[4]	P	–	–	1.2	–	–	–	–	0.35	–	–	–
Andrew and Robins, 1969	P,F	–	0.13–0.24	–	–	–	–	0.16–0.21	–	–	–	–
Werner and De Mattos, 1972	P	2.3–3.6	0.12–0.20	–	1.2–1.8	–	–	–	–	–	–	–
Gutteridge and Whiteman, 1978	F	2.3–3.0	0.14–0.23	0.8–2.4	–	–	–	–	–	–	–	–
Grof, 1982	FD[2]	–	–	–	–	–	–	0.10	0.8	0.7	–	–
	FW[2]	–	–	–	–	–	–	0.18	1.4	1.0	–	–
	FD[6]	–	–	–	–	–	–	0.09	0.7	0.6	–	–
	FD[6]	–	–	–	–	–	–	0.16	1.2	0.7	–	–

Reference		Fe	Cu	Mn	Zn	B	Mo	Cu	Mn	Zn	B	Mo
Spain et al., 1975	P	1.4-2.3	0.12-0.19	1.0-1.2	0.8-1.4	0.17-0.20	-	-	-	-	-	-
Wilson and Lansbury, 1958	F	-	0.15-0.25	1.1-2.1	1.1-1.6	0.09-0.11	-	-	-	-	-	-
Watson, 1957	P	-	0.11-0.16	1.8-2.1	0.2-0.4	-	-	-	-	-	-	-
Oliveira et al., 1978	FL[7]	4.2-5.2	0.25-0.35	1.2-2.6	0.87-1.6	0.22-0.32	0.22-0.37	-	-	-	-	-
	FS[7]	1.9-2.6	0.19-0.29	2.3-3.9	0.41-1.0	0.13-0.21	0.12-0.25	-	-	-	-	-
Watson and Whiteman, 1981a	P	-	-	-	-	-	0.13-0.28	-	-	-	-	0.24
	F	-	-	-	-	-	0.09-0.21	-	-	-	-	0.21
		Fe	Cu	Mn	Zn	B	Mo	Cu	Mn	Zn	B	Mo
Shorrocks, 1964[3]	P	60-200	18.0-20.0	50-100	20-25	25-30	0.3-0.8	6.0-12.0	5-16	7-16	2-3	0.16-0.17
Andrew and Hegarty, 1969	C	-	-	110-3125	-	-	-	-	1600[8]	-	-	-
Andrew and Thorne, 1962	P	-	2.0-5.9	-	-	-	-	4.0	-	-	-	-
	C	-	3.2-8.2	-	-	-	-	4.0	-	-	-	-
Andrew and Pieters, 1970a	P	243-454	11.0-22.0	56-181	42-73	-	0.3-4.1	-	-	-	-	-
Spain et al., 1975	P	-	5.7-10.9	83-136	32-63	16-24	-	-	-	-	-	-

1 = pot, F = Field, C = nutrient culture
2 approximate value below which a reduction in yield may occur
3 leaves from the terminal half of stems
4 photographs and descriptions
5 Personal communication. Forage agronomist, Centro Internacional de Agricultura Tropical (CIAT). Apartado Aereo 6713, Cali, Colombia.
6 C. macrocarpum
7 L = leaves, A = stems
8 Toxicity threshold

VISUAL SYMPTOMS

A key to visual nutrient deficiencies of centro is presented as an aid when laboratory data are unavailable.

I. <u>Symptoms most evident on older leaves</u>   <u>Nutrient</u>

1. Interveinal chlorosis
    (a) In severe cases interveinal chlorosis is   Mg
         not contiguous with the leaf margin.

2. Leaf tip chlorosis
    (a) Deficiency rapidly reduces size of leaf-   Zn
         lets and growth.
    (b) Increased yellowing from green color of   Fe
         older leaves to gradual shading towards
         yellow to white for youngest leaflets.
         Reduced size of leaflets as deficiency
         becomes more severe.
    (c) Interveinal chlorosis of overall leaf-   Mn
         lets, very similar to Fe deficiency.
         Only slight leaflet reduction. Younger
         leaflets are not pale yellow even with
         severe deficiency.
    (d) Also see Ca.

3. Leaf tip chlorosis with eventual margin and/or tip
    necrosis.
    (a) Mainly affects younger leaves developing   S
         into pale yellowness. Later development
         of marginal and tip necrosis, often as a
         band around most of leaflet margin. Ne-
         crosis not preceded by yellowing such as
         with K.
    (b) Growth only slightly reduced, first pale   Mo
         green and later younger leaves develop a
         pale brown papery tip and marginal scorch
         of leaflets.
    (c) Chlorosis and leaflet tip scorch begins   Cu
         at tips of younger leaves with some leaf
         distortion. Leaves are thin, become pale
         green and develop a pale brown to white
         scorch at the tips and to a lesser degree
         at the margins. When deficiency is severe,
         young leaves often die back at the tip before
         leaf expansion is completed, resulting in
         curling and cupping of the leaflets around
         the tip, without subsequent axillary growth.

4. Leaf and/or plant distortion
    (a) Young leaflet tips become wilted and fail   Ca
         to expand resulting in slightly cupped,
         deformed leaflets with tip necrosis.
         Interveinal chlorosis also evident.

(b) Stunted plants with short thick vines.   B
Leaves are much smaller and misshapen with
veins frequently prominent. Younger leaves
are noticeably thick and brittle to the
touch. Slow or ceased apical meristem
development eventually causing many apical
meristems.

(c) Also see Cu.

II.   Symptoms on both old and new foliage

1. Uniform leaflet yellowing with marked
growth reduction on leaflets and plants.   N

Note: The reader is urged to review the excellent photographs and descriptions of deficiencies by Shorrocks
(1964). We gratefully acknowledge his work in helping
with this key.

IMPROVED GENETIC POOL

During the past several years large-scale collection
and evaluation of new species and ecotypes have generated
a large pool of untested genotypes of Centrosema. The
newly recognized broad genetic diversity within and between species provides forage material that will help to
solve many nutritional and soil problems now encountered
with common and Belalto centro, and most certainly will
increase the use of other species such as C. pascuorum,
C. virginianum, C. brasilianum, and C. macrocarpum. Ad-
ditionally, plant breeding may be used to augment the
genetic diversity sought. There is a unique opportunity
for forage agronomists and soil fertility researchers to
review previous centro research in light of these rapid
new developments.

ACKNOWLEDGEMENT

This work was supported in part by a grant from the
U.S.D.A., TAD-406 7002-20108-0024.

SELECTED REFERENCES

ANDREW, C. S. and M. P. HEGARTY. 1969. Comparative responses to manganese excess of eight tropical and
four temperate pasture legume species. Australian
J. Agric. Res. 20:687-696.

ANDREW, C. S. and D. O. NORRIS. 1961. Comparative responses to calcium of five tropical and four temperate pasture legume species. Australian J. Agric.
Res. 12:40-55.

528

ANDREW, C. S. and W. H. J. PIETERS. 1970a. Manganese toxicity symptoms of one temperate and seven tropical pasture legumes. Div. Trop. Past. Tech. Paper No. 4. CSIRO. Cunningham Lab., St. Lucia, Queensland, Australia.

ANDREW, C. S. and W. H. J. PIETERS. 1970b. Effect of potassium on the growth and chemical composition of some pasture legumes. Div. Trop. Past. Tech. Paper No. 5. CSIRO. Cunningham Lab., St. Lucia, Queensland, Australia.

ANDREW, C. S. and M. F. ROBINS. 1969a. The effect of phosphorus on the growth and chemical composition of some tropical pasture legumes. 1. Growth and critical percentages of phosphorus. Australian J. Agric. Res. 20:665-674.

ANDREW, C. S. and M. F. ROBINS. 1969b. The effect of potassium on the growth and chemical composition of some tropical and temperate pasture legumes. 1. Growth and critical percentages of potassium. Australian J. Agric. Res. 20:999-1007.

ANDREW, C. S. and M. F. ROBINS. 1969c. The effect of phosphorus on the growth and chemical composition of some tropical pasture legumes. 2. Nitrogen, calcium, magnesium, potassium, and sodium contents. Australian J. Agric. Res. 20:675-685.

ANDREW, C. S. and P. M. THORNE. 1962. Comparative responses to copper of some tropical and temperate pasture legumes. Australian J. Agric. Sci. 13:821-835.

BRUCE, R. C. 1978. A review of the trace element nutrition of tropical pasture legumes in northern Australia. Trop. Grassl. 12:170-183.

BURT, R. L., P.P. ROTAR, J. L. WALKER and M. W. SILNEY, 1983. The Role of **Centrosema, Desmodium,** and **Stylosanthes** in Improving Tropical Pastures. Westview Press, Boulder, Colorado. Westview Trop. Agric. Series No. 6.

FRANCO, A. A. 1977. Contribution of the legume-Rhizobium symbiosis to the ecosystem and food production. p. 69. In J. Dobereiner, R. H. Burris, and A. Hollaender (ed.). Limitations and Potentials for Biological Nitrogen Fixation in the Tropics. Plenum Press. New York.

GROF, B. and W. A. T. HARDING. 1970. Yield attributes of some species and ecotypes of Centrosema in north Queensland. Queensland J. Agric. Anim. Sci. 27:237242.

GROF, B. and R. J. CLEMENTS. 1981. Plant nutrition. p. 57-58. CSIRO Tropical Crops and Pastures. Divisional Rept. 1979-80. Cunningham Lab., St. Lucia, Queensland, Australia.

GUTTERIDGE, R. C. and P. C. WHITEMAN. 1978. Pasture species evaluation in the Solomon Islands. Trop. Grassl. 12:113-126.

JONES, M. B. and J. L. QUAGLIATO. 1973. Response of four tropical legumes and alfalfa to varying levels of sulfur. Sulfur Inst. J. winter/spring.

KRETSCHMER, A. E. Jr. 1976. Growth and adaptability of Centrosema species in south Florida. Soil Crop Sci. Soc. Florida Proc. 36:164-168.

KRETSCHMER, A. E. Jr. 1978. Tropical forage and green manure legumes. p. 97-123. In G. A. Jung (ed.) Crop Tolerance to Suboptimal Land Conditions. Amer. Soc. Agron. Spec. Pub. 32.

OLIVEIRA, G. D. de, H. P. HAAG, J. R. SARRUGE, and M. L. V. BOSS. 1978. Nutricao mineral de leguminosas tropicais. 1. Absorcao dos macronutrientes pela centrosema (Centrosema Pubescens Benth.) Siratro (Macroptilium atropurpureum cv. 'Siratro') e soja perene (Glycine wightii Willd.) cultivadas em condicoes de campo. Anais da E. S. A. "Luis de Quiroz" 35:341-416.

SANCHEZ, P. A. and L. E. TERGAS (ed.) 1979. Pasture Production in Acid Soils of the Tropics. CIAT, Cali, Colombia.

SHORROCKS, C. M. 1964. Mineral Deficiencies of Hevea and Associated Cover Plants. Rubber Res. Inst. Malaysia. Kuala Lumpur. Kynock Press, Birmingham.

SKERMAN, P. J. 1977. Tropical Forage Legumes. FAO, Rome.

SPAIN, J. M., C.A. FRANCIS, R. H. HOWELER, and F. CALVO. 1975. Differential species and varietal tolerance to soil acidity in tropical crops and pastures. p. 308-327. In E. Bornemiza and A. Alvardado (ed.) Soil Management in Tropical America. North Carolina State Univ., Raleigh. N.C., USA.

STANDLEY, J., J. E. GILES, and G. H. PRICE. 1981. Commercial soil test results from pasture areas on the wet tropical coast of Queensland. J. Australian Inst. Agric. Sci. 113-117.

STEEL, R. J. H. and L. R. HUMPHREYS. 1974. Growth and phosphorus response of some pasture legumes sown under coconuts in Bali. Trop. Grassl. 8:171-178.

TEITZEL, J. K. and R. L. BURT. 1976. Centrosema pubescens in Australia. Trop. Grassl. 10:5-14.

WATSON, G. A. 1957. Nitrogen fixation by Centrosema pubescens. J. Rubber Inst. Malaya. 15:168-174.

WATSON, S. E. and P. C. WHITEMAN. 1981. Grazing studies on the Guadalcanal plains, Solomon Islands. 1. Climate, soils, and soil fertility assessment. J. Agric. Sci., Camb. 97:341-351.

WERNER, J. C. and H. B. de MATTOS. 1972. Studies on the fertilization of Centrosema pubescens Benth. Bol. Industr. Anim. SP (Brasil). 29:375-391.

WERNER, J. C. and H. B. de MATTOS. 1975. Fertilizer trial with four trace elements on Centrosema pubescens Benth. Bol. Industr. Anim. SP (Brasil). 32:123-135.

WHITNEY, A. S., Y. KANEHIRO, and G. D. SHERMAN. 1967. Nitrogen relationships of three tropical forage legumes in pure stands and in grass mixtures. Agron. J. 59:47-50.

WILSON, A. S. B. and T. J. LANSBURY. 1958. Centrosema pubescens: ground cover and forage crop in cleared rain forest in Ghana. Empire J. Exp. Agric. 26:351-364.

49
TROPICAL GRASSES

Jose Vicente-Chandler

There are hundreds of millions of hectares of rolling to steep lands in the humid tropics where mechanized cropping is not feasible and the soil requires the protection afforded by well-managed grasslands. Since pastures can be seeded and fertilized from the air and cattle do their own harvesting, efficient production together with conservation is feasible on these lands.

Meat is becoming an increasingly important export item and source of foreign exchange for countries in the humid tropics, and local demand for meat and milk is increasing with rising standards of living.

Discussions in this chapter are generally limited to the following widely-used tropical grasses:

Napiergrass or elephant grass (Pennisetum purpureum): A tall grass which grows into clumps by tillering abundantly, it is propagated by stem cuttings and is widely used as a cut grass although it can also be grazed if properly managed. Guinea grass (Panicum maximum): A tall clump grass propagated by clump sections or seed and used for cutting or grazing. Pangola grass (Digitaria decumbens). A sod grass propagated by stem cuttings and used mainly for grazing. Carib grass (Eriochloa polystachya) and para grass (Brachiaria mutica). These grasses often grow together, are similar in growth habit, are propagated by stem cutting and, although used mainly for grazing, can also be harvested by cutting. Stargass (Cynodon nlemfuensis var. nlemfuensis). A sod grass used primarily for grazing and propagated by stem cuttings, it can occasionally have a high cyanide content but has not been found toxic to cattle.

Numerous experiments conducted over many years in the humid region of Puerto Rico show that, when properly fertilized and cut every 40-60 days, napiergrass produces 35,000 kg/ha/yr of dry forage [14,000 kg of total digest-

[1]Agricultural Research Service, U.S. Department of Agriculture, Rio Piedras, Puerto Rico.

ible nutrients (T.D.N.)], guinea, star and congo grasses 30,000 kg/ha/yr of dry forage (12,000 kg of T.D.N.), and pangola, carib and para grasses 27,000 kg/ha/yr of dry forage (11,000 kg of T.D.N.).

Under grazing management, napier, guinea, and star-grass pastures can produce 14,000 kg/ha/yr of dry forage containing 8,000 kg of T.D.N. and carry 5.6 head of young cattle/ha gaining 1,200 kg of liveweight yearly. Pangola, congo, and carib grasses can produce 13,000 kg/ha/yr of dry forage containing 7,200 kg of T.D.N. and carry 5 head of young cattle/ha gaining 1,000 kg of live weight yearly.

Thousands of analyses and many feeding trials show that if properly managed, all these grasses are similar in palatability, composition and nutritive value. If properly fertilized and cut every 60 days, these grasses contain an average of 8.3% protein, 37% fiber, 9.9 % lignin, 45% soluble carbohydrate, 5% calcium, 0.2% phosphorus, 0.3% Mg, 1.9% K and have a digestibility of 45%.

Under grazing management, also with proper fertilization, the forage ingested by cattle grazing star, guinea, napier, pangola and congo grasses contained an average of 18% protein, 1.5% silica, 7.4% lignin, 0.2% phosphorus, 0.4% calcium and had a digestibility of 58%. These values and those for milk and meat production from pastures of these grasses with cattle receiving no concentrate feed as discussed later, are comparable to those for well managed grasses in the temperate region and refute the popular belief that tropical grasses are low in nutritive value.

Fertilization of grasses must be considered in a somewhat different manner from that of crops. In many areas grasslands are the prevailing natural vegetation and if not overgrazed, can indefinitely sustain a low level of animal production. Garcia-Molinari (1952) described the natural grasslands occurring in Puerto Rico on 6 of the 8 major soil groups of the world with rainfall ranging from 600 to 3,500 mm/yearly, thus representing vast areas of the tropics. In some cases, higher yielding grass species adapted to the environment can be introduced and productivity of the natural grasslands increased. Humphreys (1981) has described the methods of adaptation of tropical pasture plants to different environments. However, natural grasslands are generally used extensively and are rarely fertilized or limed, except for correcting adverse soil conditions such as liming very acid soils or applying small amounts of phosphorus to soils very deficient in this nutrient.

On the other hand, in many areas of the tropics grasslands can or should be considered as a crop and be fertilized and managed accordingly. These areas are generally near the population centers, and therefore to markets for milk and meat. For this reason, and because they have the necessary infrastructure, the land is

expensive and must be used intensively. Also, here
fertilizer is much more likely to be readily available.
Under the conditions, fertilization of grasslands can be
considered in a manner similar to that of other crops
except that the product (forage) is not usually sold
directly, but must first be converted by animals to the
marketable product, meat or milk. It is with this type
of grassland farming that this chapter is mainly
concerned.

In the humid tropics climate is rarely limiting,
grasses grow throughout the year, and can produce almost
twice as much forage as in temperate regions. As a result, their nutrient requirements can be very high. In
Puerto Rico, Vicente-Chandler et al. (1974) found that
moderately fertilized grasses harvested by cutting removed about 330 kg N, 50 kg P, 300 kg K, 110 kg Ca, 70 kg
Mg, and 60 kg S/ha/yr.

Thus, to a considerable extent, fertilization of
grasses in the tropics should be based primarily on information on the quantity of nutrients removed in the
forage, on losses of nutrients applied as fertilizer and,
to a much lesser degree, on the quantity of nutrients
available in the soil. Information obtained under
typical conditions on nutrient uptake by grasses and on
losses of applied nutrients should be widely applicable
throughout the tropics. This chapter is accordingly
confined mostly to work conducted in Pueto Rico with
limited references to related research in other tropical
countries.

Unless indicated, the following conditions typical
of much of the humid tropics prevail: annual rainfall
from 1,500 to 2,000 mm with lowest precipitation from
December through March, mean monthly temperatures from
21° to 27°C, and deep, porous soils (Ultisols) with a pH
of 4.8 to 5.2 and 4 to 8 meq of exchangeable bases/100 g
of soil.

FERTILIZING CUT GRASSES

Nitrogen (N). About 330 kg N are removed/ha yearly
by properly fertilized cut grasses whereas unfertilized
Ultisols provide an average of only about 100 kg N/ha
yearly to grasses (Vicente - Chandler et al, 1974).
Furthermore, about 45% of the N fertilizer applied is
lost, mostly by leaching. It therefore follows that
about 400 kg N/ha/yr should be applied in order to obtain
high yields of good quality forage.

Grasses respond much more strongly to nitrogen fertilization during seasons of fast growth than during the
drier, cooler winter months (Figure 49.1). Protein
content of the forage was higher during seasons of slow
growth owing to the concentration effect of lower yields
in the presence of an equal supply of N.

Similar responses to N fertilization by tropical
grasses harvested by cutting have been obtained elsewhere

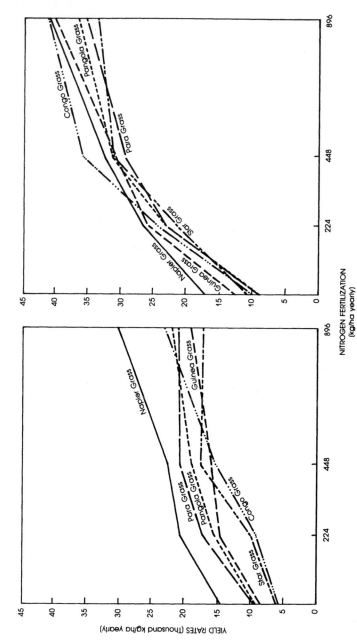

Figure 49.1 The response of grasses to nitrogen fertilization during seasons of slow and fast growth.

in the tropics. Sivalingam (1964) in Ceylon found that napier, guinea, and Brachiaria pastures responded to annual applications of up to 400 kg N/ha/yr in 3 applications, with about 60% of the applied N recovered in the forage. Salette (1970) obtained similar responses to N by pangolagrass under hot tropical conditions. In Hawaii, pangolagrass responded strongly to applications up to 640 kg N/ha with 36 to 64% of the applied N recovered in the forage. Appadurai and Arasaratnam (1969) in Ceylon found Brachiaria brizantha responded linearly to N applications up to 280 kg/ha yearly. In Colombia, pangolagrass yields increased linearly with N rates up to 200 kg/ha yearly and curvilinearly to heavier applications. Guererro et al.(1958) in Costa Rica found that napiergrass responded strongly in yield to the application of 600 kg N/ha and that such fertilization increased the protein and decreased the P content of the forage.

Vicente-Chandler, et al. (1974) found that highest yields were produced, and more fertilizer N was recovered in the forage, when N was applied immediately after each cutting, which is also the easiest practice. In Colombia, it was best to apply N after each cutting of pangolagrass, and splitting N applications or mixing nitrification inhibitors with urea increased efficiency of N fertilization of pangolagrass.

Studies by Vicente-Chandler et al. (1974) with five N compounds applied at the rate of 672 kg N/ha yearly in six equal applications to cut napiergrass showed that N sources did not significantly affect yields, although ammonium sulphate tended to be superior and urea and ammonium hydroxide resulted in lower yields of crude protein.

Vicente-Chandler et al. (1974) reported the relative efficiency of different N compounds applied in 6 equal applications yearly to a dense pangolagrass sod harvested by simulated grazing over a 2-year period. The N sources had similar effects on pangolagrass except that ammonium sulphate generally produced higher, and urea and urea-lime produced lower, yields of forage and of protein. Mixing urea or ammonium nitrate with lime did not affect their efficiency as N sources, and maintained levels of soil acidity at about pH 5.5, whereas ammonium sulphate applications increased acidity to pH 4.5 and urea and ammonium nitrate to pH 4.8.

Analytical Tools. There are no simple reliable methods of analysis to predict the quantity of N that a given soil can provide to grasses. Foliar analysis is also not very useful since N content of the forage does not increase much until near maximum growth rates are attained and can vary considerably throughout the year. A N content of about 1.2 percent (7% crude protein) in 60-day old cut grass is usually associated with near optimum economic yields. The main visual symptom of N deficiency is yellowing of the leaves.

For nitrogen, it can be concluded that: (i) grasses harvested by cutting respond strongly to applications of up to 400 kg N/ha/yr; (ii) N should be applied after each cutting; (iii) response to N fertilization is greater as the interval between cutting is lengthened although nutritive value decreases, and when rainfall is abundant; and (iv) the cheapest source of N should be used depending on price per kg of N plus the cost of application of lime required to maintain a desirable level of soil acidity.

Potassium (K). Well managed, properly fertilized cut grasses take up about 300 kg K/ha/yr, but soils can provide considerable quantities of K to grasses. Figure 49.2 shows the quantities of K taken up by pangolagrass growing on different soils over a 4 year period. Uptake of K was high during the first year and thereafter levelled off to an average rate of about 100 kg K/ha/yr. The Mollisols and Inceptisols of the semiarid region, however, provided much more K to the grass over a longer period of time.

Figure 49.3 shows the relationships between initial exchangeable K content of some of the soils, and K uptake by the grass over the first year. About 25% of K applied as fertilizer is lost, mostly by leaching. From the above it can be rationalized that, on a long term basis, otherwise-well-managed grasses will require about 300 kg of fertilizer K/ha/yr.

The effects of K fertilization on yields and composition of otherwise-well-fertilized grasses growing on a Ultisol were reported by Vicente-Chandler, et al. (1974). Yields and K content of all grasses increased rapidly with rates up to 448 kg K/ha/yr (Figure 49.4). Potassium contents of about 1.5% were associated with high yields. Recovery of fertilizer K in the forage of the different grasses ranged from 65 to 77% at the 448 kg K/ha rate. Ca and Mg contents of the forage decreased with increasing K rates.

Vicente-Chandler et al. (1974) found that potassium chloride and potassium sulphate were equally effective suppliers of K to stargrass growing on an Ultisol, and liming increased the efficiency of K fertilization as indicated by uptake of this nutrient by the grass.

The exchangeable K content of the soil provides a fairly good indication of the soil's ability to provide this nutrient to grasses over the short term. Since about 300 kg K/ha yearly in the forage is required for near optimum yields, it would seem from Figure 49.3 that about 200 kg/ha of exchangeable K in the upper 20 cm of the soil in Inceptisols (which in addition have considerable nitric acid extractable K) can provide the required amount of K to grasses. With Ultisols and Oxisols, which are much more weathered and therefore have less K reserves as indicated by extraction with acid, exchangeable

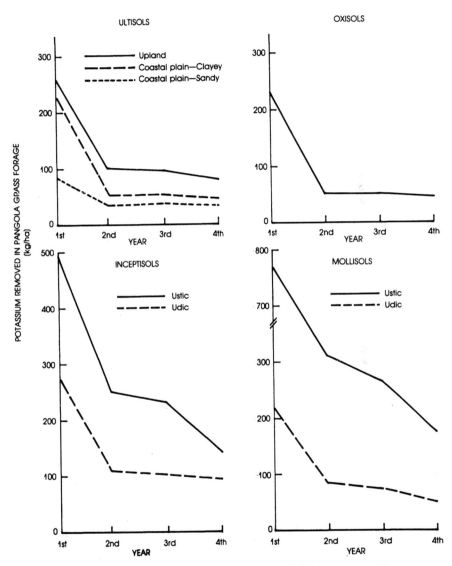

Figure 49.2 Potassium uptake from different soil groups by pangola grass over a 4-year period.

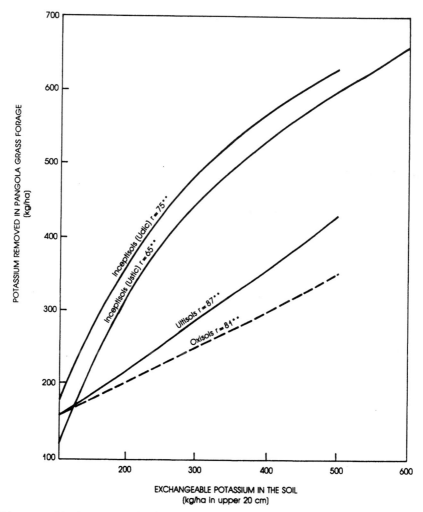

Figure 49.3   Relationship between initial exchangeable potassium content of the soil and quantities of potassium extracted from the soil by pangola grass during first year of cropping.

K levels of about 300 kg/ha are required for high
yields. Since grasses can quickly deplete the soil's K
reserves, the soil must be analyzed every 1-2 years as a
basis for fertilization.
    Analytical Tools. Potassium contents of 1.0-1.5
percent in 60-day-old cut forage indicates desirable
levels of this nutrient and therefore no need to fertilize. Brown spots and "burning" of the tips and eventually of all the older leaves indicate K deficiency.
    It can be concluded that: (i) otherwise-well-fertil-
ized grasses harvested by cutting should receive about
300 kg K/ha/yr once the exchangeable K in in the soil has
been depleted; (ii) K should be applied after each cutting; (iii) the cheapest source of K should be used; (iv)
the soils should be limed to about pH 5.5, and (v) K contents of less than 1 to 1.5% in 60-day-old forage indicate lack of this nutrient for optimum growth of grasses
as do exchangeable K levels of 200 kg/ha in the soil.

    Phosphorus (P). Most soils of the humid tropics are
low in P. Furthermore, most Ultisols and Oxisols can
bind P, rendering it unavailable to grasses and other
plants. However, over time much of this "fixed P" becomes slowly available to grasses. Furthermore, since P
is held strongly in the soil against leaching, the available P content of many soils has been built up from previous applications of fertilizer to the point where
grasses may not respond to P applications for years.
    The effects of P fertilization on yields and compo-
sition of grasses growing on two Ultisols with all other
nutrients provided in abundance were reported by Vicente-
Chandler et al. (1974). Napiergrass responded strongly
to applications of 75 kg P/ha/yr on a soil which had
received little previous fertilization with the nutrient
(Figure 49.4). Phosphorus content of the forage increased with rates up to 150 kg P/ha/yr, with a content
of about 0.17% in 60-day-old grass being associated with
high yields. This low P content in the forage may be
partly the result of heavy N fertilization.
    On the other hand, napier, guinea, and pangola-
grasses did not respond in yield or P content to P applications during 4 years on a Fajardo clay which had been
in sugarcane fertilized with P for many years. Ahmad et
al. (1968) in Trinidad also found that pangolagrass did
not respond to P applications under similar conditions.
    Vicente-Chandler et al. (1974) in 12 field experi-
ments representing 8 soil types determined the response
of pangolagrass to P applications, and the capacity of
the soils to provide this nutrient to grasses. The soils
provided an average of 20 kg P/ha/yr to the grass.
Pangolagrass responded to P applications at only 4 of the
12 sites. The grass took up an average of about 55 kg
P/ha/yr.

540

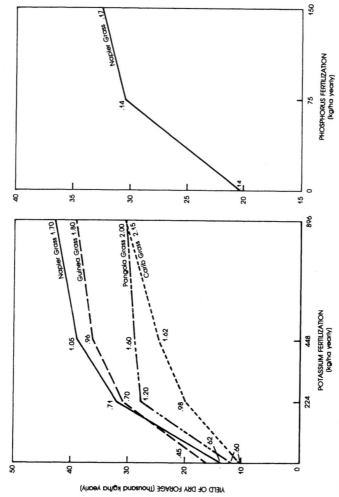

Figure 49.4 The response of tropical grasses to
fertilization with phosphorus and potassium on a typical
Ultisol. Numbers show percent potassium and phosphorus
content of forage on a dry weight basis.

Analytical Tools. Soil analyses can provide qualitative information on the available P in a given soil.
Values of "available" P in the soil much below 10 ppm
indicate a P deficiency. Foliar analysis can also be
useful in determining how much P to apply, with values of
less than about 0.18% in 60-day-old grass suggesting a
deficiency. Visual symptoms of P deficiency are rarely
seen; a deficiency of this nutrient is usually evidenced
only by reduced growth.

It can be concluded that: (i) grasses harvested by
cutting respond to applications of about 70 kg P/ha/yr if
the soil has not previously been fertilized with P. If
previously fertilized, or if soil analyses indicate sufficient available P/(more than 10 ppm), P can be reduced
correspondingly; (ii) content of less than 0.18% P in 60-
day-old forage indicates deficiency; (iii) P can be
applied once yearly, although it is usually applied in
several applications as part of a complete fertilizer.

Magnesium (Mg). Well-managed grasses take up about
70 kg Mg/ha/yr, and losses of this nutrient by leaching
are increased by heavy applications of N. Also, heavy
applications of K can interfere with uptake of Mg by the
grass. Furthermore, Oxisols and Ultisols, and especially
sandy soils, often contain little exchangeable (available) Mg. Deficiencies of Mg are not unusual.

Figure 49.5 shows that the 7 soil types studied
varied considerably in Mg supplying power, providing
after 4 years of continuous cropping an average of about
80 kg Mg/ha/yr to the grass.

The Mg requirements of grasses can be provided by
limestone which often contains 1 to 3% of Mg.

Analytical Tools. Exchangeable Mg content of a soil
is a good guide to fertilizing with Mg, with values of
less than 100 ppm indicating a possible deficiency.
Foliar analysis can also be very useful, with values of
less than 0.20% suggesting deficiencies. Visual symptoms
of deficiency other than reduced growth are rarely found
on grasses.

It can be concluded that Mg should be applied if the
soil contains less than about 200 kg exchangeable Mg/ha
in the root zone, and if limestone to be applied contains
little or no Mg. Also, a deficiency of Mg should be
suspected if 60-day-old grass contains less than 0.2%
Mg. Magnesium can be applied as ground dolomitic
limestone.

Sulfur (S) and minor elements

Serious S deficiencies occur in some areas of the
tropics. In Brazil, McClung and Quinn (1964) found that
grasses responded almost as strongly to applications of S
as to N fertilization.

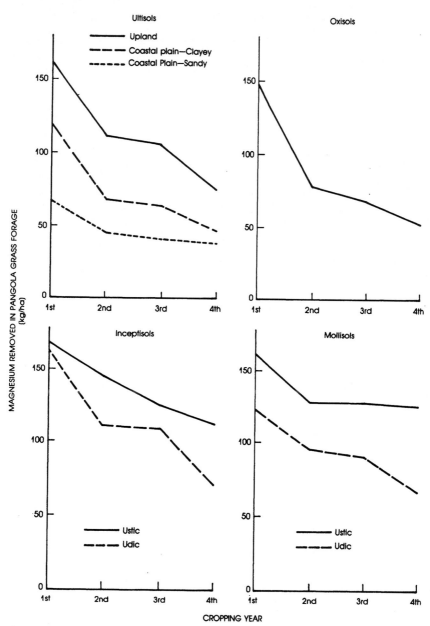

Figure 49.5 Magnesium uptake from different soil groups
by pangola grass over a 4-year period.

Virgin Catalina clay, a typical Ultisol, contains very little S in the upper 45 cm but considerable quantities at greater depths. Nipe clay, an Oxisol derived from serpentine, has an exceedingly high S content, about 600 kg of sulphate-S/ha in the upper 15 cm of soil.

Studies carried out in Puerto Rico by Vicente-Chandler et al. (1974) showed that 60 day-old pangola and guinea grasses contain about 0.15% of S, and these grasses when harvested by cutting removed about 60 kg/ha/yr.

Although grasses take up considerable S, and some soils are low in S, deficiencies are rare since grasslands are usually fertilized with such excellent sources of S as 20% superphosphate, ammonium sulphate or potassium sulphate. On S-deficient soils, where fertilizers containing S are not used, it may be necessary to apply gypsum.

Little is known of the effects of minor element applications on tropical grasses. Table 49.1 shows the minor element content of well fertilized pangolagrass growing on 7 typical soils. Levels of Zn, Fe, Co, Li and Se are normal; those for Mo are rather low. Values for Mn and Ca are high but do not represent levels toxic to grasses. With continued heavy applications of high analysis fertilizers containing few minor elements and having an acid residue, grasses growing on acid, leached soils may eventually require applications of minor elements.

Calcium (Ca). Even very acid soils usually contain sufficient Ca to meet the requirements of grasses for many years. However, liming is required to correct the natural acidity of many soils and to prevent development of high levels of acidity resulting from heavy applications of acid N fertilizers. These can leach a large quantity of bases from soils and increase soil acidity so that high contents of exchangeable Al and/or Mn become toxic to grasses. Vicente-Chandler et al. (1974) report that exchangeable bases in the upper 30 cm of a Mollisol dropped from 21.2 to 14.5 meq./100 g of soil and pH from 6.5 to 4.6 following applications of 880 kg N/ha/yr as ammonium sulphate for 3 years.

Increases in soil acidity are relatively easy to prevent by proper liming but are difficult to correct once acidity develops deep in the soil profile, since deep plowing required to mix limestone with lower soil layers is generally impractical.

The effect of liming on yields and composition of well fertilized grasses growing on two Ultisols was reported by Vicente-Chandler et al. (1974) over a 4-year period. On both soils, the grasses receiving 800 kg N/ha/yr as ammonium sulphate responded strongly in yield to applications of limestone up to about 3 tons/ha yearly, and the response was related to exchangeable Al

TABLE 49.1
Minor element content of heavily fertilized[1] pangolagrass growing on seven typical soils and cut every 60 days[2]

Soil	Mineral element content in ppm								
	Zn	Fe	Mn	Cu	Mo	Co	Li	Se	Na
Coto (Oxisol)	17	222	246	14	0.11	0.14	1.8	0.03	2,500
Colinas (Mollisol)	16	118	152	18	.08	.04	1.4	.02	800
Pandura (Inceptisol)	18	74	211	14	.08	.07	.9	.02	2,700
Los Guineos (Ultisol)	21	104	246	19	.04	.07	.9	.04	1,400
Cialitos (Ultisol)	16	129	48	16	.11	.04	1.4	.04	1,400
Humatas (Ultisol)	41	134	358	21	.08	.40	.5	.03	3,800
Mucara (Inceptisol)	20	232	232	14	.09	.43	.5	.02	7,600

[1]1600 kg/ha yearly of N, 80 of P and 400 K in six equal applications.

[2]Analysis and interpretation of results by David L. Grunes, Soil Scientist, Plant, Soils and Nutrition Laboratory, ARS, USDA, Ithaca, N.Y.

content of the soil. Figure 49.6 summarizes the results obtained with napiergrass. Whereas yields were only slightly affected by liming during the first year, they were almost doubled during the fourth year. This increasing response with time was caused by progressive sharp drops in forage yield in the unlimed and lightly limed plots, with yields of well limed plots remaining fairly constant. Yields were closely related to the exchangeable Al content of the soil.

Liming increased the Ca content of all the grasses. A Ca content of about 0.40% in 60-day-old napier and pangolagrasses and of 0.60% in guinea grass, was associated with high yields. Such levels also meet livestock requirements for this nutrient. Liming decreased the Mn content of all the grasses, but did not affect their P or Mg contents.

The effect of liming six typical soils of the humid tropics on yields of well fertilized pangolagrass was determined by Vicente-Chandler et al. (1974). The grasses did not respond to liming on soils with pH values ranging from 4.8 to 6.0, but responded strongly on soils with pH values of 4.5 or lower.

It is significant that in all these experiments about one ton of limestone was required per ton of fertilizer in order to maintain high yields and to neutralize the residual acidity of the fertilizers.

Analytical Tools. In all soils, high yields were obtained when base saturation was 50% or more, when the soils contained less than 2 m.e. exchangeable Al/100 g, equivalent to about 25% saturation of the exchange capacity, or had a pH above 5.0. A Ca content of less than about 0.4% in 60 day old napier or pangolagrasses, or of less than 0.6% in guinea or congo grasses indicate a possible need for liming. It can be concluded that: (i) soils should be limed to above 60% base saturation or above pH 5.2 at planting; (ii) about 1 ton of commercial limestone containing no less than 80% calcium carbonate equivalent, some dolomite to supply Mg, and finely ground since particles larger than 60 mesh are of little value in correcting soil acidity, should be applied for every ton of fertilizer used unless a non-acid residue N source is used.

FERTILIZING CUT GRASSES

Grasses harvested by cutting respond strongly in yield to applications of 400, 150 and 300 kg/ha/yr of N, P_2O_5 and K_2O, respectively. This suggests that a 3-1-2 (15-5-10) or similar commercial fertilizer is well suited to grasses under the described conditions.

Cost of fertilizer and value of the forage produced mainly determine how much fertilizer to use. These values vary widely, and figures are used here only as a de-

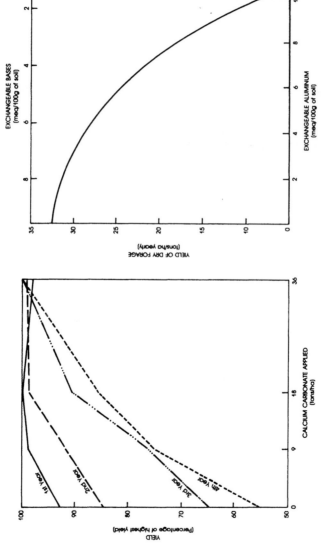

Figure 49.6   Effect of liming on yields of heavily fertilizer napier grass growing on an Ultisol as related to soil properties. A. Effect of liming on yields produced over 4 successive years. B. Relationship between yields during the fourth year and soil properties.

monstration. In all discussions related to economics in this chapter, assumed prices are as follows: 20¢/kg of fertilizer; 22¢/kg of concentrate feed; $1.00/kg of live-weight and 40¢ liter of milk.

Although the cost of fertilizer is easy to ascertain, it is difficult to determine the value of forage. One way is in terms of the beef that forage can produce. In experiments by Vicente-Chandler et al. (1974), about 14.5 kg of dry, cut grass of good quality, when eaten by young cattle, were converted to 1 kg of liveweight. Since about 20% of cut forage is wasted in feeding, 18 kg of dry grass were required to produce 1 kg of liveweight worth $1.00. Thus, a kg of dry forage can be valued at 5.5¢. Returns from different levels of fertilization can be approximated by using typical fertilizer response curves for the grasses and the above values for fertilizer and forage.

The following tabulation shows the approximate kgs of dry forage produced per kg of each increase in fertilization of grasses cut every 60 days:

Fertilizer rate (kg/ha yearly)	Kg of dry forage produced/ kg of fertilizer increment
1,000	6
2,000	5
3,000	4
4,000	1
5,000	1
6,000	1

Each kg of fertilizer (worth 20¢) would produce 33¢ worth of forage for the first ton applied, 27.5¢ for the second ton and 22¢ for the third ton. It would, therefore, be economic to apply up to 3 tons/ha of fertilizer at these prices. Furthermore, protein content of the forage is increased by fertilization, and additional protein is produced, to which a value should be ascribed. These calculations do not include savings in investments in land and overhead expenses. Similar calculations with dairy cows result in much higher returns from fertilization.

From this simple approach to what is really a complex problem, it can be concluded that it is profitable to apply up to 3 metric tons/ha of 15-5-10 or similar fertilizer yearly to cut grasses under the specified conditions.

The optimum quantity of fertilizer to apply to cut grasses depends on many other factors. Fertilization can be reduced if manure is returned to the land. About 80% of the nutrients in the forage consumed by cattle is returned in the urine and feces. Therefore, if all excreta together with rejected forage is returned to the field without loss, it is theoretically possible to reduce fertilization to 20% of optimum. However, even under the best conditions about half of the nutrients in the excreta is lost by volatilization, leaching, etc. Urine and feces should be mixed and returned to the field as soon as possible, or carefully stored. Superphosphate may be added as an absorbing agent for ammonia. Applying manure to fields is only profitable when handling costs are kept low.

More fertilizer can be used profitably as land values increase. Where land is expensive, additional forage can be produced more profitably by heavier fertilization than by planting a large area which also involves additional expenses for planting the pastures, weed control, fencing, and so on.

If the soils have little natural fertility, as is usual in the humid tropics, planting additional lands to grasses ordinarily is less profitable than fertilizing existing grasslands more heavily.

More fertilizer can be used profitably when moisture is adequate or on deep soils with a high water-holding capacity than on shallow or sandy soils which dry out after a few days. Also, more fertilizer can be used profitably during seasons of favorable weather, provided the additional forage produced is used efficiently.

It can be concluded that at the specified prices it is profitable to apply up to 3 ton/ha yearly of 15-5-10 or similar fertilizer to grasses harvested by cutting if there is sufficient moisture available . The fertilizer should be applied in 6 equal applications yearly, one after each cutting, and one ton of limestone should be applied for every ton of acid residue fertilizer used.

FERTILIZING GRASS PASTURE

Fertilizer requirements of pastures differ from those of cut grasses. Only about half as much forage, hence much less nutrients, are taken from the land under grazing than under cutting management.

A heavy return of nutrients to the soil also occurs with grazing. About 80% of nitrogen, phosphorus and potassium consumed by cattle is excreted in the urine and feces. A mature cow produces about 9,000 kg of feces and urine yearly containing 70 kg of nitrogen, 18 kg of phosphoric acid and 60 kg of potash. This high return of nutrients in the excreta suggests that pastures require little fertilizer. Grazing animals, however, are not very effective in maintaining fertility of pastures since excreta is poorly distributed and deposited in wastefully heavy concentrations on limited areas resulting in heavy leaching losses, particularly of N (Peterson, et al; 1965).

N fertilization of pastures should not be decreased because of the limited build up of the excreta N in small areas. P and K, on the other hand, can build up under heavy stocking and intensive management, as these nutrients are strongly held in the soil.

The long term build up of P and K in an Ultisol was determined by Vicente-Chandler et al. (1974). In two experiments, in which the pastures were well fertilized and intensively managed, about 20% of the K and P applied over a 14-year period accumulated in the soil in easily available form. This accumulation resulted from P and K returned in manure, rather than from applications of these nutrients in excess of the grasses needs.

It can therefore be reasoned that pastures require only about half as much fertilizer as cut grasses, and that after several years of heavy fertilization the proportion of P and K in relation to N can be reduced by about 20%.

Vicente-Chandler et al. (1974) found that with guinea grass harvested every 45 days by plucking to stimulate grazing, best results were obtained when the annual rate of fertilizer was applied in four rather than in two or eight equal applications yearly.

Pastures under actual grazing conditions may respond differently to fertilization because of variations in nutrients removed in the forage consumed by the cattle, trampling effects, and return of nutrients in the excreta.

Four grazing experiments were reported by Vicente-Chandler et al. (1974) determining the effect of fertilizer rates on carrying capacity and beef produced by napier, pangola, and stargrass pastures. The experiments were carried out on steep Ultisols limed to about pH 6.0. The 15-5-10 fertilizer was applied in four equal applications yearly. The pastures were grazed in rotation by young cattle receiving no supplementary feed. A new group of cattle was put on the pastures every year. Paired areas in each pastures were mowed, one before and one after each grazing, and the forage taken by plucking to simulate grazing was analyzed for various components. Total digestible nutrients consumed by the grazing cattle were calculated.

Napiergrass Pastures

When fertilization was increased from 675 to 2,025 kg/ha/yr, beef production increased from 638 to 1,201 kg/ha yr, TDN consumed by the cattle increased from 4,820 to 7,500 kg/ha/yr, and carrying capacity moved from 3.5 to 5.5 273-kg steers/ha (Table 49.2). Also, forage consumed by the cattle increased from 9,970 to 15,000 kg of dry matter and its protein content from 8.1 to 15.9%. Daily gains per head, averaging 0.6 kg, were not affected by fertilization. Increasing fertilization to 3,375 kg/ha/yr did not significantly increase productivity of the pastures.

TABLE 49.2
Effect of three fertilizer levels on the productivity of napiergrass pastures on a steep ultisol over a 2-year period

15-5-10 fertilizer applied kg/ha/year	Gains in weight kg/ha/year	Carrying capacity 273-kg steers/ha*	Dry forage consumed by cattle kg/ha/Year**	T.D.N. consumed by cattle kg/ha/year&	Protein content of forage consumed by cattle
675	638	3.5	9,970	4,820	8.1
2,025	1,201	5.5	7,500	7,500	15.9
3,375	1,333	6.3	9,070	9,070	17.6
LSD05	526	--	1,400	1,400	4.1

* One 273-kg steer = 3.86 kg T.D.N. daily.
** From differences in forage harvested from paired strips cut before and after grazing each pasture.
& Calculated from body weights, days of grazing, and gains in weight, following recommendations of the Pasture Research Committee (1943).

It was very profitable to increase fertilization up to 2,025 kg/ha/yr in this trial. The additional 563 kg of liveweight produced was worth $563 compared with increased fertilization costs of $270. One additional milk cow producing 3,000 liters worth $1,200 annually could be carried/ha as a result of applying 1,350 kg more of fertilizer worth $270/ha.

Pangolagrass Pastures

Pangolagrass pastures responded in terms of beef production and carrying capacity to applications of up to 2,688 kg/ha yearly of fertilizer but did not respond to heavier applications. With 2,688 kg/ha yearly of fertilizer, the pangolagrass pastures produced 976 kg of liveweight gain/ha yearly and carried the equivalent of five-273 kg steers/ha with daily gains per head of 0.6 kg.

An increase of 58 kg of beef/ha worth $578 resulted from the application of an additional 2,240 kg of fertilizer/ha yearly worth $48. The profits from this increase in fertilization are much greater in terms of increased milk production, as discussed above.

Bryan and Evans (1971) in Australia found that pangolagrass pastures produced 1,106 kg/ha of liveweight gains when 448 kg of nitrogen were applied/ha yearly compared to 699 kg/h of gains when 168 kg N were applied.

Stargrass Pastures

Stargrass pastures responded in terms of liveweight gains to the application of up to 3,136 kg of 15-5-10 fertilizer/ha yearly. The pastures produced an average of 1,032 kg of liveweight gain/ha yearly with 1,792 kg/ha of fertilizer, compared to 1,342 additional kg of fertilizer worth $268. Daily gains per head averaged 0.6 kg at all fertilizer rates.

With 3,136 kg/ha/yr of fertilizer applied, the stargrass pastures produced over 9,000 kg/ha of T.D.N./ha/yr, equivalent to a carrying capacity of six and a half 273 kg steers/ha. Milk cows on these pastures would produce about 8,000 liters of milk/ha yearly worth about $3,200 with no supplementary feed.

REFERENCES

AHMAD, N., L.I. TULLOCH-REID, C.E. DAVIS, 1963. Fertilizer studies on pangolagrass in Trinidad. Trop. Agr. (Trinidad) 46:173-178.

Pasture Research Committee (Joint Committee of the American Dairy Science Association and the American Society of Agronomy) 1943. Report on pasture investigation techniques. J. Dairy Sci. 26:353-369.

APPADURAI, R.R., R. ARASARATNAM. 1969. The effect of large applications of urea nitrogen on the growth and yield of an established pasture of Brachiaria brizantha Stapf. Trop. Agr. (Trinidad) 46:153-158.

BRYAN, W., and T.R. EVANS. 1971. A comparison of beef production from nitrogen fertilized pangola grass and pangolagrass-legume pastures. Trop. Grassl. 5:89-98.

GARCIA-MOLINARI, O. 1952. Grasslands and grasses of Puerto Rico. Agr. Expt. Sta., Univ. P.R. Bull. 102.

GUERRERO, R., H.W. FASSNDER, J. BLYDENSTEIN. 1958. Fertilizing elephant grass in Turrialba, C.R.I. Effect of increased rates of N. Turrialba 20:53-63.

HUMPHREYS, L.R. 1981. Environmental Adaptation of Tropical Pasture Plants. Macmillan Publishers, Ltd.

McCLUNG, A.D., and L.R. QUINN. 1964. Sulphur and phosphorus responses of Batatais grass (Paspalum notatum), IBEC Research Institute Rep. No. 18.

PETERSEN, R.S., W.W. WOODHOUSE, and H.L. LUCAS. 1956. The distribution of excreta by freely grazing cattle, and its effect on pasture fertility. II. Effect of returned excreta on the residual concentration of some fertilizer elements. Agron. J. 48:444-448.

SALETTE, J.E. 1970. Nitrogen use and intensive management of grasses in the wet tropics. IX Int. Grassland Congress, Australia pp. 404-407.

SIVALINGAM, T. 1964. A study of the effect of nitrogen fertilization and frequency of defoliation on yield, chemical composition and nutritive value of three tropical grasses. Trop. Agr. 159-180.

VICENTE-CHANDLER, J., F. ABRUNA, R. CARO-COSTAS, J. FIGARELLA, S. SILVA and R.W. PEARSON. 1974. Intensive grassland management in the humid tropics of Puerto Rico. Tech. Bull. 233, Agr. Expt. Sta., Mayaguez Campus. University of Puerto Rico, Rio Piedras, P.R.